CEREBRAL PALSY - CHALLENGES FOR THE FUTURE

Edited by **Emira Švraka**

Cerebral Palsy - Challenges for the Future
http://dx.doi.org/10.5772/56981
Edited by Emira Švraka

Contributors

Jennifer McGinley, María Carmen Carrascosa-Romero, Carlos De Cabo De La Vega, Deepak Sharan, Mintaze Kerem Gunel, Ozgun Kaya Kara, Duygu Turker, Cemil Özal, Yasser Awaad, Tamer Gaber, Emira Švraka, Alok Sharma, Pooja Kulkarni, Pedro Ponce

Published by InTech

Janeza Trdine 9, 51000 Rijeka, Croatia

Publishing Process Manager Iva Lipovic

Technical Editor InTech DTP team

Cover InTech Design team

First published March, 2014

Printed in Croatia

Additional hard copies can be obtained from orders@intechopen.com

Cerebral Palsy - Challenges for the Future, Edited by Emira Švraka

p. cm.

ISBN 978-953-51-1234-1

For EU product safety concerns:
IN TECH d.o.o., Prolaz Marije Krucifikse Kozulić 3, 51000 Rijeka, Croatia,
info@intechopen.com or visit our website at intechopen.com.

Meet the editor

Professor Emira Švraka, MD, PhD, is a Vice Dean for the third cycle of study and science at the Faculty of Health Studies, University of Sarajevo, Bosnia and Herzegovina. She is the author of the monographs: The Other Side of Life: Learning Difficulties in Children and Adolescents with Cerebral Palsy and Two Sides of Happiness: Quality of Life for Families of School Children with Intellectual Disabilities and co-author of the scientific book Epilepsy - Histological, Electroencephalography and Psychological Aspect, published by InTech, and leading author of University textbook Occupational Therapy. Professor Švraka is a corresponding member of the Bosnian and Herzegovinian American Academy of Arts and Sciences (BHAAAS), and president of the Cerebral Palsy Associations of FB&H since 2011 (www.cpafbih.org).

Contents

Preface

Writing a comprehensive scientific book about the cerebral palsy (CP) is a great ***challenge***. The book is international, 23 authors ; the nine chapters represent different professional approaches on the topic of cerebral palsy, from Australia, Bosnia and Herzegovina, India, Mexico, Saudi Arabia, Spain, Turkey to USA.

Many different interventions are available for persons with CP. Increasingly, it is recognized that intervention needs to be evidence-based and family-centered. Related therapies can offer improvement in some cases but do not offer a cure.

Lifelong re/habilitation (habilitation and rehabilitation) in person with cerebral palsy is the first section of this book and has four chapters about management in children and adults with cerebral palsy through the life span, providing support and services.

In families who have ***children with CP*** the "constant attendance" of the disease is present, through strict consistent long-term care of family and many other factors, such as services, support and physical aspects of the environment, which can lead to deterioration of the patient's quality of life.

Spasticity can be associated with cocontraction, clonus and hyperreflexia. Children with spastic cerebral palsy generally have a typical pattern of muscle weakness, impairment in selective motor control and sensory impairment.

Constrained-induced movement therapy (CIMT) is a treatment approach based on restraining the uninvolved upper extremity and exercising the involved upper extremity intensively. The treatment protocol is based on the principle of the limitation of the nonaffected extremity and forcing the patient to use the affected extremity during the day. The increased use of the affected extremities with CIMT is suggested to be due to an expansion in the contralateral cortical area that controls this extremity's motion and the development of new ipsilateral areas. This is reported to form the neural basis for the continuation of the use of the affected extremity after the treatment.

Many adults with CP face challenges with declining mobility and the emergence of secondary musculoskeletal conditions as they age. Decline in walking and falls are common, potentially comprising activity, participation and health-related quality of life. Current ***healthcare services for adults with CP*** lack the evidence-based knowledge needed to develop and implement best practice clinical guidelines.

Home modifications are part of the worldwide commitment for ***universal design standards***, which are encouraged by the World Health Organization. Occupational and vocational therapists are trained to analyze an individual's home, school, and work environments to prescribe and recommend modifications to meet specific needs.

Three chapters of the second section are exploring the new therapy options which could ***improve the family quality of life.***

Since hypoxic ischemic encephalopathy (HIE) is a potentially preventable cause of cerebral palsy, much interest has been focused on prevention as well as research on ***neuroprotection therapies***. The authors of fifth chapter proposed a model of *"off-label combined therapy"* based on hypothermia /antiepileptic drugs in combination with antioxidants, phospholipase A2 inhibitors, glutamate receptor antagonists or EPO using a staggered design in function of the intensity of the perinatal asphyxia and severity of the encephalopathy.

A well-planned and executed **SEMLARASS**, in the context of a multi-disciplinary team, provides the child with CP, spasticity or movement dysfunction and lever arm dysfunction, with the best hope for a dramatic, predictable and lasting functional improvement.

Stem cell therapy has been extensively studied but still needs to be standardized before it becomes a definitive treatment modality. Autologous BMMNCs are safe and feasible option but their effectiveness needs more clinical trials. Other types of stem cells need to establish safety and efficacy. Though not a cure, stem cell therapy has emerged as a novel therapeutic option to improve the quality of life.

Third section has two chapters about complementary therapies with ***new possibilities for the future.***

Future progress will depend on recognition that ***brain-computer interfaces*** (BCI) research is an interdisciplinary work: rehabilitation engineers, speech pathologists, occupational therapists, physiatrists and rehabilitation therapists are professionals who can provide access to these technologies and assistance in their use.

The results of the use of ***virtual reality (VR)*** as a treatment approach in adults and children are promising. VR implementation in children with CP positively affects brain reorganization, plasticity, motor capacity, visual perceptive skills, social participation and personal factors.

I strongly recommend this InTech international scientific book to all physicians, specialists in physical medicine and rehabilitation, pediatricians, orthopedics, physiotherapists, occupational therapists, rehabilitation nurses, psychologists, sociologists, social workers, other medical workers, teachers of all profiles, students of various studies, but, first of all, to parents of children with disabilities and to general public.

Emira Švraka, MD, PhD
Physiotherapy Department
Faculty of Health Studies
University of Sarajevo
Bosnia and Herzegovina

Section 1

Lifelong Re/habilitation in Person with Cerebral Palsy

Chapter 1

Management of Spasticity and Cerebral Palsy

Yasser Awaad, Tamer Rizk and Emira Švraka

Additional information is available at the end of the chapter

http://dx.doi.org/10.5772/58311

1. Introduction

Spasticity was defined by Lance as a "velocity-dependent increase in tonic stretch reflexes (muscle tone) with exaggerated tendon jerks, resulting from hyperexcitability of the stretch reflex" (Young RR, 1994; Francisco GE, Ivanhoe CB, 1997).

Young further added characteristics of positive and negative symptoms. Positive symptoms consist of exaggerated cutaneous reflexes, including nociceptive and flexor withdrawal reflexes, autonomic hyperreflexia, dystonia, and contractures. Negative symptoms include paresis, lack of dexterity, and fatigability (Young RR, 1994).

Treatment for spasticity was documented as early as the late 19th century, when surgeons Abbe and Bennet discussed decreasing tone in a spastic limb through sensory rhizotomies. Later, in 1898, the scientist Sherrington published experiments in which the sensory roots of spastic cats were severed to relieve spasticity (Abbott R, 1996).

The technique of sensory rhizotomies has been improved on and continues to be used today as a treatment for patients with spasticity as does neuromuscular blockage, a longstanding treatment, which has been used for over 30 years (Koman LA, Mooney JF, Smith BP, 1996).

1.1. Cerebral palsy

Cerebral palsy (CP) is a group of permanent disorders of the development of movement and posture, causing activity limitation, that are attributed to non-progressive disturbances that occurred in the developing foetal or infant brain. The motor disorders of cerebral palsy are often accompanied by disturbances of sensation, perception, cognition, communication, and behavior, by epilepsy, and by secondary musculoskeletal problems (Glinac A, Tahirović H, Delalić A, 2013).

The research *Frequency of joined disabilities of children with cerebral palsy in Tuzla canton* covers a total sample of 48 examinees, chronological age from 2-19 years, in Tuzla Canton. Research instrument was a Structural Questionnaire for the parents of children and adolescents with cerebral palsy. Research data were processed by nonparametric statistics method. Basic statistical parameters of frequency and percentages were calculated, and tabular presentation was made. After classification of examinees as per frequency of joined disabilities was done, work results have shown that speech impairment occurred with 35.4 % of children, visual impairment 33.3 %, epilepsy 29.3 %, whereas hearing impairment occurred with 2 % of children (Babajić M, Švraka E, Avdić D, 2013).

Although there are many possible causes of spasticity, this chapter will focus on children with spasticity, most of whom have diagnoses of cerebral palsy; approximately two thirds of all cerebral palsy patients suffer from spasticity (Albright AL, 1996).

A patient with spastic cerebral palsy presents with muscle imbalance, stands with bent knees and legs tightly together, and in severe cases, a scissors-type gait (Frerebeau PH, et al, 1991; Adams RD, Victor M, Ropper AH, 1997). The antigravity muscles are predominantly affected with arms in a flexed and pronated position and legs in an extended and adducted position. When the muscles are at rest they are flaccid to palpation and electromyographically silent.

Spasticity can be associated with cocontraction, clonus and hyperreflexia. Children with spastic cerebral palsy generally have a typical pattern of muscle weakness, impairment in selective motor control and sensory impairment (Mikov A, Dimitrijević L, Sekulić S, Demeši-Drljan Č, Mikov I, Švraka E, Knežević-Pogančev M, 2011).

Many children with more severe spastic CP experience *communication problems* due to disturbed neuromuscular control of speech mechanism, i. e, dysarthria, that diminish the ability of the child to speak intelligibly. However, substantial dysarthria are most often seen in children with severe CP and intellectual disability, while most children with mild and moderate CP and average cognitive level of functioning have normal or near-normal expressive language and articulation skills (Bottcher, 2010).

1.2. Etiology and epidemiology

Spasticity may result from either diffuse or localized pathology of the cerebral cortex, brain stem, or spinal cord. Possible causes of such injuries include traumatic brain injury, stroke, multiple sclerosis, spinal cord trauma, or disease and anoxic insults. The neurologic localization of the lesion causing spasticity may result in different clinical manifestations. Thus, it is important to consider whether the spasticity results from cerebral pathology, whether it is diffuse or localized, or whether it is a result of spinal cord injury.

Diffuse cerebral injury or diseases would include anoxia, toxic, or metabolic encephalopathy, where as localized cerebral injury would include tumor, abscess, cyst, arteriovenous malformations, hemorrhage, or trauma.

Spinal cord injury or disease may result as an insult to descending pathways by trauma, inflammatory or demyelinating disease, degenerative disorders, or compression such as is caused by a tumor or cyst (Albright AL, 1996; Frerebeau PH, et al, 1991; Dimitrijevic MR, 1991).

The annual incidence of spinal cord injuries in the United States is estimated to be 30 to 40 new cases per million individuals. About 3% to 5% of cases each year occur in children younger than 15 years of age (Price C, Makintubee S, Herndon W, Istre GR, 1994).

The male-to-female ratio of patients is 4:1 in the general population, but in younger age groups, the ratio is approximately 1.5:1 (Zidek K, Srinivasan R, 2003).

According to the time of insult, causes of *cerebral palsy* can be divided to prenatal (from conception till beginning of delivery), perinatal (started from beginning of labor till end of neonatal period; first 28 days of life) and postnatal (from 29th day of age until two years of age). The majority of international studies indicates that the prevalence of cerebral palsy is about 2-2,5 cases per 1000 born, although there are some reports about lower and higher prevalence rates. Majority of previous research in the world was focused on the prevalence, determination of the motor abilities, and perinatal etiological factors of the cerebral palsy. Evidences indicated that 70-80 % of cerebral palsy is caused by the prenatal factors and that the birth asphyxia has a relatively minor role with the less than 10 % (Švraka E, 2012).

Common causes of *cerebral palsy in children* that may result in spasticity are prolonged second stage labor, fetal distress, cystic degeneration of the brain, prematurity, periventricular encephalomalacia, cortical abnormalities such as porencephaly, or congenital malformations of gyri such as micropolygyria.

Through the last decades, marked improvement in the level of intensive care at Neonatal Intensive Care Units (NICU) which was reflected on an increase in the survival of very low birth-weight (VLBW) and extremely low birth-weight (ELBW) premature newborns. New risk factors have appeared among infants who previously would have died, and the incidence of neurodevelopmental impairments in survivors of NICU is higher than in normal birth-weight newborns. In particular, due to the high risk of interventricular haemorrhage and periventricular leukomalacia, an increasing prevalence of cerebral palsy has occurred in premature, low birth-weight newborns and children born with asphyxia (Švraka E, 2012).

Spasticity is present in about two thirds of cerebral palsy patients, and cerebral palsy affects anywhere from 1.5 to 2.5 per 1000 live births in the United States (Adams RD, Victor M, Ropper AH, 1997).

The number of spastic patients continues to increase due to an increased survival rate of premature births. Males and females are equally affected.

1.3. Pathogenesis and pathophysiology

There are many different types of spasticity. Because of this, more than one mechanism may be responsible for the disturbance in muscle tone and the mechanisms may vary between patients. The neuropathophysiologic processes involved in spasticity are complex and not fully understood, but there is a widely accepted hypothesis that spasticity depends on hyperexcitability of spinal alpha motor neurons, which is due to the interruption of descending modulatory influences carried by the corticospinal, vestibulospinal, and reticulospinal tracts and other possible tracts (Filloux FM, 1996).

Ia afferent fibers provide segmental input from muscle spindles to alpha motor neuron pools. They synapse on segmental inhibitory interneurons that then inhibit alpha motor neurons innervating antagonist muscles in the Ia reciprocal inhibition pathway. Ib afferents inhibit alpha motor neurons by way of the Golgi tendon organs via the Ib inhibitory interneuron in another pathway known as nonreciprocal inhibition (Young RR, 1994; Filloux FM, 1996).

Increased excitation of these afferents does not seem to be the cause of spasticity. Instead, evidence supports that reduced reciprocal inhibition of antagonist motor neuron pools by Ia afferents, decreased presynaptic inhibition of Ia afferents, and decreased nonreciprocal inhibition by Ib terminals are all possible pathophysiologic mechanisms of spasticity (Young RR, 1994).

The pathophysiology of traumatic brain injury involves a complex combination of forces that has been a subject of substantial debate (Drew LB. and Drew WE, 2004).

On occasion, *autonomic dysreflexia* may occur after an intramuscular injection, although this is relatively rare (Selcuk B, Inanir M, Kurtaran A, Sulubulut N, Akyuz M, 2004).

In some patients, autonomic dysreflexia may occur even if the level of spinal injury is below T6 (Blackmer J, 2003; Krassioukov AV, Furlan JC, Fehlings MG, 2003).

1.4. Diagnostic procedure

Examination should begin with the patient in a relaxed, lying position with the head up and arms resting to the sides because it is easier to determine the extent of spasticity in this position. The examination should include tonic stretch reflexes by manual passive stretches, elicitation of tendon jerks and clonus in a relaxed position, and tonic and phasic stretch reflexes carried out in a sitting position.

The manual passive stretch maneuver is used to assess resistance at different rates. A joint is passively moved while the muscles corresponding to that joint are lengthened and shortened. In cases of mild spasticity, the muscles will only resist when stretched at a high rate, whereas in cases of moderate spasticity, resistance is noticed at a slower rate and the clasp-knife phenomenon may be exhibited. Movement of the muscle may be difficult to impossible in cases of severe spasticity (Dimitrijevic MR, 1991).

Tendon jerks are easier to elicit in spastic patients than in patients with normal muscle tone, and reflex responses can be achieved in muscles without well-defined tendons. Percussion of the tendon reveals hyperactive tendon jerks, especially for the Achilles, patellar, biceps, and triceps tendons (Zidar J, Dimitrijevic MR, 1991).

Measurement of resistance to passive stretch, reduction in the tonic vibration reflex, and reduction of the plantar withdrawal reflex should also be evaluated. Motoneuronal overactivity should also be evaluated because any input to motoneurons produces excessive and

http://dx.doi.org/10.5772/58311

prolonged activity that can be observed in the contractions of many limb muscles (Zidar J, Dimitrijevic MR, 1991).

The amount of function the patient derives from spasticity can be evaluated by having the patient obtain and maintain standing and seated positions. To determine the degree to which the hamstring tone is affecting the alignment of the pelvis and knees, have the child sit with feet straight in front. The patient can sit in a chair to allow the examiner to assess trunk control. The side sit position exhibits a patient's ability to maintain control in an asymmetric position (Abbot R, 1991).

Modified Ashworth Scale (MAS) has been used as a diagnostic test for spasticity. Testing can be done to establish the presence of any lesion or brain or spinal cord injury. The muscle tone is graded according to the *Modified Ashworth scale (MAS),* a scale ranging from 1 to 5, in which resistance to the passive muscle stretch is measured at various velocities; MAS: 0 = No increase in muscle tone, 1 = Slight increase in the muscle tone, manifested by catch and release or minimal resistance at the end of the range of motion, 2 = more marked increase in the muscle tone through most of the range of motion, but affected parts are easily moved, 3 = considerable increase in the muscle tone, and passive movements are difficult, 4 = affected parts are rigid in flexion or extension (Albright AL, 1996).

MRI of the brain can be performed to rule out periventricular leukomalacia.

On EMG, the jerks show greater amplitudes than are normal and are followed by after-discharge of the motor units that is often slightly longer lasting than normal. The size of tendon jerks can be measured by either EMG response or by recordings of mechanical events.

H-reflex studies are electrically elicited tendon jerks and are restricted mostly to the soleus and flexor carpi radialis muscles in normal adults. In cases of upper motor neuron lesions, the H-reflex may be elicited in muscles where it is not normally seen, such as the intrinsic hand muscles, tibialis anterior, or peroneal muscles (Zidar J, Dimitrijevic MR, 1991).

A baseline EEG to establish underlying seizure activity can also be done as well as basic lab studies. *Neurophysiological studies,* such as the H-reflex study, may be performed in patients with neurodegenerative disease; an enzymatic assay should also be performed.

Video cameras are often helpful during evaluation as the patient's movements can be recorded and compared against movements during and after treatment.

Test	Use
Basic lab studies	Metabolic derangement
Enzymatic assay	Neurodegenerative disease
EEG	Underlying seizure activity
NCS	Neurodegenerative disease (Leukodystrophies)
MRI of the brain	Periventricular leukomalacia

Table 1. Useful Tests for Diagnosis

1.5. Differential diagnosis

Spasticity can be confused with rigidity when a patient is being evaluated. Stretching can distinguish rigidity from spasticity. Rigidity will relax through repeated stretching of a muscle, whereas a spastic muscle will continue to increase in resistance as the velocity of the stretch is increased (Young RR, 1994; Dimitrijevic MR, 1991).

1.6. Quality of life

In families who have children with CP the "constant attendance" of the disease is present, through strict consistent long-term care of family and many other factors, such as services, support and physical aspects of the environment, which all can lead to deterioration of the patient's quality of life (Glinac A, Tahirović H, Delalić A, 2013).

The study *Family quality of life: adult school children with intellectual disabilities in Bosnia and Herzegovina,* provides initial data for family quality of life in Bosnia and Herzegovina (B&H). It also provides suggestions for improving quality of life for families that have one or more members with intellectual disability (ID). The principle measure used was the *Family Quality of Life Survey 2006 – main caregivers of people with intellectual or developmental disabilities.* The sample consisted of the main caregivers in 35 families that have adult children 18 years and over with ID who attended classes in a specially adapted programme in the Centre for children with ID, autism and cerebral palsy (n = 16), and in the Vocational Secondary School, B&H (n = 19). Regarding diagnosis as reported by main caregivers, 15 sons or daughters had ID of unknown aetiology, eight had cerebral palsy, four had Down syndrome, four had epilepsy and another three had epilepsy as a co-morbidity, two had autism and two had Prader-Willi syndrome. One had a dual diagnosis, ID and mental illness. When asked to rate overall family quality of life, three said 'excellent', eight said 'very good', 16 said 'good', seven said 'fair' and one said 'poor'. Furthermore, when asked to rate their overall satisfaction with their family quality of life, two said 'very satisfied', 19 said 'satisfied' and 13 said 'neither satisfied nor dissatisfied' (Švraka E, Loga S, Brown I, 2011).

Spasticity results in limited functional capacity and increased inactivity. The sequelae of this inactivity may include decubiti, cardiovascular problems, thrombophlebitis, respiratory infections, fixed contractures, osteoporosis, bladder and bowel problems, and social isolation. Ultimately, these consequences of inactivity may lead to a further decrease in strength and function (Francisco GE, Ivanhoe CB, 1997).

The patient's *quality of life* may be compromised as spasticity has negative impacts on mobility, hygiene, self care, sleeping patterns, self esteem, mood, and sexual function.

It is important to evaluate the advantages and disadvantages that the patient gains from their spasticity so that treatment strategies and goals can be identified. Disadvantages may include interference with activities of daily living, inhibition of good sleep, contractures, dislocations, skin breakdown, bowl and bladder dysfunction, impairment of respiratory function, pain with stretching, and the masking of the return of voluntary movement. However, patients may rely on a certain amount of spasticity to function and the advantages they may receive include maintaining muscle tone, supporting circulatory function, assisting in activities of daily living, and preventing the formation of deep vein thrombosis.

2. Case study of two children with CP

2.1. Patient A

Patient A was a 5-year-old African-American boy with a history of developmental delay and a diagnosis of cerebral palsy of the spastic-diplegic type. He first presented at 18 months with severe spasticity in both lower extremities. Prior to treatment with botulinum toxin, the patient walked on tip toes and had hip and knee flexion. There was some scissoring of his legs. On *examination,* exaggerated deep tendon reflexes were elicited, as were sustained clonus and bilateral Babinski sign. MRI of the brain showed findings that may be secondary to previous hypoxic injury, compatible with cerebral palsy.

Prior *treatments* included physical therapy, bilateral ankle-foot orthosis, serial casting, and oral baclofen. This boy with spastic-diplegic cerebral palsy walked on tip toes until treatment with botulinum toxin injections.

Following botulinum toxin injection, at the age of 18 months, the patient's gait has improved; he is flat-footed and presently wears bilateral ankle-foot orthosis. His hygiene and positioning have also improved and he returns every 6 months to 9 months for reinjection.

Results of the study *Use of Botulinum toxin type a in children with Spastic Cerebral Palsy,* support the idea that younger children may receive more benefit from multilevel botulinum toxin type A injections, intensivephysiotherapy and appropriate orthotic management compared to older children. Younger children might have been able to maintain the functional gains because the motor pattern of very young children provides greater scope for better development and recovery. A younger child has greater potential than older child for increasing the plasticity of the central nervous system. Botulinum toxin type A injections should always be used as an adjunctive treatment to physiotherapy, occupational therapy and orthotic management. In combination with post-injection physiotherapy this treatment could provide long-term benefits (Mikov A, Dimitrijević L, Sekulić S, Demeši-Drljan Č, Mikov I, Švraka E, Knežević-Pogančev M, 2011).

2.2. Patient B

Patient B was a 7-year-old African-American boy with a history of cerebral palsy of the spastic-diplegic type. On primary examination he presented with tightness of both hamstrings and heel cords with the right more involved than the left. The patient had good toe standing, especially on the right side and good sitting balance with a kyphotic sacral-type sitting due to the tight hamstring. He uses a walker to ambulate and walks on tip toes. The EEG was abnormal, indicating the presence of epileptiform activity from the left central parietal head region and diffuse background disorganization, which indicates underlying neuronal dys-function.

Treatments before *intrathecal baclofen pump* implantation included bilateral ankle-foot orthoses, tendon releases, alcohol block, and botulinum toxin injections. Before treatment with intrathecal baclofen the patient was dependent on a care giver and used a walker to ambulate. With the intrathecal baclofen pump the patient has gained function, does not use a walker to ambulate, and performs activities of daily living independently. With the intrathecal baclofen pump the patient has gained function, does not use a walker to ambulate, and successfully performs activities of daily living.

3. Spasticity management

Traditional treatments for spasticity include physical therapy, occupational therapy and rehabilitation treatments which complete a number of crucial tasks and specific goals in the treatment of patient with CP, this will promote their sensorimotor development, improve their overall posture and position and enhance their control of movements in all their daily activities: a lot of physical therapy approaches were based on different theoretical principles though the main target is the management of abnormal muscle tone and improving the range of motion through neurodevelopment therapy, conductive education, constraint induced movement therapy, etc.

There are other modalities including electrical stimulation and cold temperature (Chiara T, Carlos J Jr, Martin D, Miller R, Nadeau S, 1998; Pease WS, 1998; Scheker LR, Chesher SP, Ramirez S, 1999; Kinnman J, Andersson T, Andersson G, 2000).

Occupational therapy is a client-centered health profession concerned with promoting health and well being through occupation. Possible problems in children with cerebral palsy are motor, sensor, cognitive, intrapersonal, interpersonal, problems of self care, productivity and leisure.

Occupational therapy, in which the patient is stretched anywhere from once daily to several times per day, but this has only a limited effect on the patient's spasticity. Rehabilitation treatment options include casting, orthotics or splints, strengthening, electrical stimulation, practice of functional tasks, sensory integration; muscle stretching, and targeted muscle training (Fetters L, Kluzik J, 1996).

Within the scope of pediatric neurorehabilitation, distinct diseases can produce specific complications. These complications; however, can also occur in association with many disorders. For example, spasticity from injury to the upper motor neuron unit can develop in many neurologic disorders in children. Several of these complications, such as autonomic dysreflexia, deep vein thrombosis, and heterotropic ossification, can be severe and potentially life-threatening (Umphred D, Dewane J, Hall-Thompson M, et al, 2001; Dobkins, BH, 2003; DeLisa JA, Gans BM, Walsh NE, Bockneck WL, Frontera WR, 2004).

http://dx.doi.org/10.5772/58311

3.1. Oral medications

Oral medications can be used to decrease spasticity; however, many have unwanted side effects such as drowsiness, sedation, confusion, and fatigue. Benzodiazepines, such as diazepam, are rarely used because of their strong sedating effects. They result in enhanced presynaptic inhibition, but because they are presumed to enhance the postsynaptic effects of GABA, they can only work if the GABA-mediated process functions. Benzodiazepines have a long half-life and an active metabolite. Benzodiazepine therapy is indicated in spinal cord injury and multiple sclerosis with possible application in traumatic brain injury, cerebral palsy, and cerebrovascular accident. Clinical effects include sedation and reduced anxiety, decreased resistance to passive range of motion, decreased hyperreflexia, and reduction in painful spasms. Side effects of all benzodiazepines include sedation, weakness, hypotension, gastrointestinal symptoms, memory impairment, incoordination, confusion, depression, and ataxia. Also, benzodiazepines are controlled substances with the potential for dependency. Diazepam is the most widely used benzodiazepine for spasticity management. The recommended initial dose is 2 mg 3 times daily with a maximum dose of 60 mg daily (20 mg 3 times daily). If nocturnal spasticity is the presenting problem the patient should be started with a single dose at night.

Like benzodiazepines, baclofen works centrally. Baclofen binds with GABA-B receptors on brain and spinal membranes, restricting calcium influx into presynaptic nerve terminals, thereby reducing spasticity [4]. The use of baclofen is indicated when spasticity is of spinal origin. The clinical effects include decreased resistance to passive range of motion, decrease in hyperreflexia, and reduction in painful spasms and clonus.

Unlike benzodiazepines and baclofen, dantrolene sodium works peripherally at the level of the muscle fiber. It has no effect on neuromuscular transmission, but works by acting directly on the skeletal muscle, hindering the release of calcium from the sarcoplasmic reticulum, thereby preventing the excitation-contraction coupling mechanism. This affects both intrafusal and extrafusal fibers by decreasing the force of muscle contraction. However, this mechanism is not selective for muscles with increased tone, and the resulting generalized muscle weakness may weaken respiratory muscles. The use of dantrolene sodium is indicated in treating spasticity secondary to cerebrovascular accident, cerebral palsy, and has possible applications for traumatic brain injury, spinal cord injury, and multiple sclerosis. Clinical effects of dantrolene sodium include decreased resistance to passive range of motion, decrease in hyperreflexia and tone, and reduction in spasms and clonus.

Another group of oral medications used in spasticity management includes clonidine and tizanidine, which are alpha 2 noradrenergic receptor agonists that release excitatory neurotransmitters and inhibit supraspinal facilitatory pathways (Young RR, 1994; Francisco GE, Ivanhoe CB, 1997).

Tizanidine is a new oral antispasticity agent that is selective in decreasing tone and spasm frequency in only spastic muscles, eliminating the unwanted side effect of generalized muscle weakness. Tizanidine is reported to have reduced symptoms of spasticity in patients with multiple sclerosis or spinal cord injury and is well tolerated in most patients. It is an imidazoline derivative similar to clonidine but without the cardiovascular effects when appropriately titrated. Tizanidine results in a direct reduction of excitatory amino acid release from spinal interneurons and inhibits facilitatory caerulospinal pathways. Its peak effect occurs 1 to 2 hours following administration and its half-life is 2.5 hours. The clinical effects of tizanidine include reduced muscle tone, spasm frequency, and hyperreflexia. Animal studies with tizanidine demonstrate antinociceptive activity under specific conditions with increased dose titration (McCarthy RJ, Kroin JS, Lubenow TR, Penn RD, Ivankovich AD, 1990).

As with other antispasticity medications, the potential side effects of tizanidine are dose related and may be mitigated by dosage titration. The potential side effects include drowsiness, dry mouth, and dizziness. Literature suggests that tizanidine may be better tolerated than other antispasticity agents as measured by the global tolerance rating scale (Lataste X, Emre M, Davis C, Groves L, 1994).

In placebo-controlled studies, tizanidine has been shown to be effective in multiple sclerosis and spinal cord injury. It is also useful for spasticity of spinal pathology when weakness is of concern. Tizanidine may also prove effective in managing spasticity of cerebral origin (Medici M, Pebet M, Ciblis D, 1989).

Secondary oral and systemic agents include tiagabine, cyproheptadine, clonidine, lamotrigine, gabapentin and carbidopa-levodopa (Gracies JM, Nance P, Elovic E, et al, 1997).

Multiple medications have been recommended, of which the most recent addition is gabapentin (Zidek K, Srinivasan R, 2003).

The use of antihypertensive pharmacologic agents in treating spasticity is unclear because randomized trials have not been performed. Nifedipine has been used in a bit-and-swallow technique; more recently, captopril also has been found to be of benefit (Esmail Z, Shalansky KF, Sunderji R, Anton H, Chambers K, Fish W, 2002).

3.2. Chemo-denervation

Chemo-denervation such as using botulinum toxin type A, has proved easier, more effective, and less painful for patients. First clinically introduced in the United States in the early 1980s, botulinum toxin is a potent neurotoxin derived from the anaerobic bacteria *Clostridium botulinum*, but when used in treatment, no serious systemic toxin effects have been reported (Francisco GE, Ivanhoe CB, 1997).

The medication is more costly than alcohol or phenol but the cost is offset by less physician time and the lack of anesthesia. The formation of antibodies has been a concern, but this can be prevented by allowing 2 months to 3 months between injections. Botulinum toxin works by acting in the neuromuscular junction, preventing the release of acetylcholine, which results in functional denervation. It can be given without EMG and anesthesia, does not cause dysesthesias, and is no more painful than an injection of saline solution. Effects are local and last 3 months to 4 months, or longer. It is contraindicated during pregnancy, lactation, in individuals with neuromuscular disorders (such as myasthenia gravis), in patients taking aminoglycosides, or in those who have a known allergy to the drug. Adverse effects are not common and are usually associated with the site of injection, such as bleeding, bruising, and soreness or redness at the injection site, or diffusion to nearby muscle groups. In patients that do not respond to botulinum toxin, possible reasons should be considered before labeling the patient as unresponsive. Reasons could be related to injection technique, improper toxin storage, or the patient's individual characteristics. Overall, botulinum toxin has proven clinically to be effective, safe, and less painful than other invasive therapies (Francisco GE, Ivanhoe CB, 1997; Keam SJ, Muir VJ, Deeks ED, 2011).

Botulinum toxin is available in serotypes A and B, which have different unit potencies, side-effect profiles, and dilution schedules. Both have been used in children with cerebral palsy, although serotype A has been used more extensively. Dosing guidelines have been suggested for botulinum toxin A for adult and pediatric patients. Adult recommendations are available for botulinum toxin B, but studies are ongoing for pediatric patients (Tilton AH, 2003; Schwerin A, Berweck S, Fietzek UM, Heinen F, 2004; Sanger TD, Kukke SN, Sherman-Levine S, 2007).

Some results suggest that botulinum toxin type A can be effective in reducing muscle tone over a longer period, but not in preventing development of contractures in spastic muscles. Mechanical and functional alterations can arise from the muscle tissue itself even though the nervous system is the site of the primary lesion. The gross mechanical changes occur in skeletal muscle secondary to spasticity and during development of contracture. Muscle stiffness can change for a variety of structural reasons, only one of which is altered fiber length. There is currently no evidence in the literature that muscle fiber length is shortened in contracture or in spastic skeletal muscle. Contracture formation results from inappropriate architectural adaptation of extremity muscles in response to upper motor neuron lesion (Mikov A, Dimitrijević L, Sekulić S, Demeši-Drljan Č, Mikov I, Švraka E, Knežević-Pogančev M, 2011).

Several studies have reported the successful use of botulinum toxin A for the treatment of drooling in children with cerebral palsy, using injection into the submandibular or parotid glands alone or in combination with other agents. In some studies, the beneficial effects have lasted for up to 4 months without serious side effects or disturbances of oral function (Jongerius PH, van den Hoogen F, van Limbeek J, Gabreels FJ, van Hulst K, Rotteveel JJ, 2004; Bothwell JE, Clarke K, Dooley JM, et al, 2002; Suskind DL, Tilton A, 2002).

Other treatments include *chemical neurolysis*, in which the nerve conduction is impaired through the use of chemical agents and therapeutic nerve block using phenol or alcohol. The goals of these treatments are to prevent muscle contractures and improve the patient's function. A common side effect is that after the nerve is injected, alcohol levels measure above the legal limit in children. Other side effects include damage to sensory and motor nerves, pain at injection site, scarring, and dysesthesias. To ensure the correct site, injection must be made using an electrical stimulator (Albright AL, 1996; Francisco GE, Ivanhoe CB, 1997).

3.3. Neurosurgical approaches

Another treatment used alleviate spasticity in children with cerebral palsy is *rhizotomy*. Studies have shown that performing selective dorsal rhizotomy at a young age can reduce the need for orthopedic surgery (Chicoine MR, Park TS, Kaufman BA, 1997).

Goals of rhizotomy are decreased tone, increased mobility, and the facilitation of care for the patient, however, the reduction in spasticity cannot be predicted and sometimes results in excessive hypotonia (Im D, McDonald CM, 1997).

The procedure is very meticulous, requiring general anesthesia and a neurophysiologist who must be present to identify which nerve is to be severed.

Other *neurosurgical approaches* include peripheral neurectomy, myelotomy, and dorsal column electrical stimulation.

It has been established that oral baclofen does not cross the blood-brain barrier effectively and that higher doses of the medication result in serious side effects (Francisco GE, Ivanhoe CB, 1997).

Intrathecal baclofen results in a greater decrease in spasticity by allowing higher concentrations of baclofen in the cerebrospinal fluid at about 1% the daily oral dosage (Im D, McDonald CM, 1997).

To be considered for intrathecal baclofen pump placement, the patient must have severe lower limb spasticity that does not respond to other less-invasive treatments. The patient must first be given a trial of 50 μg baclofen through a lumbar puncture or spinal catheter. If unresponsive, 75 μg can be tried after 24 hours and a third trial of 100 μg can be tried 24 hours after that, after which if the patient is still unresponsive he or she must be excluded from the treatment (Francisco GE, Ivanhoe CB, 1997).

Implantation lasts 1 to 2 hours and the pump is easy to refill subcutaneously. It is programmed by a computer-controlled radiotelemetry programmer that is linked to the pump's internal computer and that selects the rate and pattern of baclofen administration. Complications to intrathecal baclofen include hypersensitivity to baclofen, intolerance to the side effects of baclofen including drug tolerance, cerebrospinal fluid leakage, pump pocket seroma, hematoma, infection, and soft tissue erosion. The objective of intrathecal baclofen is to individualize the patient's dose and infusion so that the lowest dose that yields the greatest response can be achieved (Young RR, 1994; Francisco GE, Ivanhoe CB, 1997).

In comparison, intrathecal baclofen has less complications and side effects than other treatments and more generalized results in both cerebral and spinal spasticity, making intrathecal baclofen the most effective current tool for the treatment of spasticity in non-ambulant individuals. A recent systematic review showed that there was no evidence to support the clinical use of intrathecal baclofen in ambulant individuals with hypertonicity without further rigorous longitudinal studies (Pin TW, McCartney L, Lewis J, Waugh MC, 2011).

As a precaution, families are prescribed diazepam or diazepam rectal as well as oral baclofen to have at home. If there is evidence of withdrawal, one of these medications is administered, and the patient is instructed to go immediately to the emergency department. Although aggressive use of benzodiazepines and oral baclofen may be helpful, recognition and return to appropriate intrathecal baclofen dosage is essential for rapid recovery (Alden TD, Lytle RA, Park TS, Notzel MJ, Ojemann JG, 2002).

http://dx.doi.org/10.5772/58311

3.4. Orthopedic procedures

Orthopedic procedures are the most frequently performed operations for spasticity. The targets of these operations are muscles, tendons, or bones. Muscles may be denervated and tendons and muscles may be released, lengthened, or transferred. The goals of surgery may include reducing spasticity, increasing range of motion, improving access for hygiene, improving the ability to tolerate braces, or reducing pain. Orthopedic problems that may result from a spastic limb include cubital or carpal tunnel syndrome, spontaneous fracture, dislocation of the hip or knee, and heterotopic ossification.

The most common orthopedic procedure for the treatment of spasticity is a *contracture release*. In this procedure, the tendon of a muscle that has a contracture is partially or completely cut. The joint is then positioned at a more normal angle, and a cast is applied. Regrowth of the tendon to a new length occurs over several weeks. Serial casting may be used to gradually extend the joint. Following cast removal, physical therapy is used to strengthen the muscles and improve range of motion.

Spastic muscles in the shoulder, elbow, forearm, hands, and legs may all be treated with tendon or muscle lengthening. Spasticity in the shoulder muscles may cause abduction or adduction and internal rotation of the shoulder. Abduction results in difficulties with balance, which then affects walking and transferring, and adduction causes problems when reaching for an object or with hygiene and personal care. An operation known as a slide procedure may be used to lengthen the supraspinatus muscle in an abducted spastic shoulder. With adducted shoulders, the surgeon can perform a release of all 4 muscles that typically cause this deformity.

In an operation known as a tendon transfer, the orthopedic surgeon moves a tendon from the spot at which it attaches to the spastic muscle. With the tendon transferred to a different site, the muscle can no longer pull the joint into a deformed position. In some situations, the transfer allows improved function. In others, the joint retains passive but not active function. Ankle-balancing procedures are among the most effective interventions.

The goal of surgical-orthopedic treatment which is basically symptomatic improve or facilitate the movement to solve the functional or fixed contractures preventing further rehabilitation, to solve the deformation that reduces or prevents movement, sitting, causing pain as in the cases of hip luxation, or threaten respiration as in cases of severe scoliosis. Subluxation and dislocations of the hip in children with CP are most common in children and adolescents who do not walk. We must bear in mind the saying that every child and adolescent with CP has a hip disorder until proven otherwise. The occurrence of dislocation of the hips makes furniture, hygiene and often causes pain. Requires regular radiological controls hips once or twice a year in the course of growth, to hip dislocation discovered at an early stage. Subluxation and luxation of the hips treated surgically. The decision about surgery should bring those involved in the treatment of patients, carefully weighing hopper performs coarse benefits and harms of surgery. Surgery is necessary to balance the muscle forces around the hip and normalize abnormal anatomic relationships (Đapić T, Šmigovec I, Kovač-Đapić N, Polovina S, 2012).

Osteotomy and arthrodesis involves operations on the bones and are usually accompanied by operations to lengthen or split tendons to allow for fuller correction of the joint deformity.

Osteotomy can be used to correct a deformity that cannot be fixed with other procedures. In an osteotomy, a small wedge is removed from a bone to allow it to be repositioned or reshaped. A cast is applied while the bone heals in a more natural position. Osteotomy procedures are most commonly used to correct hip displacements and foot deformities. Arthrodesis is a fusing together of bones that normally move independently. This fusion limits the ability of a spastic muscle to pull the joint into an abnormal position. Arthrodesis procedures are performed most often on the bones in the ankle and foot. In triple arthrodesis, the 3 joints of the foot are exposed, the cartilage is removed, and screws are inserted into the bones, fixing the joints into position. With a short walking cast in place for 6 weeks or until the bones have fully healed, the patient may bear weight immediately after the operation (http://wemove.org/spa/spa_oss.html 2007).

The risks of developing a structural spinal deformity ranges from 24% to 36% for scoliosis and is 50% for lordosis for an average of 4 to 11 years after selective dorsal rhizotomy (Turi M, Kalen V, 2000; Johnson M, Goldstein L, Thomas SS, Piatt J, Aiona M, Sussman M, 2004).

Other principals include single event, multilevel surgery; surgery is delayed as long as possible (more than 6 years). Spasticity management is used as an adjunct to surgical intervention (Boyd R, Graham J, Nattras G, Graham K, 1999).

3.5. New treatments in spasticity management

Acupuncture and homeopathic approaches (Guo Z, Zhou M, Chen X, Wang R, 1997), herbs and hyperbaric oxygen [41-45], constraint induced training [46, 47], the Adeli suit [48], conductive education, craniosacral, and manipulation and patterning.

Context therapy is a new intervention approach that focuses on changing the task and the environment rather than children's impairments. It can be a viable treatment to achieve parent-identified functional goals for children with cerebral palsy (Darrah J, Law MC, Pollock N, et al, 2011).

A summary of management in spasticity is provided in Table 2.

Therapeutic intervention	Mechanisms	Major points
Nonpharmacologic treatments		
Physical therapy Occupational therapy	Neurodevelopmental therapy (NDT) Constraint-induced movement therapy (CIMT) Neurophysiologically based therapy/ Vojta Manual medicine Training of the muscular strength Treadmill therapy Conductive education Hand-arm bimanual intensive therapy (HABIT) Sensory integration therapy/Ayres	Different techniques are tailored depending on the individual goals Training of both manual and fine motor skills of the paretic side through activity limitation of the healthy side Reflex locomotion to encourage motor development through repetitive triggering reflex creeping and reflex turning Encouraging motor learning through active and passive mobilization, soft tissue release and manipulations Encouraging locomotion and posture through specific training of certain muscle groups

http://dx.doi.org/10.5772/58311

Therapeutic intervention	Mechanisms	Major points
		Gait training through walking on treadmill, with body weight support Systematic, intensive training of small learning steps in the motor, linguistic and cognitive domains Motivation for bimanual activity of the paretic and nonparetic side with specified tasks Everyday tasks training for coordination and sensory information enhancement
	Splints, strengthening, electrical stimulation, practice of functional tasks, muscle stretching, and targeted muscle training	Mainstays and cornerstones in spasticity management; complications, such as autonomic dysreflexia, deep vein thrombosis, and heterotropic ossification, can be severe and potentially life-threatening
Casting and orthosis	Extend joint range diminished by hypertonicity; reduce an abnormal pattern by positioning	Temporary effect
Selective posterior rhizotomy	Balancing spinal cord-mediated facilitatory and inhibitory control	Permanent effect; sometimes results in excessive hypotonia
Orthopedic surgery	Corrects deformity induced by muscle overactivity involving muscles, tendons, or bones	In moderate to severe spasticity, permanent effect
Pharmacological treatments, oral medications		
Benzodiazepine	Increases the affinity of GABA for GABA-A receptors; inhibitory effect at both the spinal cord and supraspinal levels	Short-term treatment; strong sedating effects
Dantrolene sodium	Inhibits release of calcium from sarcoplasmic reticulum in muscle; works peripherally at the muscle fibers	Serious side effects; hepatotoxicity in 1% patients, respiratory muscle weakness
Baclofen	GABA agonist; binds at the GABA-B receptor; restricts calcium influx into presynaptic nerve terminals in the spinal cord	Rapidly absorbed after oral administration; levels in the CSF are low because of low lipid solubility
Tizanidine	Centrally acting alpha-2 noradrenergic agonist; inhibits release of excitatory neurotransmitters in the spinal cord and supraspinally	Drowsiness, dry mouth, and dizziness; monitor liver function
Pharmacological treatments, chemodenervation		
Alcohol/Phenol block	Nonselective proteolytic agents; selective denervation when injecting into motor nerves or muscles	Damage to sensory and motor nerves, painful dysesthesias
Botulinum toxin injection	High affinity and specificity to the presynaptic membranes of cholinergic motor neurons	Recommended as effective treatment; no sensory disturbance
Pharmacological treatments, other		
Intrathecal baclofen pump	Using a programmable implanted pump, baclofen can be delivered intrathecally	Severe, generalized spasticity; less complications and side effects

Table 2. Management in Spasticity

Pregnancy

The patient with spasticity may expect to have a difficult pregnancy and delivery as well as difficulty managing and caring for an infant.

Anesthesia

Not applicable.

4. Conclusion

To prevent cerebral palsy in infants and, thus, the resulting spasticity, it is important that mothers receive prenatal care during pregnancy, that measures are taken to avoid premature labor, and that special consideration is given to pregnancies involving multiple gestations.

Early detection and treatment of neurodegenerative diseases may prevent the development of spasticity as well as detect the underlying diseases that could result in brain injury. If children have conditions that make them susceptible to brain or spinal cord injury or both, safety measures should be taken (i.e., helmets for patients who have frequent seizures).

The goals of and benefits to the patient are important when considering the path of treatment. In some cases, function will not return, but treatment can result in pain reduction and allow easier management of patient care. Common goals are to decrease pain, prevent or decrease contractures, improve ambulation, facilitate activities of daily living, facilitate rehabilitation participation, save caregiver's time, improve the ease of care, and increase safety. Appropriate management choices are based on therapeutic objectives. Physical and occupational therapists can play a key role in identifying these objectives. Treatments with the fewest side effects are usually given priority. Both the patient's and the caregiver's goals must be considered.

Rehabilitation multidisciplinary team could be good connection with Management. There are different approaches in rehabilitation treatment of persons with cerebral palsy, especially children and adolescents. The treatment of children with spastic cerebral palsy is a combination of intensive sensorimotor stimuli, physical therapy, occupational therapy, Vojta therapy, orthopedic procedures and/or botulinum toxin applications. It is child/family-centered management.

The ICF can guide management but does not give sufficient detail of the "hows and whys of the child activities to enable a specific treatment plan.

5. Summary

Spasticity may result from either diffuse or localized pathology of the cerebral cortex, brain stem, or spinal cord. Possible causes of such injuries include traumatic brain injury, stroke, multiple sclerosis, spinal cord trauma, or disease and anoxic insults. The neurologic localiza-

http://dx.doi.org/10.5772/58311

tion of the lesion causing spasticity may result in different clinical manifestations. Thus, it is important to consider whether the spasticity results from cerebral pathology, whether it is diffuse or localized, or whether it is a result of spinal cord injury.

Cerebral palsy (CP) is a group of permanent disorders of the development of movement and posture, causing activity limitation, that are attributed to non-progressive disturbances that occurred in the developing foetal or infant brain. The motor disorders of cerebral palsy are often accompanied by disturbances of sensation, perception, cognition, communication, and behavior, by epilepsy, and by secondary musculoskeletal problems

Spasticity can be associated with cocontraction, clonus and hyperreflexia. Children with spastic cerebral palsy generally have a typical pattern of muscle weakness, impairment in selective motor control and sensory impairment.

It is important to evaluate the advantages and disadvantages that the patient gains from their spasticity so that treatment strategies and goals can be identified. Disadvantages may include interference with activities of daily living, inhibition of good sleep, contractures, dislocations, skin breakdown, bowl and bladder dysfunction, impairment of respiratory function, pain with stretching, and the masking of the return of voluntary movement.

There are many different types of spasticity. Because of this, more than one mechanism may be responsible for the disturbance in muscle tone and the mechanisms may vary between patients. The neuropathophysiologic processes involved in spasticity are complex and not fully understood, but there is a widely accepted hypothesis that spasticity depends on hyperexcitability of spinal alpha motor neurons, which is due to the interruption of descending modulatory influences carried by the corticospinal, vestibulospinal, and reticulospinal tracts and other possible tracts.

Traditional treatments for spasticity include physical therapy, occupational therapy and rehabilitation treatments which complete a number of crucial tasks and specific goals in the treatment of patient with CP, this will promote their sensorimotor development, improve their overall posture and position and enhance their control of movements in all their daily activities: a lot of physical therapy approaches were based on different theoretical principles though the main target is the management of abnormal muscle tone and improving the range of motion through neurodevelopment therapy, conductive education, constraint induced movement therapy, etc.

Oral medications can be used to decrease spasticity; however, many have unwanted side effects such as drowsiness, sedation, confusion, and fatigue. Benzodiazepines, such as diazepam, are rarely used because of their strong sedating effects. They result in enhanced presynaptic inhibition, but because they are presumed to enhance the postsynaptic effects of GABA, they can only work if the GABA-mediated process functions.

Chemo-denervation such as using botulinum toxin type A, has proved easier, more effective, and less painful for patients. First clinically introduced in the United States in the early 1980s, botulinum toxin is a potent neurotoxin derived from the anaerobic bacteria *Clostridium botulinum*, but when used in treatment, no serious systemic toxin effects have been reported.

Another treatment used alleviate spasticity in children with cerebral palsy is *rhizotomy.* Studies have shown that performing selective dorsal rhizotomy at a young age can reduce the need for orthopedic surgery. Goals of rhizotomy are decreased tone, increased mobility, and the facilitation of care for the patient, however, the reduction in spasticity cannot be predicted and sometimes results in excessive hypotonia.

Other *neurosurgical approaches* include peripheral neurectomy, myelotomy, and dorsal column electrical stimulation.

Orthopedic procedures are the most frequently performed operations for spasticity. The targets of these operations are muscles, tendons, or bones. Muscles may be denervated and tendons and muscles may be released, lengthened, or transferred. The goals of surgery may include reducing spasticity, increasing range of motion, improving access for hygiene, improving the ability to tolerate braces, or reducing pain. Orthopedic problems that may result from a spastic limb include cubital or carpal tunnel syndrome, spontaneous fracture, dislocation of the hip or knee, and heterotopic ossification.

Abbreviations

EEG: electroencephalogram

EMG: electromyography

MRI: Magnetic Resonance Imaging

ICD codes

ICD-9:

Abnormal involuntary movements: 781.0

ICD-10:

Other and unspecified abnormal involuntary movements: R25.8

Associated disorders

Adrenoleukodystrophy

Anoxia

Cerebral palsy

Multiple sclerosis

Neurodegenerative disease

Spinal cord injury

Stroke

Traumatic brain injury

http://dx.doi.org/10.5772/58311

Major keyword descriptors

bent knees

gait disturbances

muscle imbalance

poor hygiene

poor positioning

scissors-type gait

stretch reflexes

tendon jerks

Minor keyword descriptors

bladder problems

bowel problems

cardiovascular problems

fixed contractures

osteoporosis

pain

respiratory infections

thrombophlebitis

Glossary

Adrenoleukodystrophy: demyelination of nerve cells in the brain and progressive dysfunction of the adrenal gland.

Anoxia: diminished supply of oxygen to an organ's tissues.

Cerebral palsy: Nonprogressive disorder or movement and posture that can occur anywhere from 0 to 5 years of age, caused by a brain lesion.

Clasp-knife phenomenon: characterized by a free interval of movement of the limb, followed by a sudden stop and increase in muscle resistance which melts away as the passive stretching of the limb continues.

Multiple Sclerosis: plaques form from inflammation of the white matter of the central nervous system, causing destruction of the myelin sheath, resulting in diminished or lost function.

Permuted topics, synonyms, variants

Spasticity

Related topics

Acupuncture

Autosomal dominant inherited ataxias

Baclofen

Cerebral palsy

Childhood ataxia with central nervous system hypomyelination

Childhood movement disorders

Neurodegeneration with brain iron accumulation

Hyperargininemia

Hyperbaric oxygenation for the treatment of stroke

Machado-Joseph disease

Multiple sclerosis

Nonautosomal dominant inherited ataxias

Sjogren-Larsson syndrome

Differential diagnosis

Rigidity

Author details

Yasser Awaad[1*], Tamer Rizk[2] and Emira Švraka[3]

1 Wayne State University, Oakwood Healthcare System, and the University of Michigan, USA

2 Al-Takhassusi Hospital, Habib Medical Group, Riyadh, Saudi Arabia

3 University of Sarajevo, Faculty of Health Studies, Bosnia and Herzegovina

References

[1] Abbott, R. (1996) Sensory rhizotomy for the treatment of childhood spasticity. J Child Neurol 11 (suppl1): S36-42.

[2] Abbott, R. (1991) Childhood spasticity assessment. In: Sindou M, Abbott R, and Keravel Y, editors. Neurosurgery for spasticity: a multidisciplinary approach. Wien; New York: Springer-rlag:51-6.

[3] Adams, RD; Victor, M. & Ropper, AH. (1997) Motor paralysis: cardinal manifestations of neurologic disease. In: Adams RD, Victor M, Ropper AH, editors. Principles of neurology. Vol 6. New York: McGraw Hill:54-6.

[4] Albright, AL. (1996) Spasticity and movement disorders in cerebral palsy. J Child Neurol;11(suppl 1):S1-4.

[5] Akman, MN; Loubser, PG; Fife, CE, et al. (1994) Hyperbaric oxygen therapy: implications for spinal cord injury patients with intrathecal baclofen infusion pumps. Paraplegia; 32:281-4.

[6] Alden, TD; Lytle, RA; Park, TS; Notzel, MJ. & Ojemann, JG. (2002) Intrathecal baclofen withdrawal: a case report & review of the literature. Child Nerv Syst; 18(9-10): 522-5.

[7] Babajić, M; Švraka, E. & Avdić, D. (2013) Frequency of joined disabilities of children with cerebral palsy in Tuzla canton. Journal of Health Sciences;3(3): 222-226

[8] Blackmer, J. (2003) Rehabilitation medicine: I. Autonomic dysreflexia. CMAJ; 169:931-5.

[9] Bothwell, JE; Clarke, K; Dooley, JM, et al. (2002) Botulinum toxin A as a treatment for excessive drooling in children. Pediatr Neurol;27(1):18-22.

[10] Bottcher, L. (2010). Children with spastic cerebral palsy, their cognitive functioning, and social participation: a review. Child Neuropsychology, 16: 209-228.

[11] Boyd, R; Graham, J; Nattras, G. & Graham, K. (1999) Medium-term response characterization and risk factor Analysis of botulinum toxin type A in the management of spasticity in children with cerebral palsy. Eur J Neur (Suppl 4):S37-45.

[12] Chiara, T; Carlos, J Jr; Martin, D; Miller, R. & Nadeau, S. (1998) Cold effect on oxygen uptake, perceived exertion, and spasticity on patients with multiple sclerosis. Arch Phys Med Rehabil; 79:523-8.

[13] Chicoin, MR; Park, TS. & Kaufman, BA. (1997) Selective dorsal rhizotomy and rates of orthopedic surgery in children with spastic cerebral palsy. J Neurosurg ;86:34-9.

[14] Chung, CY; Chen, CL. & Wong, AM. (2011) Pharmacotherapy of spasticity in children with cerebral palsy. J Formos Med Assoc;110(4):215-22.

[15] Collet, JP; Vanasse, M; Marois, P, et al. (2001) Hyperbaric oxygen for children with cerebral palsy: a multicenter, placebo controlled, randomized clinical trial. Lancet; 357:582-6.

[16] Crocker, MD; MacKay-Lyons, M. & McDonnell, E. (1997) Forced use of the upper extremity in cerebral palsy: a single-case design. Am J Occup Ther; 51:824-33.

[17] Darrah, J; Law, MC; Pollock, N. et al. (2011) Context therapy: a new intervention approach for children with cerebral palsy. Dev Med Child Neurol; 53(7):615-20.

[18] Dimitrijevic. MR. (1991) Clinical assessment of spasticity. In: Neurosurgery for spasticity: a multidisciplinary approach. New York: Springer-Verlag: 33-7.

[19] DeLisa, JA; Gans, BM; Walsh, NE; Bockneck, WL. & Frontera, WR. (2004) Physical medicine and rehabilitation: principles and practice. Philadelphia: Lippincott Williams & Wilkins.

[20] Dobkins, BH. (2003) The clinical science of neurologic rehabilitation. London: Oxford University Press.

[21] Drew, LB. & Drew, WE. (2004) The contrecoup-coup phenomenon: A new understanding of the mechanism if closed head injury. Neurocrit Care; 1(3):385-90.

[22] Đapić, T; Šmigovec, I; Kovač-Đapić, N. & Polovina, S. (2012) Surgery of cerebral palsy with special reference to treatment spastic luxation of the hip. Paediatrics Today; 8 (Suppl 2) : 20-30 ISSN 1840-0914 (Print) ISSN 1840-2968 (Online)

[23] Esmail, Z; Shalansky, KF; Sunderji, R; Anton, H; Chambers, K. & Fish, W. (2002) Evaluation of captopril for the management of hypertension in autonomic dysreflexia: a pilot study. Arch Phys Med Rehab; 83(5):604-8.

[24] Fetters, L. & Kluzik, J. (1996) The effects of neurodevelopmental treatment vs practice on reaching of children with spastic cerebral palsy. Phys Ther; 76:346-58.

[25] Filloux, FM. (1996) Neuropathophysiology of movement disorders in cerebral palsy. J Child Neurol; (suppl 1):S5-12.

[26] Francisco, GE. & Ivanhoe, CB. (1997) Pharmacologic management of spasticity in adults with brain Injury. In: Kraft GH, Horn LJ, editors. Physical medicine and rehabilitation 8:4. Philadelphia: WB Saunders Company: 707-31.

[27] Frerebeau, PH. et al. (1991) Clinical feature of spasticity. In: Neurosurgery for spasticity: a multidisciplinary approach. New York: Springer-Verlag: 29-32.

[28] Glinac, A; Tahirović, H. & Delalić, A. (2013) Family socioeconomic status and health-related quality of life in children with cerebral palsy: assessing differences between clinical and healthy samples. Paediatrics Today; 9(2):183-191. DOI 10.5457/p2005-114.74

[29] Gracies, JM; Nance, P; Elovic, E. et al. (1997) Traditional pharmacological treatments for spasticity part II: general and regional treatments. Muscle Nerve; 20(suppl 6):S92-120.

[30] Guo, Z; Zhou, M; Chen, X. & Wang, R. (1997) Acupuncture methods for hemiplegic spasm. J Tradit Chin Med; 17(4):284-8.

[31] Im, D. & McDonald, CM. (1997) New approaches to managing spasticity in children with cerebral palsy. West J Med; 166(4):271.

[32] Johnson, M; Goldstein, L; Thomas, SS; Piatt, J; Aiona, M. & Sussman, M. (2004) Spinal deformity after selective dorsal rhizotomy in ambulatory patients with cerebral palsy. J Pediatr Orthop; 24(5):529-36.

[33] Jongerius, PH; van den Hoogen, F; van Limbeek, J; Gabreels, FJ; van Hulst, K. & Rotteveel, JJ. (2004) Effect of botulinum toxin in the treatment of drooling: a controlled clinical trial. Pediatrics; 114(3):620-7.

[34] Keam, SJ; Muir, VJ. & Deeks, ED. (2001) Botulinum toxin A (Dysport®): in dystonias and focal spasticity. Drugs; 71(8):1043-58.

[35] Koman, LA; Mooney, JF. & Smith, BP. (1996) Neuromuscular blockage in the management of cerebral palsy. J Child Neurol ; 11(suppl1):S23-8.

[36] Kinnman, J; Andersson, T. & Andersson, G. (2000) Effect of cooling suit treatment in patients with multiple sclerosis evaluated by evoked potentials. Scand J Rehabil Med; 32:16-9.

[37] Kluger, J. (2001) The root of tranquility: is extract of kava a natural substitute for valium–or just alternative medicine's newest herb du jour? |{website:Time Website}{webURL:http://www.time.com/time/}|

[38] Krassioukov, AV; Furlan, JC. & Fehlings, MG. (2003) Autonomic dysreflexia in acute spinal cord injury: an under-recognized clinical entity. J Neurotrauma; 20(8):707-16.

[39] Lataste, X; Emre, M; Davis, C. & Groves, L. (1994) Comparative profile of tizanidine in the management of spasticity. Neurology; 44(suppl 9):53-9.

[40] McCarthy, RJ; Kroin, JS; Lubenow, TR; Penn, RD. & Ivankovich, AD. (1990) Effect of intrathecal tizanidine on antinociception and blood pressure in the rat. Pain; 40(3): 333-8.

[41] Medici, M; Pebet, M. & Ciblis, D. (1989) A double-blind, long-term study of tizanidine ('Sirdalud') in spasticity due to cerebrovascular lesions. Curr Med Res Opin; 11(6):398-407.

[42] Mikov, A; Dimitrijević, L; Sekulić, S; Demeši-Drljan, Č; Mikov, I; Švraka, E. & Knežević-Pogančev M. (2011) Use of Botulinum toxin type a in children with Spastic Cerebral Palsy. HealthMED, Journal of Society for development of teaching and business processes in new net environment in B&H. Published by DRUNPP, Sarajevo. Vol.5, No 4, p. 922-928 ISSN 1840-2291

[43] Montgomery, D; Goldberg, J; Amar, M. et al. (1999) Effects of hyperbaric oxygen therapy on children with spastic diplegic cerebral palsy: a pilot project. Undersea Hyperb Med; 26:235-42.

[44] Panteliadis, CP. (2011) Cerebral Palsy, A multidisciplinary approach. ISBN: 978-3-87185-403-3

[45] Pease, WS. (1998) Therapeutic electrical stimulation for spasticity: quantitative gait analysis. Am J Phys Med Rehabil; 77:351-5.

[46] Pin, TW; McCartney, L; Lewis, J. & Waugh, MC. (2011) Use of intrathecal baclofen therapy in ambulant children and adolescents with spasticity and dystonia of cerebral origin: a systematic review. Dev Med Child Neurol; 53(10):885-95.

[47] Pittler, MH. & Ernst, E. (2000) Efficacy of kava extract for treating anxiety: systematic review and meta-analysis. J Clin Psychopharmacol; 20:84-9.

[48] Price, C; Makintubee, S; Herndon, W. & Istre, GR. (1994) Epidemiology of traumatic spinal cord injury and acute hospitalization and rehabilitation charges for spinal cord injuries in Oklahoma, 1988-1990. Am J Epidemiol; 139(1):37-47.

[49] Sanger, TD; Kukke, SN. & Sherman-Levine, S. (2007) Botulinum toxin type B improves the speed of reaching in children with cerebral palsy and arm dystonia: an open-label, dose-escalation pilot study. J Child Neurol; 22(1):116-22.

[50] Scheker, LR; Chesher, SP. & Ramirez, S. (1999) Neuromuscular electrical stimulation and dynamic bracing as a treatment for upper-extremity spasticity in children with cerebral palsy. J Hand Surg [Br]; 24:226-32.

[51] Schwerin, A; Berweck, S; Fietzek, UM. & Heinen, F. (2004) Botulinum toxin B treatment in children with spastic movement disorders: a pilot study. Pediatr Neurol; 31(2):109-13.

[52] Selcuk, B; Inanir, M; Kurtaran, A; Sulubulut, N. & Akyuz, M. (2004) Autonomic dysreflexia after intramuscular injection in traumatic tetraplegia: a case report. Am J Phys Med Rehabil; 83(1):61-4.

[53] Stephens, K, editor. 82000) Poland: space suit technology offers new hope for cerebral palsy rehab. |{Website: Orthotic and Prosthetic Business News}{webURL:http://www.oandpbiznews.com}| [serial online]. Spring 1998;1(2). Accessed December 15, 2000.

[54] Suskind, DL. & Tilton, A. (2002) Clinical study of botulinum-A toxin in the treatment of sialorrhea in children with cerebral palsy. Laryngoscope; 112(1):73-81.

[55] Švraka, E; Loga, S. & Brown, I. (2011) Family Quality of Life: Adult school children with intellectual disabilities in Bosnia and Herzegovina. Journal of Intellectual Disability Research. The Foremost International Journal on Intellectual Disability. Volume

55 part twelve. Special Issue Part One: Family Quality of Life. Edited by Ralph Kober and Mian Wang. ISSN 0964-2633 (Print) ISSN 1365-2788 (Online)

[56] Švraka, E. (2012). Children with Cerebral Palsy and Epilepsy, Epilepsy - Histological, Electroencephalographic and Psychological Aspects, Dr. Dejan Stevanovic (Ed.), ISBN: 978-953-51-0082-9, InTech, Available from: http://www.intechopen com/books/ epilepsy-histological-electroencephalographic andpsychological- aspects/children-with-cerebral-palsy-and-epilepsy.

[57] Taub, E; Miller, NE; Novack, TA. et al. (1993) Technique to improve chronic motor deficit after stroke. Arch Phys Med Rehabil; 74:347-53.

[58] Tilton, AH. (2003) Injectable neuromuscular blockade in the treatment of spasticity and movement disorders. J Child Neurol; 18:S50-66.

[59] Turi, M. & Kalen, V. (2000) The risk of spinal deformity after selective dorsal rhizotomy. J Pediatr Orthop; 20(1):104-7.

[60] Umphred, D; Dewane, J; Hall-Thompson, M. et al. (2001) RMU model for neurological rehabilitation, Provo, UT.

[61] WE MOVE: Worldwide Education and Awareness for Movement Disorders. Orthopedic operations. Available at: http://wemove.org/spa/spa_oss.html. Accessed October 11, 2007.

[62] Young, RR. (1994) Spasticity: a review. Neurology; 44 (suppl 9):S12-20.

[63] Zidar, J. & Dimitrijevic, MR. (1991) In: Neurosurgery for spasticity: a multidisciplinary approach. New York: Springer-Verlag: 39-46.

[64] Zidek, K. & Srinivasan, R. (2003) Rehabilitation of a child with spinal cord injury. Semin Pediatr Neurol 2003;10(2):140-50.

Chapter 2

Physical Management of Children with Cerebral Palsy

Mintaze Kerem Günel, Duygu Türker,
Cemil Ozal and Ozgun Kaya Kara

Additional information is available at the end of the chapter

http://dx.doi.org/10.5772/57505

1. Introduction

Cerebral palsy (CP) refers to a group of permanent disorders of the development of movement and posture, causing activity limitations, which are attributed to non-progressive disturbances that occurred in the developing fetal or infant brain. Damage to the central nervous system cause disorders in neuromuscular, musculoskeletal and sensorial systems. These disorders result in posture and movement deficiencies. The causes of motor disorders are developmental retardation, abnormal muscle tone, muscle weakness, postural control deficiencies, sensorial problems, behavioral problems, orthopedic problems, abnormal movement patterns and reflex, activity, asymmetry and deformities. Within the scope of the assessment to be performed in terms of motor, besides the changes in the muscle tone, co-contraction capacities of the muscles, involuntary extremity and body movements, stabilization of the extremities, correction, balance and protective reactions, sitting balance, upper extremity and hand functions and sensory-perception problems; orthotics, need of mobilization tools and other aid tools, cooperation of the family and their knowledge on the disease also needs to be assessed. Modern therapy methods in CP rehabilitation aim to develop the maximum functionality and independence possible for the child by using the present neuromotor potential. The dynamic motor control approach based on changing the motor patterns and configuration of the tasks rather than the hierarchical modeling of the neurological motor development is used for rehabilitation.

2. Cerebral palsy

2.1. Definition

Cerebral palsy (CP) was first described by William Little in 1862 and initially was called Little's disease. It was described as a disorder that appeared to strike children in the first year of life, affected developmental skill progression, and did not improve over time. Little related the disorder as a lack of oxygen at the birth. After that, Sigmund Freud suggested that CP might be rooted in the brain's development in the womb and related aberrant development to factors influencing the developing fetus (Accardo, 1982). Asphyxia at the birth was thought to be the cause of CP until the 1980s, but today researches have shown that this etiology to be less likely and only one of many with potential to result in CP (Nelson & Ellenberg 1986, Moster et al., 2001). Recently, the most widely accepted consensus definition utilized for both clinical and research purposes is the one put forward by Rosenbaum et al, "cerebral palsy describes a group of permanent disorders of movement and posture, causing activity limitations, that are attributed to non-progressive disturbances that occurred in the developing fetal or infant brain. Beside the motor impairments, sensation, perception, cognition, communication, behavior, epilepsy and musculo-skeletal problems are also accompany to cerebral palsy" (Rosenbaum et al., 2007).

2.2. Incidence and etiology

Although, the exact prevalence of CP is variable and depends on definitions and case ascertainment, today, studies has shown that, CP prevalence is around 2 per 1000 live births in both developed and developing countries (even very different reasons). Conversely to perinatal mortality, the birth prevalence of CP has not declined since the 1950s, The proportion of children with CP that are born very preterm has increased with the advent of neonatal intensive care and improvements of neonatal intensive care may be necessary before its benefits can be fully realized (Blair & Stanley, 1997). CP prevalence is 1 per 1000 live births, for term children The prevalence of CP decreases significantly with increasing gestational age category: 14.6% at 22–27 weeks' gestation, 6.2% at 28–31 weeks, 0.7% at 32–36 weeks, and 0.1% in term infants. The type of CP is also changed by gestational age; in preterm infants, spastic CP is predominant and in term infants, the nonspastic form of CP is more prevalent than in preterm infants (Cans et al., 2000). For moderately preterm children (32–36 weeks' gestation) estimates are 6–10 times higher and for very preterm children (less than 32 weeks' gestation) prevalence is 10 times higher than in moderately preterm children. The birth weight changes the CP prevalence and it the highest in children weighing 1000 to 1499g (59.18 per 1000 live births), and the lowest in children weighing over 2500g (1.33 per 1000 live births). CP rates for live births show a lower prevalence for babies of birth weight less than 1000g than for those with a birth weight of 1000–1499g. Because the high numbers of babies do not live long enough to develop CP, it disappears when estimating prevalence for neonatal survivors (Cans et al., 2000). Changes in perinatal and neonatal mortality accelerated in from the 1960s, with a huge decrease up until the late 1980s, when there was an increase in the absolute number of children with CP. From

http://dx.doi.org/10.5772/57505

1990 there has been a plateauing of mortality rates but a downward trend in CP rate mainly in moderate and very low birth weight children (Cans et al., 2000).

There are a lot of conditions or risk factors associated with CP can be broken down into those occurring in the prenatal, perinatal or postnatal time periods. CP may result from one or more etiologies and can occur at any stage from before conception to infancy, with the actual cause difficult to determine in all cases (Taft, 1999, Rosembaum, 2003, Jones et al., 2007). Currently, problems occurring during intrauterine development, congenital disorders, asphyxia occurring in any gestational age and preterm birth are thought to account for the majority of cases (Naeye, et al., 1989, Moster et al., 2001). Neuroimaging studies support the current thought that prenatal causes of CP, like brain malformations intrauterine vascular malformations, and Infection are more common than birth asphyxia (Truwit et al., 1992). Although intrapartum asphyxia originally was thought to be a major contributor to CP, it accounts for only 10% to 20% of cases (Nelson & Ellenberg, 1986). The most frequent perinatal/neonatal etiologies in low-birth-weight infants are periventricular leukomalacia, periventricular hemorrhage and cerebral infarction, but in infants of normal birth weight, the most often reason is hypoxic-ischemic encephalopathy. Postnatal causes are generally result in spastic CP and represent only about 10% to 18% of cases (Pharoah et al., 1989). More than 30% of children, there are no risk factors or known etiology (Taft, 1999, Rosembaum, 2003). The 30-year survival rate is approximately 87% (Glader & Tilton. 2009).

Risk Factors Associated With Cerebral Palsy		
Prenatal	**Perinatal**	**Postnatal**
Hypoxia		
Intrauterine infections		
Consanguinity		Asphyxia
Iodine deficiency		Infantil spasms
Genetic influences	Asphyxia	Hyperbilirubinemia
Metabolic disorders	Premature birth<32 weeks	Cerebral Infarction
Plecental malformations	Low birth weight	Exposure toxins
Feotal malformation syndormes	Abnormal fetal presentation	Pulmonary problems
Toxicity	Instrument delivery	Meningitis
Multipl pregnancy	Blood incompatibility	Intraventricular hemorrhage
Intrauterin growth restriction	Infection	Neoplasmas
Thrombophilic disorders	Plecental abruption	Head trauma
Periventricular leukomalacia		Cerebral infection
Vascular accidents		
Abdominal trauma		

Table 1. Data from: Nelson & Ellenberg, 1986, Naeye et al., 1989; Nelson, 1989, Kuban & Leviton, 1994, Han, et al., 2002, Gibson et al., 2003; Jones et al., 2007, Pountney 2007.

2.3. Classification

Classification of CP is based on pathology, etiology or clinical description. Because pathology and etiology are unclear in so many cases, universal classification is currently possible only for clinical description, but reliability is elusive, partly since the term covers such a variety of clinical presentations. Classifications could include different types, distribution and severity of motor impairments and associated impairments. Because the characteristics of each factor vary widely, the combination of characteristics found in a person with CP may often be unique. Classification systems but what is important is subjective and depend on the purpose of classification (Blair, 1997). The most of the classification systems have poor reliability, since that they use terminology which is understood differently by clinicians trained in different disciplines. Using simple and everyday language and avoiding from technical terms or even pictorial representations are more beneficial to understand the classification. An early attempt to avoid technical language was followed by one that sought only to decrease reliance on it and is the subject's development (Evans, 1989, Blair, 1997, SCPE Collaborative Group, 2000).

One method to classify CP, divides CP into two major physiologic classifications, as pyramidal (spastic) and extra pyramidal (nonspastic) which are indicating the area of the brain lesion resulting predominant motor disorder. Results from defects or damage occurring in the brain's corticospinal pathways, also described as upper motor neuron damage. Spastic CP accounts for nearly 70% to 80% of all cases of CP. In pyramidal/spastic CP cognitive impairments seen in approximately 30% (Taft, 1995, Jones, 2007). Increased muscle tone is the predominant feature and hyperreflexia, clonus, extensor Babinski response, and persistent primitive reflexes are commonly accompany. Extrapyramidal (nonspastic) CP is caused by damage to outside of the pyramidal tracts in the basal ganglia or the cerebellum. It could divide into two subtypes, dyskinetic and ataxic (Jones, 2007). Dyskinetic and ataxic forms account for 15% to 20% of all cases of CP, with dyskinetic accounting for 10% to 15% and ataxic nearly 5% (Taft, 1995; Sanger, et al., 2003, Jones, 2007). These forms result with disabilities with abnormal tone regulation, postural control and coordination (Dormans & Pelligrino, 1998).

An another method is to describe the predominant motor characteristics, which include spastic, hypotonic, dyskinetic and ataxic, as well as the topographical pattern of limb involvement, such as monoplegia, diplegia, triplegia, hemiplegia or quadriplegia (Bobath, 1980, Jones, 2007, Pountney, 2007).

2.3.1. Spastic CP

The spastic child shows hypertonus of a permanent character, even at rest. The level of spasticity varies with the child's general condition, that is, child's excitability and strength of stimulation to which child is forced at any moment. If the degree of spasticity is high, the child will fix in a few typical patterns due to the severe level of co-contraction of the involved parts, especially around the proximal joints like shoulders and hips. As a result of tonic respirocal inhibition, some of the muscles may appear weak by their spastic antagonists: for example, the quadriceps by spastic hamstring and the dorsiflexors of the ankles by spastic triceps surae. But the real weakness may develop in some muscle groups because of disuse in case of long standing or prolonged immobilization. Sapasticity is of typical distribution and changes at first

http://dx.doi.org/10.5772/57505

in a predictable manner, owing to tonic reflex activity. Movements are restricted in range and require huge effort (Bobath, 1980).

2.3.2. Dyskinetic CP

The term of dystonic is now generally preferred to athetiod CP. Dystonic CP has few signs in the early months of life except possibly some variations of muscle tone, but abnormal postures and movements begin to occur in the second half of the first year. There are involuntary movements around the mouth and of the arms and legs and these become particularly prominent when attempting fine or gross motor movements. There are varied patterns of the abnormal movements and swallowing difficulties are common. Many of the children make grimacing movements of their face, which are often associated with attempted movements in other part of the body (Scrutton, 2004). Because both have dystonic phase, it may be difficult to distinguish between the child who is later going to be dystonic and child with cerebral diplegia early on in first year of life. In dyskinetic child the fluctuating tone persists, and tendon reflexes tend to be normal and may be increased in the lower extremities. The child has abnormal postures with increased tone depending on position in space, relation of head to body, contact with a surface or stimulation of the oral region. Initiating activity itself is clumsy and uncoordinated and involuntary movements may make effective voluntary activity difficult. In this type of CP bulbar problems are common, swallowing difficulties are frequent and may affect nutrition. Additionally, drooling may be important problem. Speech is usually impaired with dysarthria because of involment of the muscles (Scrutton, 2004).

Chorea: Sudden, quick, aimless, dancing movements of the head, neck and extremities. *Athetosis:* Involuntary, slow and snake-like movements. The plane, direction and timing of movements of the proximal articulars have mostly been defected. Chorea and athetosis sometimes occur at once, this is called choreoathetosis.

Ballismus: Involuntary thrusts like explosions. It is rare.

Tremor: Involuntary, rhythmic reciprocal, movements that occur due to the contraction of agonists and antagonists. These movements are generally more prominent in small articulars and extremity distal. It is rarely seen alone and is frequently accompanied by athetosis or ataxia.

Rigidity: Increase of tonus that includes both gravity and antigravity muscles (lead pipe and cogwheel indication).

Dystonia: Movements that are mostly characterized by constant muscle contractions in the trunk, neck and extremity proximal, causes contortions, repetitive movements or abnormal posture.

Diskinetic movements may occur in different ways. As it can occur as *intermittent spasms* characterised by increase in the flexor or extensor tonus due to tonic labyrinth and reflexes that affect the neck, it can also occur as *mobile spasms* that include the alternative flexion of extremities, extension, pronation and supination. Exaggerated movements, called *momentary localized contractions,* may occur with the muscle or muscle groups of anywhere in the body also being affected. Facial grimacing, exaggerated and asymmetric activation of the mimic muscles, rotating, bending movements of the hand and fingers, etc (Kerem Gunel, 2011).

2.3.3. *Ataxia CP*

The child does not have involuntary movements; however volitional movements are affected usually all over the body. In classical neurology there is an unstable, gait with wide based of support and often gross tremor in the arms and hands. The infant with ataxic CP presents as a low toned baby with tonic paresis, the opposite to spasticity. There are increased ranges of movement at all joints and postural development is delayed, as is walking. Physiological ataxia is prolonged, and speech is usually slow and of developmental patterns rather the dysarthria of acquired ataxias. Hand skills are disrupted in regulating speed, distance and power, and since the cerebellum is involved in motor learning the child may appear dysraxic (Bobath, 1980, Scrutton, 2004).

2.3.4. *Hypotonic CP*

Hypotonic CP often is included in classifications of CP because of the resulting motor delays observed. This form also is referred to as central hypotonia. To classify hypotonia as CP, myopathy or neuropathy must be excluded as potential causes. These infants are low toned, exhibit a marked reduce in overall muscle tone and will have significant delays in motor milestones. Hypotonic CP has persistent primitive reflex patterns and hyperreflexia, so it distinguishes from lower motor neuron problems which causes of hypotonia (Taft, 1999, Jones, 2007).

It is possible to see different combinations of forms of CP depending on the area of brain lesion; this can be confusing to parents when different professionals call their child's CP "mixed". However, when the types overlap on each other, it can be difficult to classify definitely the resulting disability within the typical subtypes (Kuban& Leviton, 1994, Taft, 1999, Jones, 2007)

Another classification is done according to the extremities that are affected.

Diplegia is defined as involvement of the whole body; the lower half but, is more affected then the upper half. Head control and control of the upper limbs are usually little affected and speech is nearly normal (Bobath, 1980). The child with classical diplegic CP has slightly flexed and internally rotated hips and femoral anteversion, semi-flexed knees, extended planter flexed ankles and depending on the extent of involvement and effectiveness of management, some fixed contractures potentially at all hip, knee and ankle joints. Additionally, there are some associated posturing in the upper extemities like internally rotated shoulders, flexed elbows, wrist and fingers and adducted/opposed thumbs. This pattern is often seen after 2 years of age and may be completely apparent after 3 or 4 years. Commonly, before the age of year, there will be dystonic phase when the child will have accompanying hypertonia and diagnosis of CP may be quite difficult (Scrutton, 2004). When the child gets older, usually toward the end of the first year and during the second year, spasticity becomes more clear. The most of these groups of children walk independently and these deformities develop as a result of the crouch gait which seen in many of spastic diplegic children because of spasticity in the hip adductors and flexors, hamstring and calf muscles. To compensate for tight tendo

http://dx.doi.org/10.5772/57505

Achilles, children in this group may develop hyperextension of the knee and kyphosis may develop as a sequel to tight hamstring or hyperlordosis as a compensatory balance mechanism (Pountney, 2007).

Quadriplegia is also defined as involment of the whole body, however upper parts being more involved than, or at least as equally involved as, the lower parts. Spasticity dominates in all four extremities. The children develop very minimal functional movements and they are at great risk of contractures and deformities. Distribution is usually asymmetrical. Due to the greater involvement of the upper body, head control and eye coordination poor. In general, children with quadriplegic distribution have severe CP, frequently associated with seizure and severe cognitive impairment. These children usually have feeding problems and some involvement of speech and articulation (Bobath, 1980). If their care is not good, they have tendency to develop both scoliotic and kyptotic problems in adult life. Beside these deformities, may develop dislocation of their hip joints and spinal curvature. The subluxation or dislocation of hip joint may cause significant morbidity in terms of pain and difficulty with postural control, creating limitations in sitting, standing and walking, and personal care problems which include hygiene (Bobath, 1980, Scrutton, 2004, Pountney, 2007). Children who do not walk independently, approximately 60% of this group will have hip dislocation by age 5 years (Scrutton & Baird, 1997, Scrutton et al., 2001). It is recognized that dislocation continues to occur well into adolescence offered a protocol for the surveillance of hips in young children, which recommends a baseline X ray at 30 months to determine risk (Scrutton & Baird, 1997, Miller & Bagg, 1992). The association between hip dislocation and spinal curvature is well known and children with a windswept deformity of the hip are subluxated or present as a precursor to spinal curvature. Spinal curvature occurs in up to 70% of children with bilateral cerebral palsy but it is most prevalent in those with quadriplegia. As a result of weakness of sitting stability, pain, pressure and respiration problems will occur. Scolisis is the most common curve seen; however kyphosis and hyperlordosis are also common. In many of spinal curves, rotatory elements are present and combinations of curve a and combinations of curve patterns, such as kypho-scoliosis, are present. Spinal problems explained above can occur from very young age and continue to progress well into adulthood, with individuals with the spastic form of CP at greatest risk (Lonstein, 1995, Satio et al, 1998 Pountney, 2007).

Hemiplegia is involment of upper and lower extremity on one side (Bobath, 1980). The upper limb appears to be much more involved than the lower limb, although this is partly because the less affected proximal part of body makes walking look relatively 'normal'. The lack of fine movements of the hand are very pronounced, but fine movements of the toes are equally impaired. The typical postures are similar to diplegia but affect only half the body. Bony undergrowth of the affected extremity, when present, occurs in the first two years of life and if not well managed may play a part in the development of a contracture of the tendo Achilles. What is so apparent in a unilateral disorder points to the fact that many diplegic children will have some bony undergrowth in both extremities. Although their onset of walking may be delayed, nearly all hemiplegic children walk, but they often experience underdevelopment of the affected side, which results in smaller extremities and can result shortening in the leg (Scrutton, 2004). Equines of the foot and ankle, flexion of the elbow, wrist and fingers and adducted thumb are classical deformities of the hemiplegic child. For hemiplegic children, one hand functions well, however the other has some degree of dysfunction (Uvebrant, 1988,

Scrutton, 2000). Impairment of the upper extremity results with complications in almost all forms of human activity like self-care, school or work, and engagement in play or daily life activities (Exner, 2001, Sköld et al., 2004). The hemiplegic hand can be described as slow and weak, with uncoordinated movements, incomplete finger fractionation, spasticity and commonly, impaired tactile sensibility (Uvebrant 1988, Brown & Walsh 2000, Krumlinde-Sundholm & Eliasson, 2002) additionally, Impairments in fingertip force control and timing during object manipulation and inadvertent mirror movements are also described (Gordon et al., 1999, Kuhtz-Buschbeck et al., 2000).

Recently, the clinical type of CP of children with CP is classified based on the most frequent neurologic indications. SCPE's (Surveillance CP Europe) classification system is progressing on creating an international language. The system adopted by SCPE provides a decision flow chart to aid classification into neurological and topographical categories including spastic (unilateral or bilateral), ataxic, dyskinetic (dystonic or choreoathetotic), or not classifiable. Despite careful planning of the system, there has been little work to demonstrate the validity and reliability of classification. The lack of any defined criteria for recording functional limitation in the SCPE definition was noted by lenksi et al (2001). Subsequently, SCPE, along with other research groups, demonstrated that the inclusion of a description of functional ability markedly improved the reliability of diagnosing children with CP. Consistent application of the diagnosis is of paramount importance when the prevalence of CP from different sources and places is being compared.

According to the record system that SCPE suggests, CP;

Spastic type CP is characterised by at least two:

- Abnormal posture and/or movement.
- Increased tonus (not required to be constant).
- Pathological reflexes (increase in reflexes: hyperreflexia and/or pyramidal indications, i.e. Babinski response).

Spastic CP can be bilateral or unilateral.

Spastic Bilateral CP is diagnosed if it includes extremities on both sides of the body.Spastic Unilateral CP is diagnosed if it includes extremities on one side of the body.

Ataxic type CP is characterised both of the below:

- Abnormal posture and/or movement.
- Loss of muscle control so that movements are performed with abnormal force, rhythm and accuracy

Both of the below are dominant in dyskinetic type of CP:

- Abnormal posture and/or movement.
- Involuntary, incontrollable, repetitive and sometimes stereotype movements.

Dyskinetic CP however, can by dystonic or choreo-athetoic:

http://dx.doi.org/10.5772/57505

- *Dystonic CP is active in both situations:*
- *Hypokinesis (decrease in activity, i.e. difficult movement).*

Hypertonia (tonus generally increased).Choreo-athetodic CP is active in both situations:

- *Hypokinesis (decrease in activity, i.e. severe movements).*
- *Hypertonia (tonus generally increased)* (Krägeloh-Mann, 2009, Garne, et al. 2008).

2.4. Commonly associated conditions

Because of the abnormal tone or movement associated with the disorder, nearly all children with CP have orthopedic concerns. These orthopedic concerns may be the extent of the effect of CP for some children. Many are at risk, but, for associated medical concerns as well. The most children with CP have at least one additional disability. For many children, the associated disabilities may be more significant from a functional or quality-of-life perspective than the neuromotor impairments that define the condition. It is important to be aware of potential associated disabilities and medical complications so that the child can be monitored in a proactive manner (Glader & Tilton, 2009).

2.4.1. Primary effects: Neurologic Sequel

From a neurologic view, some primary issues may arise as a result of the underlying injury causing the CP. Major manifestations include seizures disorders, intellectual and learning disabilities, neurobehavioral concerns, sensory impairment, and effects of bulbar palsy (Glader & Tilton, 2009).

2.4.1.1. Seizure disorders

The rate of epilepsy in children with cerebral palsy is, ranging from 15% to more than 60%, depending on the type of cerebral palsy and the origin of the series. In these children, epilepsy is an index of the severity of cerebral palsy. Associated disabilities like mental problems are much more common in patients with CP with epilepsy than in those without seizures. The presence of seizure seems to be a more predictive factor of mental development than the extent of the brain damage. Epilepsy associated with CP is difficult to control, although remission, even in the presence of brain damage, can occur. But, there is still controversy concerning the optimal seizure-free period needed before discontinuing antiepileptic drugs (Glader & Tilton, 2009).

2.4.1.2. Intellectual disability and learning disabilities

Nearly 65% of children with CP meet criteria for intellectual disability (Miller, 1998). There is a correlation between intellectual disability and the subtype of CP. Children with spastic quadriplegia have the highest tendency of having an intellectual disability, additionally there is some indication that the presence of epilepsy correlates more strongly with intellectual disability (Glader & Tilton, 2009).

Learning disabilities occur in children with CP, and seem to correlate with general cognitive function. There is a discussion about children with a right-sided hemiplegia have increased prevalence of language disorders based on a left-sided injury (Trauner et al., 1996). Additionally, low-birth-weight infants with CP have increased risk for educational impairments (Fennell & Dikel, 2001).

2.4.1.3. Neurobehavioral concerns

There are a lot of neurobehavioral concerns arise in children with CP. Typical problems include inattention, internalizing behavioral problems, immature adaptive skills, and undesirable behaviors. Consistent with the attention-deficit/hyperactivity disorder, inattention may be primarily neurologic. Neurobehavioral issues may indicate subclinical seizures, depression, discomfort, anxiety or fatigue. The interaction between a numerous of medical realities can result with these maladaptive behaviors such that they are of a secondary rather than primary etiology. This condition demands a different approach to treatment and knowledge of the potential associated medical concerns. It can be challenging to explain the etiology of neurobehavioral symptoms in a child with CP, particularly if communication impairment exists. The diagnosis of neurobehavioral origin is generally one of exclusion, after other explanations, like discomfort and fatigue, have been excluded (Glader & Tilton, 2009).

2.4.1.4. Hearing impairment

Hearing loss occurs in 12% of children with CP in different degrees. Most commonly, hearing loss relates to a kernicterus, very low birth weight, meningitis, or it is very important to obtain a hearing evaluation in any child suspected to have CP (Carey 2009, Glader & Tilton, 2009).

2.4.1.5. Vision problems

More than 25% of children with CP have different kinds of visual problems, and some studies place the prevalence at closer to 40%. Children with a periventricular leukomalacia seem to be particularly tend to vision problems. The range of visual impairment encountered includes retinopathy of prematurity, nystagmus, amblyopia, refractive errors, optic nerve atrophy and cerebral visual impairment (Rudank et al., 2003). All children diagnosed with CP must be evaluated by an eye specialist. A functional vision or cerebral visual impairment assessment looks for the presence of visual field cuts and behaviors, like the use of peripheral vision and gaze preference. All of these can affect a child socially and academically in term participation (Glader & Tilton, 2009).

2.4.1.6. Gastrointestinal problems

Different gastrointestinal problems are present in a child with CP. Delay in growth and malnutrition are common (Sullivan et al, 2000) and the sequel of malnutrition are important to recognize. As a result of that, endurance or ability of a child can be affected. Postoperative wounds may cause Infection.

http://dx.doi.org/10.5772/57505

Decreased oral intake may reflect underlying gastrointestinal problems, especially conditions relating to lack of motility. The child with uncontrolled gastroesophageal reflux or constipation may feel uncomfortable. Treatment of these underlying disorders can have an important effect on nutrition. Caloric enhancement and work with a dietitian to optimize caloric intake can be central to helping a child to overcome the malnutrition. The most common reason for decreased oral intake is oromotor dysfunction. Oromotor problems can be exaggerated by challenges with overall tone and poor positioning. Treatments to gain oromotor control and safety include different modifications of food textures, feeding techniques and seating. A therapist specialize in feeding can play an important role for a child challenge with oral feeding. In more severe cases, oral feeding cannot be managed safely, and assessment for direct enteral feeds (e.g, placement of a gastrostomy tube) must occur (Glader & Tilton, 2009).

Many children with CP challenge with difficulties gaining weight; on the other hand, on occasion, obesity is a problem and when it occurs, the child faces increased challenges to overall motor activity and coordination (Carey 2009, Glader & Tilton, 2009).

2.4.1.7. Sleep problems

Sleep problems are common in children with CP. There are different etiologies, include, among others, primary obstructive sleep apnea; discomfort, which requires thorough evaluation for a wide range of medical issues; and a primary neurologic complication of sleep-wake cycle abnormality or even seizures (Glader & Tilton, 2009).

3. Motor assessment

CP refers to a group of permanent disorders of the development of movement and posture, causing activity limitations, which are attributed to non-progressive disturbances that occurred in the developing fetal or infant brain (Bax et al., 2005). Damage to the central nervous system causes disorders in neuromuscular, musculoskeletal and sensorial systems (Butler et al., 1999). These disorders result in posture and movement deficiencies. The causes of motor disorders are developmental retardation, abnormal muscle tone, muscle weakness, postural control deficiencies, sensorial problems, behavioral problems, orthopedic problems, abnormal movement patterns and reflex, activity, asymmetry and deformities (Rosembaum et al., 2007). Modern therapy methods in CP rehabilitation aim to develop the maximum functionality and independence possible for the child by using the present neuromotor potential (Hamamci & Dursun, 1995). Evaluation is very important in understanding and efficiently treating motor function problems that are the major factor influencing functional independence in CP (Livanelioğlu & Kerem Günel, 2009).

In an assessment of a child with CP, whose physiotherapy and rehabilitation needs were determined, the physiotherapist should be search the functional status, active neurophysiologic and biomechanical mechanisms and accompanying problems effect the situation. The clinical type, severity of the disease, chronologic age, age of initiating physiotherapy, existence and severity of abnormal reflexes, cognitive problems appearing together, hearing disorders, visual impairment, sensory-perception problems, general state of health and the socio- cultural and economic status of the family should be considered while deciding on suitable physiotherapy methods (Stranger & Oresic, 2003).

The actual question that needs to be answered within scope of the information obtained, as a result of the assessment, is what is important in the child's life. What needs to be provided is not only motor development abilities such as sitting, crawling, walking, muscle tonus regulation, balance and coordination training. The acquisitions shall be ensured to be able to be used in daily life (Bower & McLellan, 1992).

While clinical observation is one of the most important parts of the assessment, it completes standardized tests and contributes information which carries at least the same significance. By assessing the child, according to the parameters listed below, the physiotherapist shall present a general table of the child. The child must be calm and trust the physiotherapist during the observations conducted in terms of motor, sensory, cognitive, emotional and social/family. The parent or the guardian undertaking the care of the child shall be with the physiotherapist during the observation. The child must not be hungry, nor should be observed right after eating. The room where the observation will be done should be quiet, at an agreeable temperature and not contain unnecessary toys and equipment; if possible it should be a room covered with material that is appropriate for the child to move on the ground, with walls painted in warm colours and should not be too small. Firstly, what the child can do on his/her own should be observed while examining the functional movements, fine and gross motor skills during the observation (Mayston, 2008).

Within the scope of the assessment to be performed in terms of motor, besides the changes in the muscle tonus, co-contraction capacities of the muscles, involuntary extremity and body movements, stabilization of the extremities, correction, balance and protective reactions, sitting balance, upper extremity and hand functions and sensory-perception problems; orthotics, need of mobilization tools and other aid tools, cooperation of the family and their knowledge on the disease also needs to be assessed. The assessment of the motor function should be based on the normal process of a normal motor function development but it should also be sensitive towards special problems. For motor development reflex development, proper posture, sufficient extremity movements, appropriate muscle tonus, sensory development and cognitive functions within an integral neurologic and musculoskeletal system is required. Full completion of the motor development is required for the functional independence and social and emotional development of the child. Therefore it is required to know the normal development of a child. By knowing the normal development, the developmental problems that may occur in the child due to any reason can be better understood (Tsorlakis et al., 2004).

3.1. Muscle tone assessment

The methods used for assessing spasticity take place within a wide range that extends from clinical scales to more complex systems based on Electromyographic Analysis (EMG). Collecting comprehensive history and observations are very important in assessing the effect of spasticity on functions. The muscle groups, in which the spasticity exists, and their interaction with postural reactions' effect on functions should be researched. Although assessing functional activities and daily-life activities does not directly determine the severity of spasticity, it could present an idea on the reflection of the changes of the spasticity on the functional condition (Kerem Günel, 2011). One method of assessing spasticity in the clinic is to determine the amount of resistance that the spastic muscle presents during a passive movement of the relevant extremity. Ashworth has, accordingly, defined a 5–point scale. This scale evaluates the resistance that occurs during the passive movements of the extremities with points between 0-4. Although the Modified Ashworth Scale (MAS) is a subjective method in our day, it is widely used as an easily applied method that does not require any tool in assessing spasticity (Bohannon & Smith, 1987, Clopton et al., 2005).

Modified Ashworth Scale for Grading Spasticity

Grade	Description
0	no increase in muscle tone
1	slight increase in muscle tone, manifested by a catch and release or by minimal resistance at the end of the range of motion when the affected part(s) is moved in flexion or extension
1 +	slight increase in muscle tone, manifested by a catch, followed by minimal resistance throughout the remainder (less than half) of the ROM
2	more marked increase in muscle tone through most of the ROM, but affected part(s) easily moved
3	considerable increase in muscle tone, passive movement difficult
4	affected part(s) rigid in flexion or extension

The Tardieu Scale is another scale that assesses spasticity with passive movements, as does the AS and MAS. This scale presents spasticity's nature that depends on speed. Passive straining is performed at the speed of the extremity segments falling with gravity and slower and faster than this speed. The Modified Tardieu Scale (MTS) has added the assessment positions and spasticity angles of the extremities to the original scale (Boyd & Graham, 1999, Gracies et al., 2010). The MAS, Pendulum test and MTS for measuring the spasticity of children with CP was compared and MTS was determined to be most appropriate measurement method (Mutlu et al., 2007).

Spastic muscles limit articular movements in antagonist directions. Therefore, in addition to assessing the movement of the articular with a goniometre can also be used as an objective method although it presents conflicting results in terms of reliability. Assessments, which are not widely used in the clinic and are used more in assessments researches, are methods such as the dynamic flexometre, pendulum test, electrophysiologically assessing the H reflex and M response and the biomechanical analysis of response of the spastic muscle to angular and speed differences, etc. (Mutlu et al., 2008, Akbayrak et al., 2005).

The Barry Albright Dystonia Scale is a highly reliable rating scale developed in order to asses the dystonia in children with CP and traumatic brain injuries. The scoring is "none"; 0, "slight"; 1, "mild"; 2, "moderate"; 3, and "severe";4. Each region has specific descriptors for a scoring. Generally if dystonia is present less than 10% of the time it is "slight", if it does not interfere with function or care it is "mild", if it makes functional movements harder it is "moderate", and if it prevents function it is "severe" (Albright, 1996).

3.2. Assessment of functional level and gross motor functions

The most widely-used test battery that measures the functional motor level in order to determine the motor development level of children with CP is the Gross Motor Function Measurement (GMFM). With GMFM, physiotherapists can define the motor function level of the child; obtain aid in specifying the targets of the treatment, follow-up the post- treatment development and present objective information regarding the child to relevant colleagues, other inter-discipliner professionals and families. It was developed in 1989 by Russell et al. by considering the motor function level of a 5-year old child with normal motor development. The GMFM measures how much of the action is achieved rather than measure the quality of the motor performance. The purpose is to determine the capacity and change. It is comprised of sections of supine-facedown positions and turning, sitting, crawling and standing on knees, standing on feet, walking and running and jumping (Russel et al., 1989, Russel et al., 2000).

The Gross Motor Function Classification System (GMFCS) is a classification system developed for children with CP. The GMFCS has been developed by Palisano et al. based on the actions the child can perform from sitting to walking. It is a practical system that can be used in clinics for the rehabilitation team to classify a child with CP, observe the efficiency of the applications and follow-up on the patient in inter-intra discipliner applications. Initially, children with CP aged below 12 were divided into five levels by considering their independency in gross motor functions such as sitting, walking, mobilization and transfer activities and the tools-equipment, tools that assist in walking that they use. As motor functions of children differ according to age, functions have been defined as under 2-years old, between 2-4 years old, between 4-6 years old and between 6-12 years old for each level. This system was extended in order to include the age ranges of between 12-15 and 15-18 years old in 2007 (Palisano et al., 1997, Palisano et al., 2007).

Commonly Used Tests:
Gross Motor Function Classification System –GMFCS (Palisano et al, 1997,2007)
Gross Motor Function Measure- GMFM 88 or 66 (Russel et al, 2003)
The Functional Indepence Measure for Children- WeeFIM (Mshall et al., 1993)
Pediatric Evaluation Disability Inventory- PEDI (Vos-Vromans et al., 2005).
The Neurological, Sensory, Motor Developmental Assessment –NSMDA (Burns et al., 1989)
Bayley Scales of Infant Development (Bayley 1983)
Trunk Control Measurement Scale –TCMS (Heyrman et al., 2011)
Gait Analysis

Table 3. Data from: Palisano et al. 1997,2007, Russel et al 2003, Mshall et al. 1993, Vos-Vromans et al. 2005, Burns et al. 1989, Heyrman et al. 2011.

Trunk Control can evaluate by Trunk Control Measurement Scale (TCMS). This scale consists of 15 items measuring two main components of trunk control: (a) a stable base of support (static sitting balance), and (b) an actively moving body segment (dynamic sitting balance). The subscale static sitting balance (items 1–5) evaluates the ability of the child to maintain a stable trunk posture during movements of upper and lower limbs. The section dynamic sitting balance is further divided into two subscales: selective movement control and dynamic reaching. The subscale selective movement control (items 6–12) measures selective trunk movements in the sagittal (flexion/extension), frontal (lateral flexion) and transverse (rotation) plane within the base of support. The subscale dynamic reaching (items 13–15) evaluates the performance of three reaching tasks, requiring active trunk movements beyond the base of support. All items are scored on a two-, three- or four-point ordinal scale and administered bilaterally in case of clinical relevance. Maximum scores on the three subscales are 20, 28 and 10, respectively, resulting in a total score from 0 to 58. A higher score indicates a better performance (Heyrman et al., 2011).

3.3. Functional level assessment related by motor performance

The Functional Independence Measure for Children (WeeFIM) has been developed by using the Functional Independence Measure (FIM) developed for adults by "Uniform Data System" in 2003. It is a useful, short, comprehensive measurement method that determines the development, educational and social functional limitations of children that have CP and other development disorders. WeeFIM contains a total of 18 articles in 6 fields; self-care, sphincter control, transfers, locomotion, communication and social and cognitive. Whether or not the child is aided, performs on time or if they required an aiding device while performing the function in each article of these fields is scored from 1 to 7. 1 point is given if they perform the mission with aid, 2 for independently performing and 7 if they perform on the right time and safely. Accordingly, the child can score 18 the least (fully dependant) and 126 the most (fully independent) (Mshall et al., 1993, Ottenbacher et al., 1999)

Also The Pediatric Evaluation Disability Inventory (PEDI), is a comprehensive clinical assessment tool that assesses the functional ability and performance of disabled children. It has been developed especially to assess the function of small children and is a distinguishing measurement method that can be used for children below 7,5 years old and also older children. PEDI is comprised of three main sub-sections; functional abilities, help of caretakers and modifications. Each of these sections assesses self-care, mobility and social function areas. The functional abilities part comprises of 197 articles and measures the functional abilities of the child. In this section the "self-care" sub-section comprises of 73, the "mobility" sub-section comprises of 59 and the "social functions" sub-section comprises of 65 articles. The section regarding help of the caretakers comprises of 20 articles and measures the disability condition of the child according to the amount of aid required in order to perform the functional activity. The modifications section also comprises of 20 articles and shows the environmental modifications and tools that the child uses during his/her daily life. Each sub-section of PEDI can be used independently (Vos-Vromans et al., 2005).

3.4. Developmental evaluations

The Bayley Infant and Child Development Assessment Scale was developed by Nancy Bayley. It was revised in 1993 and the Bayley Infant and Child Development Evaluation Scale-version 2 (Bayley-II) was published. It evaluates the developmental condition of the child according to age in general. Bayley-II is used to evaluate children aged 1-42 months and to follow their development in the USA. Its standardization has been done on 1700 children in the USA and it has been used in studies and clinical applications for over 40 years (Bayley, 1993). It is one of the best tests for evaluating child development. The test is valid and reliable for determining the child's development (Blauw-Hospers & Hadders – Algra, 2005).

The Bayley and Bayley-II evaluation scales are the mental development scale (MDS) and psychomotor development scale (PDS). MDS evaluates the cognitive and language development while PDS evaluates gross and fine motor development. The MDS and PDS scales are the basic limitations of Bayley and Bayley-II (Johnson & Marlow, 2006). This limitation has been revised in Bayley-III. Thus, the composite scores of cognitive, language and motor parameters can be calculated separately. The structure of Bayley-III, the new version of Bayley-II, provides more useful information in understanding the development in the early stage, increases our capacity to identify early developmental problems and ensures focusing on special areas specific to the weakness with early intervention programs. In terms of research, it provides a better understanding of early development in the high-risk group and provides more sensitive results for clinical studies (Anderson et al., 2010). The primary purpose of Bayley-III is to identify children with a developmental delay and provide information for an intervention plan. It identifies the developmental delay risks of children aged between 1 and 42 months, and helps professionals in the identification of future applications (Bayley, 2006).

The test has been divided into 5 subgroups: cognitive, language, motor, social-emotional and adaptation. It has been created for the evaluation of the separate parts of the total development for each child. The marking system of the subtests produce scores that can be used to define an end point for each subtest in different age groups. It includes a registry form; cognitive, receptive language, expressive language subtests; and fine and gross motor subtests. The cognitive scale evaluates the sensory-perceptual sensitivity, discrimination and the resultant abilities; early period object recognition-memory retention, learning, problem-solving skills, initiation of verbal communication and vocalization, and generalized evidence of early ability and classification (Burns, 1992).

The neurological, sensory, motor developmental assessment (NSMDA) has been designed to examine the characteristics of motor development in the early childhood period and is a standard motor development test used to evaluate infants or children between the ages of 1 and 6 years (Burns et al., 1989). It is preferred for characterizing the motor development and identifying any problematic motor development areas during long-term follow-up of of premature children. It is also used in the prediction of general developmental results such as to help in the diagnosis of CP, compare motor results in different problems, and predict the motor development and cognitive performance of premature children (Burns et al., 1989, Spittle et al., 2008, Burns et al., 2009, MacDonald & Burns, 2005).

NSMDA evaluates the development of the motor performance of the child especially in certain periods and defines normal or abnormal features, motor performance, and abnormal or dysfunctional motion components at various ages. The test shows whether the motor development and mobility components of the infants and children are within normal limits, suspect or abnormal. Test parameters evaluate the age-appropriate motor skills, muscle tone, deep tendon reflexes, movement patterns, postural reactions and balance, and the tactile, proprioceptive, visual and vestibular sensory systems (Burns et al., 1989). NSMDA has been shown to be valid and reliable from birth until the age of 2 years. NSMDA was mentioned as an adequate and differentiating set of tests in a review on the clinometric characteristics of various test sets evaluating neuromotor development in first year of life (Spittle et al., 2008).

3.5. The evaluation of gait

Various methods such as observation, measurement of time-distance characteristics, video recording systems and kinetic and kinematic analysis performed with computer-aided systems are used in the evaluation of gait in children with SP (Livanelioğlu & Kerem Günel, 2009). While gait can be evaluated by observation in the clinical environment, it is digitally evaluated in gait analysis laboratories. Functional observational gait evaluation can be performed with the Gilette Functional Evaluation Survey (Novacheck et al., 2000), Physician Rating Scale (Maathuis, 2005,) and Functional Mobility Scale (Graham et al., 2004). The gait substep of GMFM can also be used in the functional evaluation of the gait (Ross & Engsberg, 2007). Although many gait problems can be understood by the visual examinations performed by experienced clinicians or analysis performed with video records, gait analysis technology is necessary to interpret the problem numerically, to record and reevaluate later, and objectively reveal the effectiveness of the treatment is necessary in complex walking problems (Kawamura et al., 2007).

Gait analysis is a systematic measurement used to identify and evaluate human movement. The numerical evaluation, identification and interpretation of gait is possible with gait analysis. Modern gait analysis laboratories are based on four disciplines: Visual inspection, quantitative analysis, biomechanical analysis and electromyography (EMG). While visual examination evaluates the body motions in repetitive gait, quantitative analysis provides the kinematic parameters, time and distance characteristics of the joints during the gait. Biomechanical analysis and EMI provide information about the muscle activity during gait and its effect on gait. A detailed gait analysis includes all of these methods (Kawamura et al., 2007). Gait is evaluated in each of the three planes as sagittal, coronal and transverse with 3D gait analysis (Deluca, 1991, Patric et al., 2001). The pelvic tilt, flexion-extension of the hip and knee, and the plantar flexion and dorsiflexion of the ankle are commonly evaluated in the sagittal plane while pelvic obliqueness, abduction-adduction of the hip, knee varus-valgus, foot inversion-eversion are evaluated on the coronal plane with 3D gait analysis. The internal and external rotations of the pelvis, femur, knee, tibia, and ankle are evaluated in the transverse plane. Kinetic (strength, pressure, moment and torque) and kinematic (changing place, linear velocity, acceleration) analyses and the analysis of the time-distance characteristics are performed with 3D gait analysis. Angular sizes are recorded in kinematic analysis (Schwartz, 2009).

4. Therapy approaches

The dynamic motor control approach based on changing motor patterns and configuration of the tasks rather than the hierarchical modeling of the neurological motor development is used for the rehabilitation. This approach allows looking at the child from the functional aspect in environments such as the home and school and performing a realistic evaluation. The aim of CP rehabilitation should be the development of current abilities of the child and to keep these abilities at an optimal level. It is important to minimalize the effects of functional limitations and disorders and protect the patient from the disability to prevent secondary disorders and maximize the motor functions to the extent permitted by the existing deficit. Therefore, the type of treatment where the optimal functions can be gained should be determined. The main goal of treatment should be reaching maximum functional capacity. However, correction of abnormal posture and patterns, prevention of the deformities that may develop, mobilization, development of existing skills, and teaching new skills are also among the targets of treatment (Miller, 2007, Camper et al., 2000).

The rehabilitation program of children with CP is determined according to the age and functional condition of the child and the general treatment approach shows significant changes according to the age (Camper et al., 2000). Combinations of different approaches may be used in each age group. The environment where the therapy will be conducted can be the home of the child as well as clinics and hospital departments, inpatient rehabilitation hospitals, rehabilitation centers or school-based therapy environments. The therapy plan should be consistent with the patient's age as well and have a specific purpose and relevant measurable short-term targets. These short-term targets allow the therapist, family and the child to be informed about the process. Another aspect of the treatment is family training. The exercise plan should be taught to the child and family, the functionality of the child in the home environment should be evaluated, and the long-term expectations of the family should be taken into account (Miller, 2007, Camper et al., 2000). It is important that the rehabilitation process be started as early as possible due to the plasticity characteristics of the central nervous system (Bluw-hospers & Hadders-Alga).

The current function of the child and prognosis for acquiring new skills should be taken into account when deciding on the effectiveness of the physiotherapy. While the treatment program is planned, the growth spurts of the child and procedures such as surgery should be taken into account. Whatever the functional level of the child, the family should also be included in the process through home programs. Deciding on the intensity of the treatment is important in the management of the process. The use of intensified therapy after orthopedic interventions and neurosurgery, during growth spurts that can affect the motion biomechanics of the child, and when a specific task is being focused on has critical importance. Physiotherapy in adolescents and adults with CP should be supported with recreational activities. Thus, motivation can be provided for the acquisition of new skills (Styer- Acevedo, 1999).

http://dx.doi.org/10.5772/57505

An extensive framework of methods is included within the concept of physiotherapy and rehabilitation in CP. Some of the approaches based on neurophysiological grounds have been developed over time and remain valid while some are no longer preferred (Damiano, 2004). In addition to neurophysiological approaches, there are also strength training, hydrotherapy, restrictive obligatory motion treatment, electrotherapy applications, hippotherapy and alternative-complementary therapy approaches that are used together with such approaches or independently.

4.1. Neurodevelopmental therapy

Today, the Bobath approach, initially, aims to observe the existing performance of the child with CP, analyses it, interpret it and then enable the child to reach the maximum level of independency within the limitations of the child's potential assessment and result (DeGangi & Royeen, 1994). Neurodevelopmental therapy (NDT) was developed by Bertha Bobath, physiotherapist, Karel Bobath, neuropsychiatrist. Bobath's approach was shaped in order to involve scientific theories that were and empirical experiments that were developed and has a structure that is open to development and is dynamic. Thus, it has been developed until our day since its first application and has undergone some changes. According to the Bobath's, the motor problem is one of the most important problems and delay or disorder of normal motor development or not being able to establish postural control against gravity due to function problems in the central nervous system is the most significant factor that causes motor problems (Tsorlakis, 2004, Bly, 1991). The NDT method, which regards all problems occurring in the child as a result of the injury in the central nervous system, has focused on working on memory, perception, sense, postural control and abnormal patterns, reflexes and sensory motor components in the muscle tonus. It is used to facilitate special gripping techniques movement patterns, balance responses and normal muscle tonus and also to decrease abnormal movement patterns, reflexes and spasticity. During the years when the NDT was first developed, the child was more passive in this approach; however, as it has received the name of "living concept" it is observed that the child is more active now (Kerem Günel, 2009).

The effect of the family is very important and the family must act like a part of the rehabilitation team within the scope of NDT (Butler & Darrah, 2001). There have been debates on whether NDT principles affect motor development in terms of reflex and hierarchic model of motor control is focused on only neural explanation. For instance, in the motor control model, the central nervous system is regarded as one of the systems that affect only motor behavior. Motor control is also affected by cognitive and environmental factors. However, physiological components and environmental contents are accepted as non-neural explanations in the child's progress (Fetters & Kluzik, 1996).

Implementing clinical practices with applications based on evidence is increasingly becoming more important today. Although NDT is the most commonly used method in child rehabilitation by physiotherapists all over the world, research that presents its effects are deemed to be insufficient. There are many reasons for this. Research presenting NDT's effect was organized by AACPDM (American Cerebral Palsy Association) and as a result the difficulties and the evidence encountered were studied (Butler & Darrah, 2001)

The most prominent difficulty is that all problems, diagnosed on research that includes low incidence and high heterogeneity conditions, become complex with the change of children along with their growth and development process. Despite these obstacles, due to different practices and understandings in applying NDT and its ongoing and wide effect in CP treatment, it is important to collect information on NDT. Researchers indicate that NDT is clinically significant but that no statistical assessment can exactly present its result, partially due to the difficulties mentioned above. In researches where the NDT's clinical effect is attempted to be presented by practice, the NDT structure changes with time and in these researches NDT practices are usually performed with other therapy techniques and medical treatments (Law et al., 1997).

The primary target of NDT is to change the central nervous system's neural based motor responses. Various aspects of the motor response have been assessed with measurement methods used in conducted researches. These are qualitative movement or physiological motor function (i.e. involuntary muscle tonus changes, spasticity, etc), reflex activity, weight transfer, postural control, trunk rotation, combined reactions, upper extremity movements and walking parameters. As a result of these researches, generally, it has been indicated that a better motor response occurred and that there were positive changes in terms of physiological motor function, movement time, step length for walking, speed and foot angle after the NDT practices (Bobath, 1971). Nonetheless, the evidence of this development in physiological motor functions and qualitative movement is not consistent. One other very important target of NDT is to prevent or slow down deformities. The measurements of articular movement width, orthosis or surgical suggestions after the NDT practice are used for researching the degree of contractures. It has been indicated that NDT provides advantages in protecting the dynamic articular movement width in the ankle and knee (Kluziket al., 1990). In other words, when the articular limitation was repetitively and immediately assessed after 20-25 minute NDT sessions, it decreased further. To decrease spasticity, provide normal movement experience, support functional independence during daily activities and thus indirectly support motor learning, physiotherapists use special grips and positioning within the scope of NDT. Dynamic articular movements and the child's active participation during the movement can be clinically descriptive. When considering principles of evidence-based practices in studies, it is possible to mention that the studies were conducted with groups that have low-level work force and insufficient number of cases and are heterogeneous. When the results are considered, it can be emphasized that NDT practices have positive results on postural tonus, functional independency and dynamic articular movements; however, NDT cannot be proved to be superior to other practices and further studies are strongly required. These efforts should involve randomized studies in more comprehensive groups whereby only NDT is applied in homogeneous groups by making use of reliable and valid evaluation analyses where age, sex, severity and type of disease, socio-economic and cultural structure of family are kept under control and which indicate long-term effects (Mayston, 2008, Butler & Darrah, 2001, Herndon et al., 1987).

http://dx.doi.org/10.5772/57505

4.2. Strength training

Disorders affecting muscle strength and motor control in children with CP are indicated among the main reasons of the motor performance disorder (Giuliani, 1991, Damiano & Abel, 1998, Engsberg et al., 2000). Muscle weakness is a common disorder in children with CP and is associated with insufficient or reduced motor unit discharge, inadequate coactivation of antagonist muscles, secondary myopathy and impaired muscle physiology. Studies have shown the usefulness of strength training in children with CP and revealed the relationship of muscle strength with activity (Scianni et al., 2009). Strength exercises increase muscle strength, flexibility, posture and balance in CP. They also increase the level of activity in daily life and develop functional activities such as walking and running (McBurney et al., 2003).

Isotonic, isometric and isokinetic exercises can be used to increase muscle strength and motor function recovery. Strengthening methods commonly used in all age groups are functional activities, gravity and body weight (Finlay et al., 2012, Berry et al., 2004, Damiano et al., 1995, Fowler et al., 2001). A sufficient level of loading is necessary to increase the strength of the muscles when choosing among different strengthening methods. Strength training requires effort against progressive resistance. Progressive resistance exercises that develop muscle performance and motor skills by increasing the force production capacity are important for individuals with CP. Increase in muscle strength and joint range of motion are provided with resistance exercise training (Mockford & Caulton, 2008). Damiano concluded that resistance exercises that involve the lower and upper extremities in children with spastic diplegia increase the strength capacity (Damiano et al., 2010). An increase in the spasticity was not observed with resistance exercise training (Dodd et al., 2002). Training conducted with manual resistance, fitness equipment, free weights, Gymball, theraband, running band, static bike, leg press and isokinetic devices are examples of resistance exercises (Finlay, 2012).

Studies reveal the presence of weakness compared to their peers in the affected extremities of children with CP even if they have a high functional level, and this weakness increases with neurological involvement (Damiano et al., 1995, Wiley & Damiano 1998). In addition, Thompson reported that children with CP show lower strength-generating capacity in all lower extremity muscle groups, except for the hip extensors, compared to their healthy peers. When the gross motor function classification system (GMFCS) levels are taken into account, the muscle strengthening trainings are most commonly used in levels I-III. Children at this level have better selective control and less coactivation and are therefore considered to tolerate the specific progressive exercise training better. Muscle strengthening in children on level IV and V is controversial due to the problems with motor control. Hydrotherapy is the most popular muscle-strengthening method in children at this level (Finlay, 2012).

Previous reviews provide contradictory results about the effect of muscle strengthening interventions. Although the investigators have proven that strength training in children with CP increases their motor abilities, it has not been proven to create a positive change in their functional capacities. The transfer of gains obtained with increased strength to functional activities requires time. It is stated that changes in muscle power should be associated with functional results. The increase in function is not parallel to the increase in isometric muscle strength in studies conducted in children. Strength training in studies includes open chain or isokinetic exercises without weights and isotonic exercises with weights (Damiano et al., 1995, Damiano et al., 1995). The strengthening effect is specific to the mode of exercise. The transfer of exercises without weights to conditions with weights is quite limited as activities with weights involve different and more complex muscle activity patterns. When strength exercises involve close kinetic chain exercises that are more associated with function, the transfer of the strength to functional motor performance improves. The person puts weight on his feet and the body mass rises and falls with the concentric and eccentric activation of the lower extremity muscles in these exercises. These movement characteristics are used in many activities that involve the lower extremity, such as standing up and walking (Blundell, 2003).

Functional strengthening training, increase the power of the weak antagonist and responsible spastic agonist and aim to provide functional benefits in children with CP (Damiano, et al., 1995). Functional exercises are a combination of aerobic and anaerobic capacity and strength training; they develop the physical fitness, activity intensity and quality of life in ambulatory children (MacPhail, et al., 1995).

A treadmill can provide functional exercise and is a dynamic approach that can be used to support the motor development of individuals with CP. The normal walking rate and distance of individuals with CP increase with treadmill exercises (Cheng, et al., 2007, Dodd & Foley, 2007). Treadmills that support the body weight can be an option in individuals with CP who have no gait ability (DiBiasio & Lewis, 2012).

Strength training is significant only when the aim is the development of a specific motor skill or function. The functional gains of children without voluntary muscle control capacity from a strength training program are therefore restricted. Surgical interventions such as muscle-tendon lengthening, selective dorsal rhizotomy, botulinum toxin injection, and intrathecal baclofen pump implantation can increase the muscle length or improve muscle control in these children. Thus, strength training can be more effective and longer lasting effects can be ensured (Miller, 2007).

Training the same muscle groups on different days is appropriate in children. Strength training should be modified in the presence of muscle pain or when muscle pain and tension develops with exercise. The child should be able to comprehend and consistently produce maximum or almost maximum effort for strength training. Although strength training can be implemented for children aged 3 years or older, it is therefore more realistic for children aged 4-5 years (Miller, 2007). Despite the lack of an evidence regarding the harm of strength training, it should not be forgotten that excessive physical effort may trigger seizures in children with a relevant history.

4.3. Constrained Induced movement therapy

Constrained Induced movement therapy (CIMT) is a treatment approach based on restraining the uninvolved upper extremity and exercising the involved upper extremity intensively. The treatment protocol is based on the principle of the limitation of the nonaffected extremity and forcing the patient to use the affected extremity during the day (Taub et al., 1999).

Children with hemiplegic CP develop strategies and techniques during their growth and development to perform their daily tasks with one hand. They discover that performing tasks with the unaffected extremity is more effective and efficient even when there is only a mild disorder in the affected extremity (Kuhtz et al., 2000). DeLuca introduced the term developmental disregard to describe a child with hemiplegia who may disregard, or learn not to use, the affected limb during the development of motor function. Although the behavior mechanism in children with CP is similar to the consolidation of the unaffected extremity and not using the affected extremity seen in adults, Eliasson has reported that learned disuse could be a different condition in these children. As development in children continues, the term "learned disuse" is replaced with "developmental disuse". A hemiplegic child cannot experience normal motor function of the extremity, and the opportunity, experience and environment that allows the child to learn how to use the affected extremity must therefore be created during therapy. CIMT makes this possible (DeLuca, 2002). Studies that started with adults have spread to the pediatric field. The frequency and intensity of the application were decreased and applications modified for pediatric use in most of the studies (Charles et al., 2009, Pierce, 2002). A limited number of controlled studies have been performed on CIMT and obligatory use in hemiplegic CP (Hoare et al., 2007, Masciento et al., 2009). The restriction was ensured with various gloves, splints or material in these studies and the duration of using the splints varied greatly. Although studies vary regarding the restriction duration within the day, the concentrated repetitive training of the involved extremity 3-6 hours a day with the aim of shaping motor behavior has been shown to be effective (Masciento et al., 2009, Charles et al., 2006, Sakzewski et al., 2001). CIMT involves providing verbal feedback for small progresses in accomplishment of the task choosing specific tasks in order to address the motor deficiencies of the child, helping the child in case he cannot complete the motion alone and during the realization of the motion stages, and systematically increasing the degree of difficulty of the performed task (Hoare et al., 2007).

The increased use of the affected extremities with CIMT is suggested to be due to an expansion in the contralateral cortical area that controls this extremity's motion and the development of new ipsilateral areas. This is reported to form the neural basis for the continuation of the use of the affected extremity after the treatment (Morris & Taub, 2001).

Charles et al. reported improvement of hand function and two-point differentiation with CIMT (Charles & Gordon, 2005, Charles et al., 2001), De Luca reported CIMT to increase dependent reaching, grabbing, weight transferring in both upper extremities and the quality of the involved upper extremity (DeLuca et al., 2003). Charles reported that modified CIMT increases the efficiency of the movement in the affected extremity in a study with decreased intensity and they described the method as "child friendly" (Charles et al., 2006). Gordon et al emphasized that both younger and older children benefited from CIMT in the same way and CIMT was useful at any age (Gordon et al., 2006). Taub et al reported very good progress in the functional use of the involved extremity in patients with the use of CIMT (Taub et al., 2007, Taub et al., 2004). Cope et al showed evidence of cortical reorganization in hemiplegic children in a pilot study (Cope et al., 2008). Although all these studies provide important data showing that CIMT is useful for the hemiplegic upper extremity, the advantages and disadvantages of the method are still being discussed.

Disadvantages such as reduction in children's self-confidence along with decreased motivation when the child finds the method difficult have been suggested with CIMT use in some studies. Although it is said there are no medical complications related to the use of splints, there are also studies indicating friction by the splint on the healthy side and mild contractures in the joints of the long-term restricted extremity. An attempt is made to eliminate these negative factors by decreasing the splint use duration, opening the splints at certain intervals and checking the skin integrity, and getting the children do play games, get involved in outside activities or join social activities such as trips or camps during CIMT. Charles, Ries, Naylor, Eliason and Brandao emphasized in their studies conducted with young children that this problem can be overcome by decreasing the restriction duration, extending the application duration, and including more games and entertaining activities in the treatment program (Charles et al., 2005, Cope et al., 2008, Ries & Leonard, 2006, Naylor & Bower, 2005, Eliason et al., 2005, Brandao et al., 2009).

4.4. Electrical stimulation

Electrotherapy is the name given to any treatment or evaluation applied to the body from outside by using electrophysical agents. Electrical stimulations within a wide range from low-level stimulations such as to decrease pain, TENS or threshold electrical stimulation where there is no muscle activation to neuromuscular electrical stimulation where active muscle contraction is observed can be used (Wright et al., 2012). The rehabilitation of the pediatric group is different than the adult group. The communication and cooperation skills as well as the histological and physiological features of this group are different. Electrotherapy applications in children therefore have basic differences than those in adults and have special applications (Palisano et al., 2006).

The main electrical stimulations used to increase muscle strength in children with CP are neuromuscular electric stimulation (NMES) and threshold electrical stimulation (Kerr et al., 2004). NMES is the application an electrical current intensive enough to cause muscle contraction. Electrodes are placed on the skin over the targeted muscles in order to reveal the contraction. Two strengthening mechanisms are aimed for. The first of these is the loading principle; muscle strength increase is ensured with an increase in the cross-sectional area of the muscle. The selective development of type II fibers ensures the development of synaptic efficiency in the muscle in the second mechanism (Reed, 1997).

NMES is used to help physiotherapy in order to increase strength, normal joint motion, motor control and co-contraction and also to temporarily decrease spasticity. The results of studies on the effect of NMES on spasticity and function vary. Although NMES is most commonly used to create reciprocal relaxation in the antagonists of the spastic muscle, it can be used for the same purpose by tiring the spastic muscles (Arya et al., 2012).

The use of NMES on children with spastic CP in order to develop gait parameters and functional results has recently increased and it has become a popular technique in physio-

http://dx.doi.org/10.5772/57505

therapy and rehabilitation. However, it has not been completely proven on which muscle NMES is effective on in CP. The use of non-invasive NMES in CP has significant advantages such as not being a surgical procedure and having relatively mild side effects (Arya et al., 2012).

Studies on the use of electrical stimulation in CP are limited and have provided various results. According to the result of a meta-analysis by Cauraugh et al., electrical stimulation minimizes activity limitation during the disturbance and the gait (Cauraught et al., 2011). Although NMES is used for the treatment of many clinical problems, the contraction required by the activity the patient is participating in during stimulation is not task-specific (Kerr et al., 2004).

The use of neuromuscular stimulation for a functional target is also known as functional electrical stimulation (FES) (Reed, 1997). FES can be defined as the stimulation of the nerve and muscle electrically in order to produce the requested joint motion. FES can be used to develop underlying motor control by increasing repetition of the specific task movement (Kapadia et al., 2013). FES can develop motor control and decrease spasticity in hemiparetic patients. FES is accepted to increase afferent input and activate neuronal plasticity (Pierber et al., 2011).

Threshold electrical stimulation is defined as the application of low-level, subcontraction electrical stimulation in the home environment during sleep. It is thought that the increased blood flow will result in increased muscle mass as long as trophic hormone secretion is high (Dali et al., 2002).

Another type of current used in physiotherapy is high-voltage pulsed galvanic stimulation. This weak current has an extremely short pulse duration and causes minor electrochemical pain during the stimulation (Noreau et al., 2008).

The use of electrical stimulation is recommended for muscle disuse atrophy, following cast use, long-term orthosis use and postoperatively in children with CP. The use of electrical stimulation in children younger than the age of two and in obese patients has been reported to be contraindicated.

4.5. Hippotherapy

Horse-assisted rehabilitation practices are ancillary treatment methods that use the repetitive rhythmic movement of the horse as their basis. The motivation of the child with CP and his participation in the treatment in these applications, performed mostly in a natural environment, are usually positive as he is in continuous interaction with a living creature. The method positively contributes to the physical, emotional, cognitive and social aspects of the children. Horse-assisted rehabilitation practices can be divided into two as Recreational Horse Riding Treatment (RHRT) and hippotherapy (Şik et al., 2012).

RHRT is performed with trained horses and a horse trainer and focuses on the progressive protection of balance and posture, only using the slow and rhythmic gait of the horse. The success of defeating the emotional fear and anxiety and going through the riding phases enable the child to notice his own value and increases self-respect. The method provides motivation to teach something new to the child in the cognitive sense (Şik et al., 2012). Hippotherapy consists of a physiotherapist or occupational therapist using the movements of the horse as a therapy tool or method. Hippotherapy is an individual therapy that uses an interdisciplinary team approach. The basis of the method is the horse gait providing a marked, soft, rhythmic and repetitive movement model similar to the mechanics of the human gait (Winchester et al., 2002). The movements of the horse have a dynamic effect on the body of the child. Pelvic movements of the horse during the walk enable the pelvis and body of the child to move close to a normal gait. There are also views that when this rhythmic movement is combined with the neutral body temperature of the horse of 38 degrees, it decreases the hypertonicity in the child with CP and provides comfort. Adapting to the movements of the horse activates the muscles and joints and this can increase strength and provide a range of motion within time. The movement of the horse generally provides various inputs and these help to develop joint stability, weight transfer and postural balance responses in children with CP (Zadnikar & Kastrin, 2011, Bertoti, 1988, Quint & Toomey, 1998).

Hippotherapy mainly aims to provide balance and proper body posture in various positions, develop the child's sensory-motor and cognitive-motor skills, and gradually increase the stretching and movement capacity of the child while the horse is moving at a slow pace. When applied together with the neurodevelopmental treatment approach, it helps development of rough motor functions regarding balance, posture and mobility and brings muscle tone to a normal level in children with CP (Miller, 2007, Zadnikar & Kastrin, 2011).

There is a limited number of studies in the literature investigating the effectiveness of RHRT or hippotherapy on rough motor functions and postural control in children with CP. Hippotherapy is reported to develop rough motor functions and improve postural control, and contribute to balance, strength, coordination, muscle tone, joint range of motion, weight transfer and body posture in children with CP in the majority of these studies. Hippotherapy is also reported to have a positive effect on providing symmetry and functional motor skills in children with cerebral palsy. Besides, it contributes to psychological self-confidence, self-esteem, motivation, attention span, spatial awareness, concentration, and ability to speak (Miller, 2007, Şik et al., 2012, Zadnikar & Kastrin, 2011, Quint & Toomey, 1998).

4.6. Aquatherapy

Aquatherapy is one of the physiotherapy methods used for children with CP (Blohm, 2011). Aquatherapy can decrease spasticity, develop the tolerance to multisensor stimulators and increase the circulation due to the effect of hydrostatic pressure. The purpose of this therapy is to develop the ability of performing daily activities. Compared with the motions performed on land, water facilitates positioning by decreasing the gravity effect, decreases the pressure force applied to the joints, and therefore helps the children who cannot perform certain activities on land to move more fluently and actively. Additionally, the viscosity and flow characteristics of water increase body stabilization and help increase strength with the resistance they provide (Dumas & Francesconi, 2001, Thorpe, 2005, Hillier, et al., 2010). Aquatherapy also contains many elements of physiotherapy performed on land such as resistance exercise, aerobic exercise, endurance and motor skills. It also involves adapting to the water, functional independence, movement control in water, rotation, swimming and

http://dx.doi.org/10.5772/57505

breathing activities. It is considered to provide psychosocial benefits (Ennis, 2011, Bumin, et al., 2003).

The aquatherapy techniques used today are the Halliwick, Bad Ragaz and Watsu methods. The Halliwick method commonly used in children with CP is divided into four stages: (1) adapting to water, (2) rotation, (3) control of movement in water, (4) movement in water. The essence of this method consists of 10 points focusing on postural control while learning how to swim. The disabled individual first learns how to ensure balance in a supine stable position and then how to maintain balance in an unstable position. In other words, the Halliwick method is a motor learning program where the individual learns how to secure his balance (Bumin et al., 2003).

Watsu shiatsu is an aquatherapy method that combines stretching, joint mobilization and dancing. The movements of the individual are continuously supported during the session (http://www.watsu.org.nz/.). Another therapy method is Bad Ragaz where the individual is supported with floating devices and the therapist provides manual resistance to the individual's active movements. The therapist also applies facilitation that will provide proprioceptive input in order to activate the weak muscles. The Bad Ragaz method uses the principles of proprioceptive neuromuscular facilitation (PNF) (www.wcpt.org/apti/terminology).

5. Adulthood and cerebral palsy

Today, survival in childhood and adulthood CP continues to improve. The largest epidemiological database is a series of 47 000 people registered as using services in the State of California between 1983 and 2002 (Strauss et al., 2004). According to studies, survival has increased in the last 20 years by 3.4% per year, even in the most disabled group. There are different factors playing a role like improvement in nutrition, increased quality of care, and improvement of society's attitude to people with CP with a consequent provision of high quality medical care (Kent, 2013). Individuals with mild CP have nearly normal life expectancy (Hutton & Pharoah, 2006). In a study, (Strauss et al., 2004) there was a mild decline in ambulation in late adulthood and few who walked well initially maintained the skill over the following 15 years Additionally, there was also some evidence of a reduction in the abilities of upper extremity functions, possibly related to upper limb contractures. Speech, self-feeding and the communication in the wider community were all well continuous. Disabled people may suffer from intercurrent illness that is suboptimally managed resulted by communication difficulties, discrimination, or poor access to services compared to able-bodied peers, and this is more evident in younger age groups (Cannell et al., 2011, Kent, 2013).

5.1. Survival in adults with cerebral palsy

Survival in childhood and adulthood CP continues to improve although adverse prognostic factors include immobility, reduced upper limb function, and gastrostomy feeding. The largest epidemiological database is a series of 47 000 people registered as using services in the State of California between 1983 and 2002 (Strauss et al., 2004, Kent, 2013). Even in the most disabled group, survival has increased in the last 20 years by 3.4% per year. Improvement in nutrition, increased quality of care, and improvement of society's attitude to people with CP with a consequent provision of high quality medical care all appear to be playing a role (Kent, 2013). Individuals with mild CP have an almost normal life expectancy (Hutton and Pharoah, 2006). The management of aging in individuals with physical impairment is a new medical challenge. In a study of patients over 60 (Strauss et al., 2004) there was a mild decline in ambulation in late adulthood and few who walked well initially maintained the skill over the following 15 years. There was also some evidence of a reduction in the ability to self-dress, possibly related to upper limb contractures. Speech, self-feeding and the ability to communicate in the wider community were all well preserved. People with disabilities may suffer from intercurrent illness that is suboptimally managed because of communication difficulties, discrimination, or poor access to services compared to able-bodied peers, and this is more evident in younger age groups (Cannell et al., 2011, Kent, 2013).

5.2. Medical complications of cerebral palsy in adulthood

Medical problems in CP include those directly associated with the condition which may be present on a lifelong basis; these can be anticipated and will need monitoring. Second are the predictable complications of the condition such as worsening spasticity, where the main aim of treatment is to prevent deformity, improve nursing care, facilitate therapy, and increase tolerance of bracing (Kent., 2013).

Mechanisms of deterioration may include physical growth and weight gain, spasticity, and deformity leading to biomechanical disadvantage and muscle weakness. Abnormal compensations may break down with a loss of energy or fitness, for instance hip hitching for foot clearance, use of hip adductors to pull through in the presence of hip flexor weakness, or excessive use of lateral trunk flexion for gait progression. Spasticity is also a possible mechanism for accelerating problems with osteoarthritis. The onset of osteoarthritis is common in the general population and their various predisposing factors. In people with cerebral palsy, including abnormal gait and congenital joint malalignment, it may be anticipated that the effect of arthritis will manifest more rapidly and have a greater impact; people with mobility impairments use more energy to mobilize, have skeletal malalignment, deformity, contracture may contribute to pain and joint changes, and there is evidence of onset of musculoskeletal alteration in performance at an earlier age in people with CP (Kent 2013).

Deterioration in ambulation is a frequent presenting complaint. Most individuals remain in the same functional class of ambulation through adolescence and early adulthood. A large longitudinal study of 7550 children at 10 years and 5721 adults at 25 years (Wu et al., 2004, Day et al., 2007) showed that, although most improved their ambulatory capacity, 25% of those who walk at 10 years lose the ability by the age of 25. Of those using a wheelchair on an occasional basis, a third will be expected to lose their ability to walk by the age of 25, while the rest will remain ambulant for the next 15 years. In aging in the able-bodied, physical work capacity reduces, muscle strength generally is maintained into middle age, and complex performance activities can show a grade change because of coordination and integration of multiple functions, and these can be anticipated in those with a disability.

http://dx.doi.org/10.5772/57505

Other risks of deterioration include physiological burnout (Pimm, 1992) as a result of fatigue, reduced muscle power, dexterity, and mobility, intercurrent illness, and injury. Long bone fractures and prolonged immobilization and cognitive and depressive factors may also be important, although these can be interrelated (Ando & Ueda, 2000, Jahnsen et al., 2003; Strauss et al., 2004, Opheim et al., 2009). How individuals perceive their body influences how they manage everyday life and use coping mechanisms. In spite of this, life satisfaction in CP is similar to that in the general population, certainly in the Swedish and German populations (Sandstrom, 2007; Hergenr€oder & Blank, 2009). Pain and loss of function are more distressing than the overall level of functioning (Andren & Grimby, 2004) and should be treated appropriately.

The most common cause of death in CP is respiratory related (Reddihough et al., 2001). If people survive into their 40s and 50s cardiovascular disease (Krigger, 2006) and neoplastic disorders become more significant (Poulos et al., 2006). It is thought that there is an increase in mortality from cancer, stroke, and heart disease, partly due to lack of early detection and poor surveillance; breast cancer mortality is around three times the national rate. The incidence of cardiovascular and cerebral vascular conditions is two to six times higher than in the general population. The prevalence of poor health in patients with CP is not known as many do not present to health services. In a widely quoted study it was found that individuals had problems with kyphoscoliosis (26%), lower extremity contractures (71%), poor nutrition, i.e., underweight (60%), skin/hair problems (31%), bladder (56%) and bowel dysfunction (53%), and overall health problems that would warrant health service intervention (59%) (Thomas et al., 1989). The exact incidence of complications reflects the type, distribution, and age of subjects within the studies. The literature tends to support the view that individuals with CP "adjust to their own normality." In one study of the health of a group of women with a mean age of 37.5 years, around 68% were able to walk and 50% were independent in activities of daily living despite over a third of them having some degree of learning disability and 40% a seizure history.

Eighty-four percent complained of any sort of pain, 59% of hip and back deformities, 56% had bowel problems, 49% bladder problems, 43% had poor dental health, and 28% gastro-oesophageal reflux (Turk et al., 1997). In a similar vein, in a Swedish study 84% lived in their own home, 24% worked full time, and 64% could walk with or without aids. As many as 35% reported deterioration in walking ability and 9% had stopped completely. The prevalence of specific problems was 77% with spasticity, 80% with some contractures, and 18% with daily pain. In spite of this 60% regarded themselves as active and 54% were unlimited in their community mobility (Andersson & Mattsson, 2001).

Pain is an important underreported symptom. Common causes include osteoarthritis, soft tissue rheumatism, overuse injuries, fractures, and postural deformity. In one study 27% of the adults with CP had chronic pain compared with 15% in the general population (Loge & Kaasa, 1998; Jahnsen et al., 2004a, b). In adults with CP, however, pain did not increase with age, which is different from the general population (Gajdosik & Cicirello, 2001). The most frequent site was back pain, both in adults with CP and in the general population. Pain in different body parts was associated with those exposed to special strain in the different types of CP, for example a high prevalence of neck and shoulder pain in persons with dyskinesia. More pain was significantly associated with being female, having a high fatigue score, low life satisfaction, and low and deteriorated physical function. Pain was associated with both overuse and inactivity. In a systematic review of studies including those that included adults with CP, psychosocial factors were shown to be significantly associated with pain and dysfunction in all disability groups. The psychosocial factors most closely associated with pain and dysfunction across the samples included: (1) catastrophizing cognitions; (2) task persistence, guarding, and resting coping responses; and (3) perceived social support and solicitous responding social factors. Psychosocial factors are significant predictors of pain and functioning in persons with physical disabilities. It is probable that psychosocial interventions are as helpful in patients with CP as in the general population, but this needs further research (Jensen et al., 2011).

In one series 76% of community-living adults self identified more than one muscular skeletal complaint and 55% of these had sustained fractures (Murphy et al., 1995). Pain limits activity (Turk et al., 1997), and two-thirds can anticipate having moderate to severe, 24% constant, and 56% daily pain. In another study 32% reported dissatisfaction with pain management (Engel et al., 2006). Other causes of chronic pain in musculoskeletal disorders include hip dysplasia (Hodgkinson et al., 2001) leading to postural problems and back pain. A wide variety of chronic pain syndromes include back pain, spinal stenosis, and degenerative disk disease. Individuals coped well with pain considering its duration and persistence (Castle et al., 2007). Joint pain may be related to primary or secondary osteoarthritis, as well as spasticity, contractures, or co-contraction leading to gait disturbance, for example spastic equinus and hyperextension of the knee.

Gait analysis and appropriate use of focal treatment of spasticity orthotics can be helpful in management. In one study 64% of ambulators and 91% of nonambulators had contractures, 27% had pain in weight-bearing joints, and 21% had muscle pain and spasm (Murphy et al., 1995). Similar findings were pain (59%) and joint deformities (19_57%) which were observed in a cohort of 25–36-year-olds, many of whom had lost contact with follow-up services (Hilberink et al., 2007).

Cervical myelopathy is an important reversible cause of deterioration, particularly in those with dyskinetic CP. It has been postulated that cervical instability, disk herniation, spondylosis, osteophytes, and stenosis of the spinal canal may all lead to this condition (Fletcher & Marsden, 1996; Amess et al., 1998).

Metabolic bone disease is more prevalent in institutional environments and vitamin D supplements may need to be considered in such groups. Osteoporosis is also common in people who are immobile, have never been mobile, those with neuroendocrine abnormalities, and those who have used anticonvulsants (King et al., 2003). Osteoporosis can give rise to low impact fractures of either the vertebrae or long bones such as the femur.

Falls risk can also increase the prevalence of fractures. One series gave 30% as having suffered from a fracture (Murphy et al., 1995). Soft tissue rheumatism includes tenosynovitis and elbow or hip bursitis. Some of this is related to problems associated with abnormal forces across joints worsened by spasticity and deformity. The use of crutches can be associated with ulnar neuropathies, and self-propulsion of wheelchairs can also lead to shoulder and elbow problems.

In hemiplegia,musculoskeletal overuse injuries may affect the unaffected side. One study reported that 10% of a mixed CP sample and 20% of people with dyskinesia had carpal tunnel symptoms (Murphy et al., 1995).

6. Conclusion

The main problem in CP is locomotion. Problems in muscle tone, muscle strength, balance, and reflex development affect the motor development and the muscle contracture, joint limitation, postural disorders added in later years decrease motor performance. The evaluation of these problems in detail and the use of the appropriate approaches are of vital importance.

Author details

Mintaze Kerem Günel, Duygu Türker, Cemil Ozal and Ozgun Kaya Kara

Hacettepe University, Faculty of Health Sciences, Department of Physiotherapy and Rehabilitation, Ankara, Turkey

References

[1] Accardo P J, Freud on Diplegia: Commentary and Translation. American Journal of Diseases of Children. 1982;136:452–456

[2] Akbayrak T, Armutlu K, Kerem Gunel M, Nurlu G. Assessment Of The Short-Term Effect Of Antispastic Positioning On Spasticity. Pediatrics International 2005;47(4): 440–445.

[3] Albright A. Spasticity And Movement Disorders In Cerebral Palsy. J Child Neurol 1996;11(1):1-4.

[4] Anderson PJ, De Luca CR, Hutchinson E, Roberts G, Doyle LW. Underestimation Of Developmental Delay By The New Bayley-III Scale. Archives of pediatrics & adolescent medicine 2010; 164 (4): 352-356.

[5] Andersson C, Mattsson E . Adults with cerebral palsy: a survey describing problems, needs, and resources, with special emphasis on locomotion. Dev Med Child Neurol. 2001; 43: 76–82.

[6] Ando N, Ueda S. Functional deterioration in adults with cerebral palsy. Clin Rehabil. 2000; 14: 300–306.

[7] Andren E, Grimby G. Dependence in daily activities and life satisfaction in adult subjects with cerebral palsy or spina bifida: a follow-up study. Disabil Rehabil.2004; 26: 528–536.

[8] Aquatic Bodywork Association New Zealand. What is Watsu? Available at: http://www.watsu.org.nz/. Accessed July. 2013

[9] Arya B K, Subramanya K, Mahadevappa M, Kumar R. Electrical Stimulation Devices For Cerebral Palsy: Design Considerations, Therapeutic Effects And Future Directions. In:Yue W., Chattopadhyay S, Lim T C, Acharya U R, (Ed). Advances in Therapeutic Engineering. Boca Raton, FL: CRC Press, Taylor & Francis. 2012.

[10] Bax M, Goldstein M, Rosenbaum P, Leviton A, Paneth N, Jacobsson B, Damiano DL. Proposed Definition And Classification Of Cerebral Palsy. Dev Med Child Neurol 2005; 47(8):571-576

[11] Bayley N. Bayley Scales Of İnfant Development. II Manual. San Antonio 1993: Harcourt Brace.

[12] Bayley N. Bayley Scales of Infant and Toddler Development-Third Edition. San Antonio 2006, TX, USA: Harcourt Assessment.

[13] Berry ET, Giuliani CA, Damiano DL. Intrasession and Intersession Reliability of Handheld Dynamometry in Children with Cerebral Palsy. Pediatric Physical Therapy 2004;16:191- 198.

[14] Bertoti D B. Effect Of Therapeutic Horseback Riding On Posture In Children With Cerebral Palsy. Phys Ther 1988; 68: 1505–12.

[15] Blair E, Stanley F J. Mental Retardation and Developmental Disabilities Research Reviews. Special Issue: Cerebral Palsy. 1997; 3(2):184–193

[16] Blauw-Hospers CH, Hadders-Algra M. A Systematic Review Of The Effects Of Early İntervention On Motor Development. Developmental Medicine and Child Neurology 2005; 47 (6): 421-432.

[17] Blohm D. Effectiveness Of Aquatic İnterventions For Children With Cerebral Palsy: Systematic Review Of Current Literature. J Aquatic Phys Ther 2011;19(1): 19-29.

[18] Blundell S W, Shephard R B, Dean C M, Adams R D. Functional Strength Training In Cerebral Palsy: A Pilot Study Of A Group Circuit Training Class For Children Aged 4–8 Years. Clinical Rehabilitation 2003;17: 48-57

[19] Bly L. A Historical And Current View Of The Basis Of NDT. Pediatric Physical Therapy 1991;3(3):131-135.

[20] Bobath K. The Normal Postural Reflect Mechanism And İts Deviation In Children With Cerebral Palsy. Physiotherapy 1971;57(11):526–532.

[21] Bobath K. A Neurophysiological Basis For Treatment Of Cerebral Palsy. Second edition.1980

[22] Bohannon RW. Smith MB. Interrater Reability Of A Modified Ashworth Scale Of Muscle Spasticity. Phys ther 1987;67(2): 206-207

[23] Bower E, McLellan DL. Effect Of Increased Exposure To Physiotherapy On Skill Acquisition Of Children With Cerebral Palsy. Dev Med Child Neurol 1992;34(1):25–39

[24] Boyd RN, Graham HK. Objective Measurement of Clinical Findings In The Use of Botulinum Toxin Type A for The Management of Children with Cerebral Palsy. European Journul of Neurology 1999; 6: 23-35.

[25] Brandäo M B., Mancini M C, Vaz D V, Bueno A M, Furtado S R C, Coelho Z A C. Effects Of Constraint-Induced Movement Therapy In Children With Hemiplegia: A Single Case Experimental Study. Revista Brasileira de Fisioterapia 2009; 13(6): 527-534.

[26] Brown J K. Walsh E G. Neurology of the upper limb. In: Neville B, Goodman R, editors. Congenital Hemiplegia. Clinics in Developmental Medicine No. 150. London: Mac Keith Press. 2000; p 113–149.

[27] Bumin G, Uyanik M, Yilmaz I, Kayihan H, Topçu M. Hydrotherapy for Rett Syndrome. J Rehabil Med 2003; 35(1): 44-45.

[28] Burns Y, Ensbey R, Norrie M. The Neurosensory Motor Developmental Assessment Part I & Part II. The Australian Journal of Physiotherapy 1989; 35: 141-157.

[29] Burns Y. NSMDA: Physiotherapy Assessment For Infants And Young Children. Brisbane 1992, Australia: Copyright Publishing Co Pty Ltd.

[30] Burns Y R, Danks M, O'Callaghan MJ, Gray PH, Cooper D, Poulsen L, et al Motor Coordination Difficulties And Physical Fitness Of Extremely-Low-Birthweight Children. Developmental medicine and child neurology 2009; 51 (2): 136-142.

[31] Butler C, Chambers H, Goldstein M. Evaluating Research In Developmental Disabilities: A Conceptual Framework For Reviewing Treatment Outcomes. Dev Med Child Neurol 1999; 41(1):55–59

[32] Butler C, Darrah J. AACPDM Evidence Report: Effects Of Neurodevelopmental Treatment (NDT) For Cerebral Palsy. Dev Med Child Neurol 2001;.43(11):778–790.

[33] Camper KS, Linden DWV, Palisano R J. Physical Therapy for Children. 2. ed. Philadelphia 2000: WB Saunders Company. p.: 534.

[34] Cannell MB, Brumback BA, Bouldin E D et al. Age Group Differences In Healthcare Access For People With Disabilities: Are Young Adults At Increased Risk? JAdolescHealth. 2011; 49: 219–221.

[35] Cans C, De-la-Cruz J, Mermet M A. Epidemiology of Cerebral Palsy. Paediatrics And Child Health, 2008; 18(9): 393-398

[36] Carey W. Crocker A. Elias E. Feldman H. Coleman W. Developmental-Behavioral Pediatrics, 4th Edition, Saunders, Philadephia, 2009; 658-660

[37] Castle K, Imms C, Howie L. Being In Pain: A Phenomenological Study Of Young People With Cerebral Palsy. Dev Med Child Neurol.2007; 49: 445–449.

[38] Cauraugh J H, Naik S K, Wen Hao H, Coombes S A, Hol KG. Children With Cerebral Palsy: A Systematic Review And Metaanalysis On Gait And Electrical Stimulation. Clinical Rehabilitation 2011; 24: 963-978

[39] Charles J R, Gordon A M. A Critical Review Of Constraint-Induced Movement Therapy And Forced Use In Children With Hemiplegia. Neural Plasticity 2005; 12(2-3): 245-61.

[40] Charles J R, Wolf S L, Schneider JA, Gordon A M. Efficacy Of A Childfriendly Form Of Constraint-Induced Movement Therapy In Hemiplegic Cerebral Palsy: A Randomized Control Trial. Developmental Medicine and Child Neurology 2006; 48(8): 635-642.

[41] Charles J R, Wolf S L. Invited Commentary. CIMT in Children with Cerebral Palsy. Physical Therapy 2009; 89:1142-1143.

[42] Charles J R, Lavinder G, Gordon, A M. Effects Of Constraint-Induced Therapy On Hand Function In Children With Hemiplegic Cerebral Palsy. Pediatric Physical Therapy 2001; 13(2): 68-76.

[43] Cheng R J, Liu C F, Lau T W, Hong R B. Effect of Treadmill Training with Body Weight Support on Gait and Gross Motor Function in Children with Spastic Cerebral Palsy. American Journal of Physical Medicine & Rehabilitation 2007; 86(7):548-555

[44] Clopton N, Dutton J, Featherston T, Grisby A, Mobley J, Melvin J. Interrater And İntrarater Reliability Of The Modified Ashworth Scale In Children With Hypertonia. Pediatr. Phys. Ther 2005;17: 268-274.

[45] Cope S M, Forst H C, Bibis D, Liu X C. Modified Constraint-Induced Movement Therapy For A 12-Month-Old Child With Hemiplegia: A Case Report. American Journal of Occupational Therapy 2008; 62(4):430-7.

[46] Dali C, Hansen F J, Pedersen S A, Skov L, Hilden J, Bjørnskov I, Strandberg C, Christensen J, Haugsted U, Herbst G. Threshold Electrical Stimulation (TES) In Ambulant Children With CP: A Randomized Double-Blind Placebocontrolled Clinical Trial. Dev Med Child Neurol 2002; 44: 364–369.

http://dx.doi.org/10.5772/57505

[47] Damiano DL, Kelly LE, Vaughn CLB. Effects of Quadriceps Femoris Muscle Strengthening on Crouch Gait in Children With Spastic Diplegia. Physical Tehrapy 1995;75(8): 658-671.

[48] Damiano D L, Kelly L E, Vaughan C L. Effects Of A Quadriceps Femoris Strengthening Programme On Crouch Gait In Children With Cerebral Palsy. Phys Ther 1995; 75(8): 658-667.

[49] Damiano D. Physiotherapy Management In Cerebral Palsy:Moving Beyond Philosophies. In: Scrutton D, Damiano D, Myston M (Ed). Management Of The Motor Disorders Of Children With Cerebral Palsy. Mac Ketith Pres. London 2004; 161-169 .

[50] Damiano DL, Abel MF. Functional Outcomes Of Strength Training In Spastic Cerebral Palsy. Arch Phys Med Rehabil 1998; 79: 119–25.

[51] Damiano DL, Arnold SA, Steeled KM, Delp SL. Can Strength Training Predictably Improve Gait Kinematics? A Pilot Study on the Effects of Hip and Knee Extensor Strengthening on Lower- Extremity Alignment in Cerebral Palsy. Physical Therapy 2010; 90(2):269-279

[52] Damiano DL, Vaughan CL, Abel MF. Muscle Response To Heavy Resistance Exercise In Children With Spastic Cerebral Palsy. Dev Med Child Neurol 1995; 37: 732-739

[53] Day SM, Wu WV, Strauss DJ et al. Change In Ambulatory Ability Of Adolescents And Young Adults With Cerebral Palsy. Dev Med Child Neurol 49:2007; 647–653.

[54] DeGangi G, Royeen C. Current Practice Among Neurodevelopmental Treatment Association Members. Am J Occup Ther 1994;.48(9):803-809.

[55] Deluca PA. Gait Analysis In The Treatment Of Ambulatory Child With Cerebral Palsy. Clin. Orthop. Relat. Res 1991; 264:65-75.

[56] DeLuca S C, Echols K, Ramey S, Taub E. Pediatric Constraint Induced Movement Therapy For A Young Child With Cerebral Palsy: Two Episodes Of Care. Physical Therapy 2003; 83: 1003-1013.

[57] DeLuca S. Intensive movement therapy with casting for children with hemiparetic cerebral palsy: a randomised controlled trial. Dissertation, The University of Alabama at Birmingham, 2002.

[58] DiBiasio P, Lewis C. Exercise Training Utilizing Body Weight Supperted Treadmill Walking With Young Adult With Cerebral Palsy Who Was Non-Ambulatory. Physiotherapy Theory and Pract 2012; 28(8): 641-652.

[59] Dodd J K, Foley S. Partial Body-Weight-Supported Treadmill Training Can Improve Walking In Children With Cerebral Palsy: A Clinical Controlled Trial. Developmental Medicine & Child Neurology 2007;49(2): 101-105.

[60] Dodd KJ, Taylor, NF, Damiano DL. A Systematic Review Of The Effectiveness Of The Strength-Training Programs In People With Cerebral Palsy. Arch Phys Med Rehabil 2002, 83, 1157–64.

[61] Dormans J P. Pellegrino L. Caring For Children With Cerebral Palsy .. Baltimore: Brooks Publishing. 1998; 8-21

[62] Dormans J P. Pellegrino L. Caring For Children With Cerebral Palsy .. Baltimore: Brooks Publishing. 1998; 8-21

[63] Dumas H, Francesconi S. Aquatic Therapy In Pediatrics: Annotated Bibliography. Phys Occup Ther Pediatr 2001;20(4): 63-78.

[64] Eliasson A C, Krumlinde-sundholm L, Shaw K, Wang C. Effects Of Constraint Induced Movement Therapy In Young Children With Hemiplegic Cerebral Palsy: An Adapted Model. Developmental Medicine and Child Neurology 2005; 47(4):266-75.

[65] Engel JM, Jensen MP, Schwartz L (2006). Coping With Chronic Pain Associated With Cerebral Palsy. Occup Ther Int 13: 224–233.

[66] Engsberg JR, Ross SA, Hollander KW et al. Hip Spasticity And Strength In Children With Spastic Diplegia Cerebral Palsy. J Appl Biomech 2000; 16: 221–33.

[67] Ennis E. The Effects Of A Physical Therapy-Directed Aquatic Program On Children With Autism Spectrum Disorders. J Aquatic Phys Ther 2011; 19(1): 4-10.

[68] Evans P. Johnson A. Mutch L. Alberman E A. Standard Form For Reporting Clinical Findings in Children With A Motor Deficit of Central Origin. Dev Med Child Neurol 1989; 31(1):119-27.

[69] Exner C E. Development Of Hand Skills. In: Case-Smith J, editor. Occupational Therapy for Children. 4th Ed. St. Louis: Mosby. 2001; p 289–327.

[70] Fennell E B, Dikel TN: Cognitive And Neuropsychological Functioning In Children With Cerebral Palsy. J Child Neurol 2001; 16:58-63

[71] Fetters L, Kluzik J. The Effects Of Neurodevelopmental Treatment Versus Practice On The Reaching Of Children With Spastic Cerebral Palsy. Phys Ther 1996;.76(4): 346-358.

[72] Finlay H, Ainscough J, Craig J. et al. Current Clinical Practice In The Use Of Muscle Strengthening In Children And Young People With Cerebral Palsy - A Regional Survey Of Paediatric Physiotherapists. APCP Journal 2012; 3(1): 27-41

[73] Fletcher NA, Marsden CD. Dyskinetic Cerebral Palsy: A Clinical And Genetic Study. DevMed Child Neurol.1996; 38: 873–880.

[74] Fowler EG, Ho TW, Nwigwe AI, Dorey FJ. The Effect of Quadríceps Femoris Muscle Strengthening Exercises On Spasticity in Children With Cerebral Palsy. Physical Therapy2001; 81(1): 1215-1223

[75] Gajdosik CG, Cicirello N. Secondary conditions of the musculoskeletal system in adolescents and adults with cerebral palsy. Phys Occup Ther Pediatr. 2001; 21: 49–68.

[76] Garne E. Dolk H. Krägeloh-Mann I. Holst Ravn S. Cans C. SCPE Collaborative Group. Cerebral Palsy And Congenital Malformations. Eur J Paediatr Neurol. 2008; 12(2): 82-88

[77] Gibson C S. MacLennan A H. Goldwater P N. Dekker G A. Antenatal Cause Of Cerebral Palsy: Associations Between Inherited Thrombophilias, Viral And Bacterial Infection, And Inherited Susceptibility To Infection. 2003; 58(3):209-20.

[78] Giuliani CA. Dorsal Rhizotomy For Children With Cerebral Palsy: Support For Concepts Of Motor Control. Phys Ther 1991; 71: 248–59.

[79] Gordon A M, Charles J, Wolf S L. Efficacy Of Constraint-Induced Movement Therapy On İnvolved Upper-Extremity Use In Children With Hemiplegic Cerebral Palsy Is Not Age-Dependent. Pediatrics 2006; 117(3): 363-73.

[80] Gordon AM. Charles J. Duff S V. Fingertip Forces In Children With Hemiplegic Cerebral Palsy. II: Bilateral Coordination. Dev Med Child Neurol 1999; 41:176–185.

[81] Gracies JM, Burke K, Clegg NJ, Browne R, Rushing C, Fehlings D, et al. Reliability of the Tardieu Scale for Assessing Spasticity in Children With Cerebral Palsy. Arch Phys Med Rehabilitation 2010; 91:421-428.

[82] Graham HK, Harvey A, et al.The Functional Mobility Scale FMS.J Pediatre Orthop 2004; 24:514-520.

[83] Hamamci N, Dursun E. Serebral Palsi Rehabilitasyonu ve Guillain Barré Rehabilitasyonu. In: Oğuz h. (ed). Tibbi Rehabilitasyon. İstanbul: Nobel Tip Kitabevleri; 1995. p633-652

[84] Han T R. Bang M S. Lim J Y. Yoon B H. Kim I W. Risk Factors Of Cerebral Palsy In Preterm Infants. American Journal of Physical Medicine and Rehabilitation. 2005; 81: 297-303.

[85] Hergenröder H, Blank R. Subjective Well-Being And Satisfaction With Life In Adults With Spastic Cerebral Palsy: A Pilot Study Of A Randomized. Dev Med Child Neurol. 2009 May;51(5):389-96

[86] Herndon WA, Troup P, Yngve DA, Sullivan JA. Effects Of Neurodevelopmental Treatment On Movement Patterns Of Children With Cerebral Palsy. J Pediatr Orthop 1987;7(4):395-400.

[87] Heyrman L Molenaers G, Desloovere K, Verheyden G, De Cat J, Monbaliu E. Feys H. A Clinical Tool To Measure Trunk Control In Children With Cerebral Palsy: The Trunk Control Measurement Scale. Research in developmental Disabilities 2011;32(6): 2624–2635.

[88] Hilberink SR, Roebroeck ME, Nieuwstraten W et al. Health Issues In Young Adults With Cerebral Palsy: Towards A Life-Span Perspective. J Rehabil Med.2007; 39: 605–611.

[89] Hillier S, McIntyre A, Plummer L. Aquatic Physical Therapy For Children With Developmental Coordination Disorder: A Pilot Randomized Controlled Trial. Phys Occup Ther Pedi 2010; 30(2): 111-124.

[90] Hoare B, Imms C, Carey L, Wasiak J. Constraint-Induced Movement Therapy In The Treatment Of The Upper Limb In Children With Hemiplegic Cerebral Palsy: A Cochrane Systematic Review.Clinical Rehabilitation 2007; 21: 675–685

[91] Hodgkinson I, Jindrich ML, Duhaut P et al. Hip pain in 234 non-ambulatory adolescents and young adults with cerebral palsy: a cross-sectional multicentre study. Dev Med Child Neurol .2001; 43: 806–808.

[92] Hutton JL, Pharoah PO. Life Expectancy In Severe Cerebral Palsy. Arch Dis Child. 2006; 91: 254–258.

[93] Jahnsen R, Villien L, Aamodt G et al. Musculoskeletal Pain In Adults With Cerebral Palsy Compared With The General Population. J Rehabil Med. 2004b; 36: 78–84.

[94] Jahnsen R, Villien L, Egeland T et al. Locomotion Skills In Adults With Cerebral Palsy. Clin Rehabil 2004a; 18: 309–316.

[95] Jahnsen R, Villien L, Stanghelle JK et al. Fatigue In Adults With Cerebral Palsy In Norway Compared With The General Population. Dev Med Child Neurol 2003;45: 296–303.

[96] Jensen MP, Moore MR, Bockow TB et al. Psychosocial Factors And Adjustment To Chronic Pain In Persons With Physical Disabilities: A Systematic Review. Arch Phys Med Rehabil.2011; 92: 146–160.

[97] Johnson S, Marlow N. Developmental Screen Or Developmental Testing? Early human development 2006;82 (3): 173-183.

[98] Jones M W, Morgan E, Shelton C E, Thorogood C, Cerebral Palsy: Introduction and Diagnosis (Part I). J Pediatr Health Care. 2007; 21: 146-152.

[99] Kapadia N M, Nagai M K, Zivanovic V, Bernstein J, Woodhouse J, Rumney P, Popovic MR. Functional Electrical Stimulation Therapy for Recovery of Reaching and Grasping In Severe Chronic Pediatric Stroke Patients. Journal of Child Neurology 2013; 0: 1-7.

[100] Kawamura C, Filh MCM, Barreto M M, Asa SKP, Juliano Y, Novo NF. Comparsion Between Visual And Three-Dimensional Gait Analysis In Patiens With Spastic Diplegic Cerebral Palsy. Gait Posture 2007; 25: 18-24.

[101] Kent R. M. Cerebral palsy. Handbook of Clinical Neurology, 2013; 110 (3rd series. Neurological Rehabilitation M.P. Barnes and D.C. Good, Editors, Elsevier B.V.

[102] Kerem Gunel M. Physiotherapy For Children With Cerebral Palsy. In: Zeljka Petelin Gadze.(Ed).Epilepsy In Children-Clinical And Social Aspects. Rijeka:Intech; 2011.p213-134

[103] Kerem Gunel M. Rehabilitation Of Children With Cerebral Palsy From A Physiotherapist's Perspective. Acta Orthop Traumatol Turc 2009;43(2):173-181.

[104] Kerr C, McDowell B, McDonough S. Electrical Stimulation In Cerebral Palsy: A Review Of Effects On Strength And Motor Function. Dev Med Child Neurol 2004; 46: 205–213.

[105] King W, Levin R, Schmidt R et al. Prevalence Of Reduced Bone Mass In Children And Adults With Spastic Quadriplegia. Dev Med Child Neurol.2003; 45: 12–16.

[106] Kluzik J, Fetters L, Coryell J. Quantification Of Control: A Preliminary Study Of Effects Of Neurodevelopmental Treatment On Reaching In Children With Spastic Cerebral Palsy. Physical Therapy 1990; 70(2):65-76

[107] Krägeloh-Mann I. Cans C. Cerebral Palsy Update. Brain Dev, 2009; 31(7): 537-544

[108] Krigger KW (2006). Cerebral Palsy: An Overview. Am Fam Physician 73: 91–100.

[109] Krumlinde-Sundholm L. Eliasson A C. Development Of The Assisting Hand Assessment: A Rasch-Built Measure Intended For Children With Unilateral Upper Limb Impairments. Scan J Occup Ther. 2003; 10: 16–26.

[110] Kuban K. Leviton A. Cerebral Palsy. The New England Journal of Medicine. 1994; 330: 188-195.

[111] Kuhtz-Buschbeck J P. Krumlinde-Sundholm L. Eliasson A C. Forssberg H. Quantitative Evaluation Of Mirror Movements In Children With Hemiplegic Cerebral Palsy. Dev Med Child Neurol. 2000; 42: 728–736.

[112] Glader L, Tilton A. Cerebral Palsy. In: Carey W. Crocker A. Elias E. Feldman H. Coleman W. (ed.) Developmental-Behavioral Pediatrics, 4th Edition, Saunders, Philadephia, 2009; 658-662

[113] Law M, Russell D, Pollock N, Rosenbaum P, Walter S, King G. A Comparison Of İntensive Neurodevelopmental Therapy Plus Casting And A Regular Occupational Therapy Program For Children With Cerebral Palsy. Dev Med Child Neurol 1997;39(1):664- 670.

[114] Livanelioğlu A, Kerem Gunel M. Serebral Palside Fizyoterapi. Ankara: Yeni Özbek: 2009

[115] Loge JH, Kaasa S. Short Form 36 (Sf36) Health Survey: Normative Data From The General Norwegian population. Scand J Social Med.1998; 26: 250–258.

[116] Lonstein JE. The Spine In Cerebral Palsy. Current Orthopaedics 1995; 9:164-177

[117] Maathuis KBB, Van der Shans CP, Van iperen A, Riedman HS, Geertzen JHB. Gait In Children With Cerebral Palsy Observer Reliability Of Physcian Raiting Scale And

Edinburgh Vizual Gait Analysis Interval Testing Scale. J Pediatre Orthop 2005; 25:268-272

[118] MacDOnald J, Burns Y. Performance On The NSMDA During First And Second Year Of Life To Predict Functional Ability At The Age Of 4 In Children With Cerebral Palsy. Hong Kong Physiotherapy Journal 2005; 23: 40-45.

[119] MacPhail H E A, Kramer J F. Effect Of Isokinetic Strengthtraining On Functional Ability And Walking Efficiency In Adolescents With Cerebral Palsy. Dev Med Child Neurol 1995; 37(9):763–775.

[120] Masciento L R, Gloria A E, Habip E S. Effects Of Constraint-Induced Movement Therapy As A Rehabilitation Strategy For The Affected Upper Limb Of Children With Hemiparesis:Systematic Rewiev Of The Literature. Revista Brasileira de Fisioterapia 2009;13(2): 97-102.

[121] Mayston M. Bobath Concept: Bobath@50: Mid-Life Crisis--What Of The Future? Physiother Res Int 2008;13(3):131-136

[122] McBurney H, Taylor NF, Dodd KJ, Graham HK. A Qualitative Analysis Of The Benefits Of Strength Training For Young People With Cerebral Palsy. Dev Med Child Neurol 2003; 45: 658–63

[123] Miller, F. Physical Therapy of Cerebral Palsy.Wilmington 2007: Springer. 354-391.

[124] Miller F., Bagg M R. Age And Migration Percentage As Risk Factors For Progression In Spastic Hip Disease. Developmental medicine and child neurology. 1992; 37: 449-455

[125] Miller G. Cerebral Palsies: An Overview. In Miller G, Clark GD (eds): The Cerebral Palsies: Causes, Consequences, and Management, Boston, Butterworth-Heinemann, 1998; 1-35.

[126] Mockford M. Caulton JM. Systematic Review of Progressive Strength Training in Children and Adolescents with Cerebral Palsy Who Are Ambulatory. Pediatr Phys Ther 2008; 20(4):318-3

[127] Morris D M, Taub E. Constraint-Induced Therapy Approach To Restoring Function After Neurological Injury. Top Stroke Rehabil 2001; 8: 16–30.

[128] Moster D, Lie R T. Irgens L. Bjerkedal T, Markestad T. The Association Of Apgar Score With Subsequent Death And Cerebral Palsy: A Population-Based Study In Term Infants. Journal of Pediatrics. 2001; 138: 798-803.

[129] Msall M E, Digaudio K M, Duffy L C. Use Of Functional Assessment In Children With Developmental Disabilities. Phys Med Rehabi. Clin North Am 1993; 4: 517- 27.

[130] Murphy K P, Molnar G E, Lankasky K (1995). Medical And Functional Status Of Adults With Cerebral Palsy. Dev Med Child Neurol 37: 1075–1084.

[131] Mutlu A, Livanelioglu A, Kerem Gunel M. Reliability Of Ashworth And Modified Ashworth Scales In Children With Spastic Cerebral Palsy. BMC Musculoskeletal Disorders 2008;9:1471-2474.

[132] Mutlu A, Livanelioglu A, Kerem Gunel M. Reliability Of Goniometric Measurements In Children With Spastic Cerebral Palsy. Med Sci Monit 2007;13(7):323-329.

[133] Naeye R L. Peters E. Bartholomew M. Landis R. Origins Of Cerebral Palsy. American Journal of Disease of Children. 1989; 143: 1154-1161

[134] Naylor C E. Bower E. Modified Constraint Induced Movement Therapy For Young Children With Hemiplegic Cerebral Palsy: A Pilot Study. Developmental Medicine and Child Neurology 2005; 47: 365-369.

[135] Nelson K B, Ellenberg J H. Antecedents Of Cerebral Palsy-Multivariant Analysis. N Engl J Med. 1986; 315:81-86

[136] Nelson K B. Causes of Cerebral Palsy. Current Opinions in Pediatrics.1989; 11: 487-491.

[137] Nelson K B, Ellenberg J H. Antecedents Of Cerebral Palsy: Multivariate Analysis Of Risk. New England Journal of Medicine.1986; 315: 81–86

[138] Noreau L, Lepage C, Bernard P M. Recent Modality In Physical Therapy In Treatment Of Cerebral Palsy. Phys. Ther 2008; 78(5): 458- 469.

[139] Novacheck T, Stout J, Tervo R. Reliability And Validity Of The Gilette Functional Asessment Questioner As An Outcome Measure In Children With Walkinh Diasbilities. Journal of pediatric orthopaedics 2000; 20(1): 75 -81

[140] Opheim A, Jahnsen R, Olsson E et al. Walking Function, Pain, And Fatigue In Adults With Cerebral Palsy: A 7-Year Follow-Up Study. Dev Med Child Neurol.2009;51: 381–388.

[141] Ottenbacher KJ, Msall ME, Lyon N, Duffy LC, Granger CV, Braun S. Measuring Developmental And Functional Status In Children With Disabilities. Dev Med Child Neurol 1999;.41(3):186-194.

[142] Palisano R, Rosenbaum P, Walter S, Russell D, et al. Development And Reliability Of A System To Classify Gross Motor Function In Children With Cerebral Palsy. Dev Med Child Neurol 1997;39(4):214-223.

[143] Palisano R J, Campbell S K, Harris S R. Evidence- Based Decision Making In Pediatric Physical Therapy. In: Campbell SK, Vander Linden D W Palisano R J (ed). Physical therapy for children. 3rd ed. Missouri: Saunders Elsevier, 2006: 0-32

[144] Palisano R, Rosenbaum P, Backett P, Livingston M. Gross Motor Function Classification System Expended And Revised. Dev Med Child Neurol 2007;39 : 214-223.

[145] Patrick JH, Roberts P, Cole GF.Therapeutic Choise In Locomotor Management Of The Child With Cerebral Palsy- More Luck Than Judgement?. Arch. Dis. Chil 2001; 85: 275-279.

[146] Pharoah P, Cooke T, Rosenbloom L. Acquired Cerebral Palsy. Arch Dis Child. 1989; 64: 1013-1016

[147] Pieber K, Herceg M, Wick F, Grim-Stieger M, Bernert G, Paternostro-Sluga T. Functional Electrical Stimulation Combined With Botulinum Toxin Type A To Improve Hand Function In Children With Spastic Hemiparesis – A Pilot Study. Wien Klin Wochenschr 2011; 123: 100–105.

[148] Pierce S, Daly K, Gallagher K. Constraint Induced Therapy For A Child With Hemiplegic Cerebral Palsy: A Case Report. Archive of Physical Medicine and Rehabilitation 2002;83:1462-1463.

[149] Pimm P (1992). The Progression Of Cerebral Palsy In Adulthood. Educational and Child Psychology 9: 3–4.

[150] Poulos AE, Balandin S, Llewellyn G et al. Women With Cerebral Palsy And Breast Cancer Screening By Mammography. Arch Phys Med Rehabil . 2006;87: 304–307.

[151] Quint C, Toomey M. Powered Saddle And Pelvic Mobility: An Investigation Into The Effects On Pelvic Mobility Of Children With Cerebral Palsy Of A Powered Saddle Which Imitates The Movements Of A Walking Horse. Physiotherapy 1998; 84: 376–84.

[152] Reddihough DS, Baikie G, Walstab JE. Cerebral palsy in Victoria, Australia: Mortality And Causes Of Death.J Paediatr Child Health.2001; 37: 183–186.

[153] Reed B. The Physiology Of Neuromuscular Electrical Stimulation. Pediatr Phys Ther. 1997; 9: 96–102.

[154] Ries J D. Leonard R. Is There Evidence To Support The Use Of Constraint-Induced Therapy To Improve The Quality Or Quantity Of Upper Extremity Function Of A 2 1/2-Year-Old Girl With Congenital Hemiparesis? If So, What Are The Optimal Parameters Of This İntervention? Phyical Therapy 2006; 86: 746-752.

[155] Rosembaum P, Panet N, Leviton A, Goldstein M, Bax M. A Report: The Definiation And Classification Of Cerebral Palsy. Dev Med Child Neurol 2007 49: 7-14

[156] Rosembaum P. Cerebral Palsy: What Parents and Doctors Want to Know. British Medical Journal. 2003; 326, 970-974.

[157] Rosenbaum P, Paneth N, Leviton A, et al. A Report: The Definition And Classification Of Cerebral Palsy. Dev Med Child Neurol. 2007; 109:8-14

[158] Ross S A, Engsberg JR. Relitionship Between Spasticity, Strenght, Gait And The GMFM -66 İmpersons With Spastic Diplegia Cerebral Palsy. Arch Phys Med Rehabil 2007; 88(9): 1114-1120

[159] Rudank S L. Fellman V. Laatikainen L. Visual Impairment In Children Born Prematurely From 1972-1989. Ophthalmology 2003; 110:1639-1645

[160] Russell D J, Rosenbaum P L, Cadman D T, Gowland C, et al. The Gross Motor Function Measure: A Means To Evaluate The Effects Of Physical Therapy. Dev Med Child Neurlo 1989;32(3):341-352.

[161] Russell D J, Avery L M, Rosenbaum P L, Raina P S, Walter S D, Palisano R J. Improved Scaling Of The Gross Motor Function Measure For Children With Cerebral Palsy: Evidence Of Reliability And Validity. Phys Ther 2000;80(9):873-885.

[162] Rutherford O M. Muscular Coordination And Strength Training: Implications For Injury Rehabilitation. Sports Med 1988; 5: 196–202

[163] Saito N. Ebara S, Ohotsuka K et al. Natural History Of Scolipsis In Spastic Cerebral Palsy, Lancet. 1998; 351: 1687-1692

[164] Sakzewski L, Ziviani J, Abbott D F, Macdonell R A, Jackson G D, Boyd R N. Participation Outcomes In A Randomized Trial Of 2 Models Of Upper-Limb Rehabilitation For Children With Congenital Hemiplegia. Archives of Physical Medicine and Rehabilitation 2001;92 (4): 531-9.

[165] Sandstrom K. The lived body _ experiences from adults with cerebral palsy. Clin Rehabil. 2007; 21: 432–441.

[166] Sanger T D. Delgado M R. Gaebler- Spira D. Hallett M. Mind J W. Classification And Definition Of Disorders Causing Hypertonia In Childhood. Pediatrics. 2003; 111: 89-97.

[167] Schwartz M. Kinamatics of Normal Gait. In: GAGE, J. (Ed). The Indentification And The Treatment Of Gait Problems In Cerebral Palsy. London. Mac Keith Pres. 2009; p.: 99-133.

[168] Scianni A, Butler JM, Ada L, Teixeira- Salmela LF. Muscle Strengthining Is Not Effective In Children And Adolescents With Cerebral Palsy: A Systematic Review. Australian Journal of Physiotheraphy 2009. 55: 81-87.

[169] SCPE Collaborative Group. Surveillance Of Cerebral Palsy In Europe: A Collaboration Of Cerebral Palsy Surveys And Registers. Dev Med Child Neurol 2000; 42:816-24.

[170] Scrutton D, Baird G, Smeeton N. Hip Dsyplasia In Bilateral Crebral Palsy: Indidence And Natural History in Children Aged 18 Months To 5 Years. Developmental medicine and child Neurology. 2001; 43: 586-600.

[171] Scrutton D. Baird G. Surveillance Measures Of The Hips Of Children With Bilateraly Cerebral Palsy. Archives of disease in childhood. 1997; 56: 381-384

[172] Scrutton D. Damiano D. Myston M. Management Of The Motor Disorders Of Children With Cerebral Palsy. Mac Keith Pres. 2004; p 12-14

[173] Scrutton D. Physical Assessment And Aims Of Treatment. In: Neville B, Goodman R, Editors. Congenital Hemiplegia. Clinics in Developmental Medicine No.150. London: Mac Keith Press. 2000; 65–80.

[174] Sköld A. Josephsson S. Eliasson A C. Performing Bimanual Activities – The Experiences Of Young Persons With Hemiplegic Cerebral Palsy. Am J Occup Ther. 2004; 58: 416–425.

[175] Spittle AJ, Brown NC, Doyle LW, Boyd RN, Hunt RW, Bear M et al. Quality Of General Movements Is Related To White Matter Pathology In Very Preterm Infants. Pediatrics 2008; 121 (5):184-1189.

[176] Spittle AJ, Doyle LW, Boyd RN. A Systematic Review Of The Clinimetric Properties Of Neuromotor Assessments For Preterm Infants During The First Year Of Life. Developmental medicine and child neurology 2008; 50 (4): 254-266.

[177] Stanger M. Oresic S. Rehabilitation Approaches For Children With Cerebral Palsy: Overview. J Child Neurol 2003;18(1):79-88

[178] Strauss D, Ojdana K, Shavelle R et al. Decline In Function And Life Expectancy Of Older Persons With Cerebral Palsy. Neurorehabilitation.2004; 19: 69–78.

[179] Styer-Acevedo J. Physical Therapy For The Child With Cerebral Palsy. In: Tecklin JS.Pediatric Physical Therapy (3rd ed). Lippincott Williams&Wilkins. Philadelphia 1999:107-162.

[180] Sullivan P. Lambert B. Rose M. Et al. Prevalence And Severity Of Feeding And Nutritional Problems In Children With Neurological Impairment: Oxford Feeding Study. Dev Med Child Neurol. 2000; 42: 674-680

[181] Şik B Y, Çekmece Ç, Dursun N, Dursun E, Balikçi E, Altunkanat Z, Gülcü MA. Hippoterapi Serebral Palsili Çocuklarin Rehabilitasyonunda Yararli mıdır? .Turkiye Klinikleri J Med Sci 2012; 32(3): 601-8.

[182] Taft L T. Cerebral Palsy. Pediatrics in Review. 1995; 16: 411-418.

[183] Taft, L. T. Accentuating The Positive For Children With Cerebral Palsy. Exceptional Parent, 1999; 29: 64-66.

[184] Taub E, Griffin A, Nick J, Gammons K, Uswatte G, Law C R. Pediatric CI Therapy For Stroke-Induced Hemiparesis In Young Children. Developmental neurorehabilitation 2007; 10 (1): 3-18.

[185] Taub E, Ramey S L, DeLuca S C, Echols K. Efficacy Of Constraint-Induced Movement Therapy For Children With Cerebral Palsy With Asymmetric Motor Impairment. Pediatrics 2004; 113(2): 305-12.

[186] Taub E, Uswatte G, Pidikiti R. Constraint Induced-Movement Therapy: A New Family Of Techniques With Broad Application To Physical Rehabilitation - A Clinical Review. Journal of Rehabilitation Research and Development 1999; 36: 237-251.

[187] Pountney T. Cerebral Palsy. In , Pountney T (ed). Physical therapy For Children.Butterworth Heinemann Elsevier. USA. 2007: 90-108

[188] Thomas A, BaxM, Smyth D. The Health and Social Needs of Young Adults With Physical Disabilities. 1989. MacKeith Press, Blackwell Scientific Publications Ltd, Oxford.

[189] Thorpe D, Reilly M, Case L. The Effects Of An Aquatic Resistive Exercise Program On Ambulatory Children With Cerebral Palsy. J Aquatic Phys Ther 2005;13(2):21-24

[190] Trauner D A. Ballantyne A. Friedland S. Et al. Disorders Of Affective And Linguistic Prosody In Children After Early Unilateral Brain Damage. Ann Neurol. 1996; 39:361-367

[191] Truwit C L. Barkovich A J. Koch T K. Ferriero D M. Cerebral Palsy: MR Findings In 40 Patients. American Journal of Neuroradiology 1992; 13: 67-78.

[192] Tsorlakis N, Evaggelinou C, Grouios G, Tsorbatzoudis C. Effect Of Intensive Neurodevelopmental Treatment In Gross Motor Function Of Children With Cerebral Palsy. Dev Med Child Neurol 2004;46(11):740–745

[193] Turk M, Geremski C, Rosenbaum P (1997). The Health Status Of Women With Cerebral Palsy. Arch Dis Child 78: S10–S17.

[194] Uvebrant P. Hemiplegic Cerebral Palsy Aetiology And Outcome. Acta Paedtr Suppl. 1988; 34: 1–100.

[195] Vos-Vromans D C, Ketelaar M, Gorter JW. Responsiveness Of Evaluative Measures For Children With Cerebral Palsy: The Gross Motor Function Measure And The Pediatric Evaluation Of Disability Inventory. Disabil Rehabil 2005;27(20):1245-1252.

[196] Wiley ME., Damiano DL. Lowe-Extremity Strength Profiles In Spastic Cerebral Palsy. Dev Med Child Neurol 1998; 40: 100-107

[197] Winchester P, Kendall K, Peters H, Sears N, Winkley T. The Effect Of Therapeutic Horseback Riding On Gross Motor Function And Gait Speed In Children Who Are Developmentally Delayed. Phys Occup Ther Pediatr 2002; 22: 37–50.

[198] World Confederation for Physical Therapy. Terminology In Aquatic Physical Therapy. Available at: http://www.wcpt.org/apti/terminology. Accessed july. 2013.

[199] Wright A P. Durham S, Ewins D J, Swain IS. Neuromuscular Electrical Stimulation For Children With Cerebral Palsy: A Review. Arch Dis Child 2012; 97:364–371.

[200] Wu YW, Day SM, Strauss DJ et al. (2004). Prognosis for ambulation in cerebral palsy: a population-based study. Pediatrics 114: 1264–1271.

[201] Zadnikar M, Kastrin A. Effects Of Hippotherapy And Therapeutic Horseback Riding On Postural Control Or Balance In Children With Cerebral Palsy: A Meta-Analysis. Dev Med Child Neurol 2011; 53(8): 684-91.

Chapter 3

Mobility in Ambulant Adults with Cerebral Palsy — Challenges for the Future

Jennifer L. McGinley, Dina Pogrebnoy and
Prue Morgan

Additional information is available at the end of the chapter

http://dx.doi.org/10.5772/58344

1. Introduction

Cerebral palsy (CP) is a lifelong cause of disability, with an incidence of around 2 per 1000 people (Himmelmann 2013). It is the most common cause of physical disability in childhood, and children are commonly treated by specialized pediatric services from a broad range of disciplines. Adults with CP are a growing community who are now recognised as outnumbering children 3:1 in some countries (Access Economics 2008). With advances in healthcare, it is now usual for those who walk during childhood to have a relatively normal life expectancy (Strauss, Brooks, Rosenbloom and Shavelle 2008). Over recent times, the need to consider CP as a lifespan, rather than childhood condition has been highlighted, with particular interest in the reported difficulties that emerge during the adult years (Tosi et al. 2009; Morgan and McGinley 2013).

Throughout childhood it is common for much emphasis and resources to be directed at gaining and optimizing walking ability, through a variety of established rehabilitation, medical and surgical interventions. Although around 60% of individuals are able to walk independently or with aids when entering adulthood (Himmelmann 2013), it is common for walking to deteriorate in early and middle adulthood with many young to middle-aged adults describing worsening or loss of walking ability (Morgan and McGinley 2013). The lifespan health challenges faced by this group have been previously poorly described, are arguably still poorly understood, but are now the focus of urgently needed new research directions in CP (Tosi et al. 2009). Many adults with CP commonly develop secondary conditions, such as osteoarthritis (OA), pain, fatigue or falls. Maintaining the ability to walk or maintaining flexible mobility options is important to enable societal participation, maintain employment and retain independence. This chapter provides an overview and exploration of current knowledge of mobility decline in adults with CP, and considers service provision in the context of lifelong rehabilitation.

2. Current knowledge of mobility decline and associated factors in adults with CP

2.1. Mobility decline and associated factors

Accumulating reports provide clear evidence that many adults with CP experience a decline in walking ability during adulthood. In childhood, understanding of the maturation and change in functional mobility over time has been greatly enhanced by the development of the Gross Motor Functional Classification System (GMFCS) and the reporting of typical 'curves' to reflect motor performance over time (Palisano et al. 1997; Hanna et al. 2009). The recent extension of this data to extend to age 21 suggests that deterioration in function may begin in more impaired children from around age 8 (Hanna et al. 2009). Such similar curves are not yet available for adults, although trends in data from the large studies by Strauss and colleagues suggest a similar profile of decline (Strauss, Ojdana, Shavelle and Rosenbloom 2004; Strauss et al. 2008). A recent systematic review of 16 studies reporting gait decline in adults found variation in the incidence of decline, but with most studies reporting decline affecting around 30% or more of the study participants (Morgan and McGinley 2013). Figure 1 illustrates the proportion of adults who experience decline in walking ability from a selection of studies.

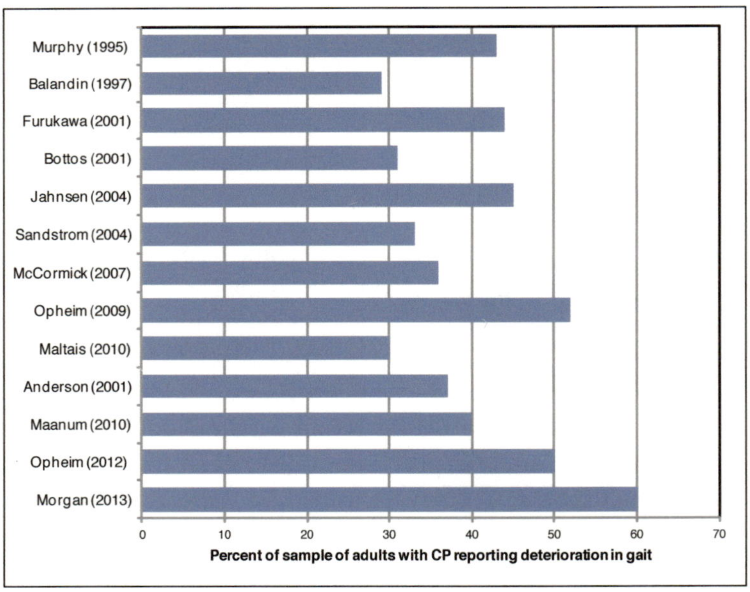

Figure 1. The proportion of adults with CP who experience decline in walking from a range of studies of adults with CP.

http://dx.doi.org/10.5772/58344

A number of studies have now provided insights into factors associated with a higher risk of decline. Unsurprisingly, the most recognised predictor of long term walking function stability is ability during childhood, with age of walking debut associated with likelihood of decline (Bottos et al. 2001; Jahnsen et al. 2004). Similarly, those individuals with poorer gait function requiring the use of gait aids during childhood (GMFCS level III) are more likely to report a deterioration in walking ability, or stop walking entirely (Jahnsen et al. 2003; Opheim, Jahnsen, Olsson and Stangelle 2009). Individuals who are older also appear to be at greater risk of decline. Jahnsen et al (Jahnsen et al. 2004) found that the risk of deterioration progressively increased across age bands from 25 to 45 years, with over 70% of those aged over 45 years reporting a decline in walking. The age of deterioration appears associated with the motor impairment, with the median age of deterioration reported in people with bilateral CP to be 37, compared to 52 in those with unilateral symptoms (Opheim et al. 2009). At even older ages, the potential for more rapid decline may be increased, with around 40% of those walking independently and climbing stairs unassisted at age 60 losing this ability prior to age 75 (Strauss et al. 2004).

As well as worsening walking function, many previously ambulant adults with CP report stopping walking altogether at a wide range of ages. Often this occurs relatively early in adulthood, in the early 20's and 30's (Morgan and McGinley 2013). For some individuals, it may be that this primarily reflects the inevitable consequence of decline in function, with loss of ability or confidence. Alternatively, for others it may simply reflect their personal choices in response to their current or changing symptoms, environment, life activities, or vocational pursuits. For example, a previously ambulant adult may decide to use a wheelchair to travel greater distances, to save energy for other activities, or for greater time efficiency (Jahnsen et al. 2004). Similarly, individuals may choose strategies to prevent falling and remain safe, for example, electing to use a wheelchair or scooter (Horsman, Suto, Dudgeon and Harris 2010). Conversely, others may choose to stay walking because they view wheelchairs or scooters as compromising their independence (Horsman et al. 2010), and may commence using a gait aid to optimize this choice.

The nature of primary motor symptoms and secondary medical conditions also appear to influence an individual's likelihood of decline. Adults with bilateral syndromes, quadriplegia or choreoathetosis are more prone to report problems than those with hemiplegia with milder deficits, who are likely to walk well until their 50's or beyond (Morgan and McGinley 2013). Similarly, higher levels of pain, higher pain intensity, and higher levels of fatigue are also associated with self-reported deterioration (Morgan and McGinley 2013). Many studies have explored the perceived causes of gait decline from the self-reported perspective of the individual. Reduced balance is the reason most commonly reported as a possible cause, nominated by over 60% of adults in the studies by Jahnsen, Opheim and colleagues (Jahnsen et al. 2004; Opheim et al. 2009). Perceived difficulty with balance is also consistent with the observation by Bottos and colleagues (Bottos et al. 2001) that many individuals experience "a rising sense of near-falling" as they age, due to progressive postural instability associated with increased additional weight and other aging factors. Falls or marked insecurity when walking were also cited as causal factors leading to cessation of walking by nine of the 13 people in one study who stopped walking (Bottos et al. 2001). Figure 2 illustrates many of the common factors self-reported as perceived causes of gait deterioration.

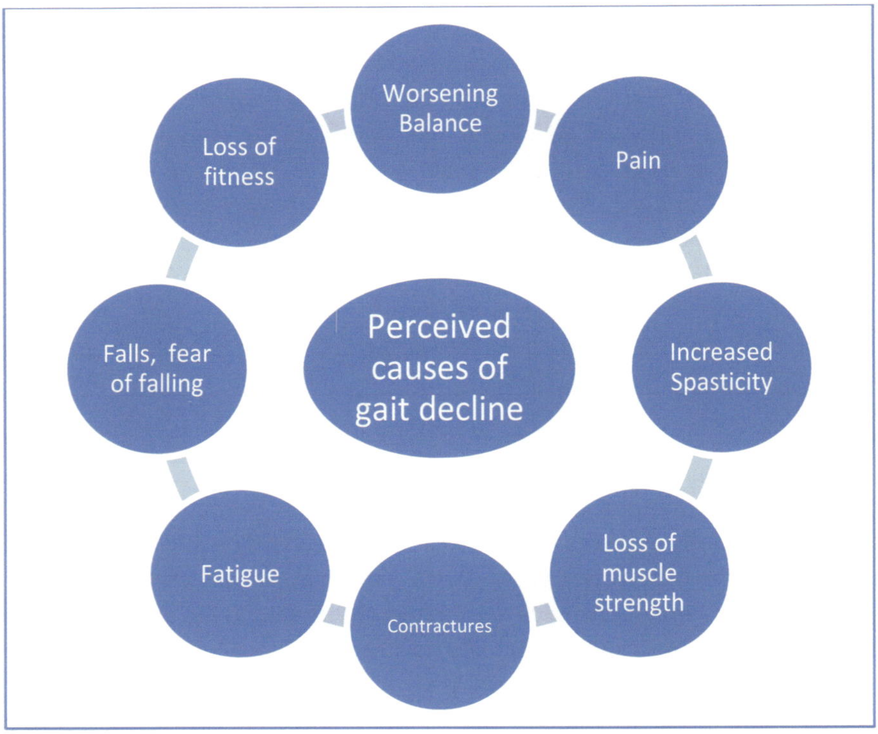

Figure 2. Key self reported factors from adults with CP potentially associated with a decline in mobility

2.2. The nature of gait decline

Gait decline is a multidimensional construct and a great deal of further research is needed to gain a better understanding of the nature of changes experienced by individuals as they age with CP. Adequate walking function is a complex task that requires an individual to successfully navigate varied environments in order to complete personal goal-directed tasks. Physiological changes in any of the systems underpinning motor control or dynamic balance may impact on walking, such as changes in strength, sensation, vision, vestibular function or cognition. Additional symptoms due to secondary conditions such as fatigue, pain, or shortness of breath may also impact on function, as may changes in self-efficacy such as fear of falling or loss of confidence.

A wide range of factors may also influence an individual's perception of their own ability. One individual with mild disability may self-report decline due to a reduced ability to walk long distances, or the need to hold a rail during stair walking. A more impaired individual at GMFCS III may reduce their walking activity and rely on a wheelchair to achieve vocational

http://dx.doi.org/10.5772/58344

or recreational demands, yet not perceive a change in walking ability. The accumulation of literature does not offer a comprehensive understanding of how walking changes for adults with CP. Reported common changes include the need for greater assistance, or aids to walk, or the use of a rail on stairs, walking shorter distances or walking only indoors. Other reports suggest that it is common for walking to become slower, unsteadier, and more difficult in varied outdoor terrains. Symptoms such as pain, fatigue, shortness of breath or concern about falling also may be limiting factors (Morgan and McGinley 2013). The nature of walking decline is also challenging to interpret from the literature due to the diversity of outcome measures used in different studies. Measures of walking may capture what the person "does in their daily environment" (performance), what they "can do in their daily environment" (capability), or what they "can do in a controlled standardized environment" (capacity) (Holsbeeke, Ketelaar, Schoemaker and Gorter 2009). Although highly related, these measures are not interchangeable, as capacity and capability may differ and often exceed everyday actual performance. Future longitudinal studies with measures reflecting all three constructs are needed to distinguish and detail objective changes in gait across a range of environmental contexts. These measures could be considered in conjunction with the individual's perceived (self-reported) capabilities, self-efficacy and personal mobility preferences.

Evidence-based knowledge about decline in mobility is markedly limited, and the causes of gait decline are likely to be multifactorial. Current attempts to understand the causes of gait decline are limited by the absence of longitudinal studies that measure gait and provide detailed participant characteristics, preferably from large population based samples. It is also likely that the outcomes of the current generations of adults have been influenced by their treatment during childhood, and may thus vary across age ranges. For example, over the last one to two decades, botulinum toxin therapy, gait analysis to guide surgical planning and single-event multilevel surgery (SEMLS) have been introduced and become standard practice in many specialized tertiary centres. Access to and uptake of these recent treatments may lead to different or better outcomes than in past times. Current and relatively recent cohorts of children may therefore enter adulthood with better limb and joint alignment or different gait ability than cohorts from two or more decades past. Many of the adults who report decline are in their mid 30's to 40's and have likely to have undergone historically different treatment plans to those in current practice. A recent study of outcome after childhood SEMLS surgery guided by gait analysis suggests that gait patterns appear to be largely stable for periods up to 10 years after surgery (Gannotti, Gorton, Nahorniak and Masso 2010). Although promising, it remains uncertain yet whether apparent biomechanical stability of gait patterns will also be accompanied by stable self-reported appraisals of walking. For example, although an individual's gait pattern may remain stable with respect to the kinematic pattern, increasing pain, loss of confidence or falls, or even a change in vocational or recreational demands may lead to an individual walking less, using a gait aid, or preferring to adopt alternative mobility choices. Well designed studies are therefore urgently required to clarify the longer term outcomes of contemporary standard interventions. The greatest and clearest insights will be provided by longitudinal studies of defined cohorts, who have well-detailed treatment records and existing measures of gait biomechanics and function using high quality outcome measurement tools in conjunction with qualitative methods.

3. Fear of falling and falls in CP across the lifespan

3.1. Falls frequency in adults with CP

Difficulty with balance or falls is frequently self-reported by adults with CP as a main reason for changes in mobility throughout adulthood (Opheim et al. 2009; Morgan and McGinley 2013). Identification, early prevention, remediation and monitoring of falls and falls risk factors is now recognised as a standard component of best practice health care in older adults. Surprisingly, there is very little information on falls in adults ageing with CP. Mosqueda (Mosqueda 2004) alarmingly reported that 40% of a cohort of adults with CP (mean age 44 years) fell monthly, and 75% fell at least every two months. More recent literature (Opheim, Jahnsen, Olsson and Stanghelle 2012; Morgan and McGinley 2013) also reported frequent falls experienced by adults with CP, with some reporting two or more falls each week. Falls rates were high, with 80% of ambulant individuals in the study by Opheim falling five or more times in the past year (Opheim et al. 2012) and 68% of those in the prospective study by Morgan falling during a six month period (Morgan and McGinley 2013). For many adults ageing with a disability, falls have been an accepted 'way of life', perhaps regarded as a natural consequence of impaired mobility throughout childhood and beyond. However, whereas it may be considered acceptable and common to fall in childhood and adolescence, the physical and social consequences of falling in adulthood become more serious over time, particularly with the potential occurrence of co-morbidities such as osteoarthritis or osteoporosis. For other ambulant adults with CP, falling may be a new development, as a result of the onset of newly acquired mobility and balance decline. Evidence suggests that falls frequency in adults with CP may be as high or higher than other commonly occurring neurological disorders such as Parkinson's disease or stroke (Mackintosh et al. 2005; Pickering et al. 2007).

3.2. Falls consequences in adults with CP

The common consequences of falls in older adults include minor soft tissue injuries (55%), with fractures or lacerations and fractures requiring hospitalization occurring less frequently (5-10%) (Nevitt, Cummings and Hudes 1991). Much less is known about the injuries incurred by falls experienced by adults with CP. For this group, falls have been reported as less likely to result in minor soft tissue injuries, possibly due to the greater integrity of soft tissue compared to older adults (Morgan and McGinley 2013). However, the rate of serious injuries is equivalent, around 10% (Morgan and McGinley 2013), with fractures and serious lacerations requiring medical attention evident. Furthermore, the functional consequences and outcome of serious injuries in adults with CP may be significant. For example, a fractured wrist may result in an inability to use a gait aid, dress and shower independently, or use a wheelchair for longer distances. It is likely that the consequences of falls in the older adult – fractures, soft tissue injuries, fear of falling, plus costs and resources associated with hospital admissions, carer requirements, rehabilitation and supported accommodation (Watson, Clapperton and Mitchell 2010) – may be similar to those experienced by ambulant adults with CP.

It is recognized that individuals with other neurological conditions and older adults typically reduce their activities as a consequence of a fall (Pieterse et al. 2006). Similarly, older people

http://dx.doi.org/10.5772/58344

who fall frequently implement changes post-fall such as seeking assistance with shopping (Murray, Hill, Phillips and Waterston 2005), or bathing. Adults with CP appear less likely to change their activities as a result of a fall (Morgan and McGinley 2013), possibly due to habituation to long-standing falls, or reduced focus on risk taking behavior.

3.3. Fear of falling

Fear of falling is increasingly recognised as a serious consequence of falls in the healthy older adult. In older adults, falling predicted poorer physical health, greater negative emotions and less physical activity due to self-imposed restriction (Ruthig et al. 2007). For adults ageing with a physical disability, any potential limitation in physical activity is undesirable. Eighty-two percent of adults with multiple sclerosis who reported fear of falling, admit to subsequent physical activity restriction (Peterson, Cho and Finlayson 2007). The Falls Efficacy Scale-International (FES-I) is an instrument to assess level of 'concern' about falling, a term closely related to fear (Yardley et al. 2005), and has been found to be predictive of falls in longitudinal research (Delbaere et al. 2010). Recent research has indicated that moderate fear of falling is experienced by ambulant adults with CP (Opheim et al. 2012; Morgan and McGinley 2013), according to the FES-I, at a level equivalent or higher than that reported in elderly people who were treated for fall related fractures (Nordell, Andreasson, Gall and Thorngren 2009), and ambulant adults with stroke or Parkinson's disease (Faria, Teixeira-Salmela and Nadeau 2009; Allen et al. 2010). Interestingly fear of falling appears not to be related to falls frequency (Morgan and McGinley 2013) or the presence of recent mobility decline in adults with CP. This may reflect longstanding awareness of balance and mobility dysfunction that is different to the more recently acquired balance decline in older adults or those who have acquired health conditions.

3.4. Falls risk assessment

Recommended protocols now exist for the assessment of falls and balance dysfunction in older adults with the inclusion of a battery of objective measures (e.g. timed up and go (TUG), 10 metre walk test), and the use of standardised assessment tools to identify level of falls risk. Performance on these measures can assist prescription of targeted rehabilitation, and implementation of falls risk reduction strategies. Although these tools are not typically used in CP management, they have been extensively used in other older adult and disabled populations to describe and define falls risk (Berg, Wood-Dauphinee, Williams and Gayton 1989; Podsiadlo and Richardson 1991; Russell et al. 2008).

A recent study used the FROP-Com (Falls Risk in Older People – community) risk assessment tool to appraise falls risk in adults with CP (Morgan and McGinley 2013). The FROP-com was developed as a tool to evaluate falls risk in community dwelling older people (Russell et al. 2008). The majority of adults with CP who fell were considered at 'mild risk' of future falls, according to their FROP-com scores (Russell et al. 2008), considerably underestimating their ongoing falls risk. Currently available tools such as the FROP-com, developed to identify multifactorial risk factors typically present in older people such as multiple medications, footwear and continence, appear to lack sensitivity in identifying falls risk factors in adults with CP. For example, in the study by Morgan and McGinley (Morgan and McGinley 2013), no adults with CP demonstrated problems with footwear, in contrast to 70% of people who fell and 50% of people who did not fall in an elderly cohort (Murray et al. 2005). The evidence to date suggests that ambulant adults with CP who fall do not have higher falls risk factors (as identified by current risk 'tools') than those who do not fall (Morgan and McGinley 2013).

3.5. Interventions to reduce falls risk in adults with CP: An evidence gap

Effective falls prevention has the potential to prevent injury, improve quality of life, and decrease the likelihood of subsequent fear of falling and activity restriction. Published clinical practice guidelines on the prevention of falls in older adults have summarised effective interventions to address single or multifactorial causation. For example, strength and balance re-training, tai chi, medication review and management of Vitamin D deficiency may be advocated to address risk factors in older adults experiencing falls (Campbell and Robertson 2007). Similar interventions have been trialed and evaluated in Parkinson's disease and stroke (Pickering et al. 2007; Batchelor et al. 2012) but not as yet in adults ageing with CP. A systematic review reported that structured exercise programmes may increase habitual physical activity levels in people with CP, however none to date have evaluated the impact on falls (Bania, Dodd and Taylor 2011). Although it is tempting to assume that effective interventions applied to older adults or adults with acquired neurological dysfunction may apply to adults with CP, there is no evidence to support this proposal. Many adults with CP begin to fall, or increase their falling behaviour as a result of age-associated mobility decline. Falls consequences can have significant impact on physical, social and economic outcomes. Current falls risk assessment tools appear to have limited application to adults ageing with CP. Adults with CP who seek health services to address mobility decline are typically not provided with comprehensive falls prevention or falls risk reduction strategies. Practitioners working with adults with CP need to consider falls management as an essential component of care.

4. Common musculoskeletal disorders that impact on mobility

Adults living with CP are at risk of developing or worsening secondary musculoskeletal conditions as they get older. In ambulant adults with CP these secondary conditions often contribute to functional decline, consequently reducing independence with activities of daily living, participation in the community, social interactions and psychological wellbeing. The emergence and changing nature of secondary musculoskeletal conditions reflects the contemporary recognition of CP as a condition in which the brain lesion itself is static and non-progressive but accompanied by secondary musculoskeletal problems, which typically *do progress*. The functional impact of CP therefore can be changing and dynamic across the lifespan, and thus evidence based knowledge to develop and evaluate interventions that limit or prevent secondary health conditions are urgently needed. Some of the most commonly reported musculoskeletal conditions include pain, osteoarthritis and fatigue. Specific evidence of the prevalence and impact of these and other musculoskeletal symptoms is gradually building, yet currently limited in detail, often poorly differentiates between adults who are

http://dx.doi.org/10.5772/58344

ambulant and those who are more functionally impaired, and remains primarily based upon cross-sectional samples of convenience.

Pain is very common in the lifelong experience of living with CP, with estimates of prevalence of any type or site of pain ranging to over 80% (Turk, Geremski, Rosenbaum and Weber 1997; Schwartz, Engel and Jensen 1999; Jahnsen, Villien, Stangelle and Holm 2004). Acute and chronic pain both occur, with reportedly nearly one third of adults with CP incurring chronic pain (Jahnsen et al. 2004). Pain is the most consistent musculoskeletal disorder reported by adults living with CP (Turk et al. 1997) and is proposed to be directly linked to age and increased inactivity (Vogtle 2009), with deterioration of functional skills found to be significantly associated with chronic pain (Jahnsen et al. 2004). Notably, pain was also the physical symptom most frequently associated with CP by a survey of rehabilitation physicians (Hilberink et al. 2007). Reported pain locations varied widely but commonly include the back and neck, along with the hips, knees and feet. Pain often affects multiple body areas, with a large longitudinal study finding that it was typical to experience pain at multiple sites, with a median of three locations identified (Opheim, Jahnsen, Olsson and Stanghelle 2011). Pain was reported by many individuals to be worsened by overexertion and fatigue and improved by rest, physiotherapy, or participation in exercise (Schwartz et al. 1999; Jahnsen et al. 2004; Opheim et al. 2011).

The relationship between pain presence, location and an individual's level of function is not yet well understood, although data from two studies suggests no association between pain and GMFCS level (Sandstrom, Alinder and Oberg 2004; Hilberink et al. 2007). This may be expected as it is likely that many individuals experience pain, but perhaps due to different profiles of physical symptoms. For example, back pain may be related to a severe postural deformity or scoliosis in a non-ambulant person, or to excessive movement and joint load in an ambulant person. Adults with CP who walk tend to have excessive pelvic tilt and a larger range of lumbar rotation during gait, which is likely to contribute to or possibly exacerbate low back pain, potentially leading to lumbar spondylolysis (Harada et al. 1993; Opheim et al. 2011). The accumulating toll of weight bearing on joints with abnormal alignment, including cavus feet, knee deformities and displaced hips can in the long term also lead to the onset of pain in the affected joints during ambulation. In unilateral CP, it is also suggested that asymmetry in motor control could lead to overuse of the non affected side, ineffective recruitment of available muscles and asymmetrical joint loading further contributing to back pain (Opheim et al. 2011). Of interest, however, a study by Opheim and colleagues (Opheim et al. 2011) did not find there to be a correlation between the number of pain sites and psychological health in this population. This may be reflective of life long experience of pain from childhood onwards, possibly resulting in better coping strategies in comparison to general public.

Other musculoskeletal disorders in adults with CP can be classified as deformities and they too have the potential to impact on an individual's ability to walk. Relatively common musculoskeletal deformities include subluxations or dislocations of the hip, structural abnormalities of the foot/feet, patella alta, pelvic asymmetry/obliquity and contractures of various muscle groups (Gajdosik and Cicirello 2001). Hip subluxation or dislocation is an acquired condition resulting from muscle imbalance, persistent bony mal-alignment and altered patterns of weightbearing, affecting an estimated 18% to 59% of individuals with CP (Root 2009). It is unknown what proportion of adults with CP who walk have abnormal hip joint structure, but it is well recognised as a common factor causing many adults to stop walking due to the development of painful joint degeneration. A number of small case series studies have emerged that outline surgical interventions for hip joint misalignment and degeneration, with individuals reporting improved pain and walking function after surgery (Root 2009; Shroeder et al. 2009). A follow up study of 16 ambulatory adults 10 years after hip arthroplasty found that this surgery can provide long-term pain relief and improved function, albeit with a higher complication rate than in non-CP individuals (Shroeder et al. 2009). Patella alta is another relatively common condition in ambulatory adults with CP, often associated with longstanding anterior knee pain or a crouch gait pattern. Stress fractures and pain can occur due to the underdeveloped and poorly aligned patella, in conjunction with altered tendon structure. Surgical interventions including distal femoral extension osteotomy and patella tendon surgery are now common during adolescence (Novacheck, Stout, Gage and Schwartz 2009), but the long term outcomes of such interventions in adulthood are unknown.

Osteoarthritis is also a common cause of pain for patients with CP and typically has an earlier onset in this population group when compared to the non CP population (Gajdosik and Cicirello 2001). It is thought that the altered joint loading patterns in childhood due to delayed weight bearing and altered muscle activity can lead to poor joint integrity and irreversible damage to the articular cartilage of the joint surface, consequently developing early onset OA, predominantly seen in the hip, knee and feet (Carter and Tse 2009). No large sample population data exists to indicate how common OA is, and who is affected. It is likely that the joint distributions and severity of OA may vary between ambulant and non-ambulant adults, and may also relate to the movement disorder type. Some studies report the incidence of hip OA to be as high as in 59% of non-ambulant adults living with CP (Boldingh et al. 2005). Individuals with more severe CP are thought to develop OA at a higher rate, as reduced weight-bearing and restricted range of motion does not provide sufficient cyclic loads to different areas of the hip that is needed to maintain cartilage (Carter and Tse 2009). Another study found clinical evidence of OA in 27% of young adults with CP between the age of 15 and 25, occurring more commonly in those who could walk (Cathels and Reddihough 1993). Population based studies to investigate the occurrence of OA with radiological confirmation are needed to determine the prevalence of this common debilitating secondary condition.

Fatigue is also common in adults with CP and potentially associated with both decline in walking or adoption of alternate mobility choices. Adults with CP reported significantly more physical fatigue than the general population, with fatigue levels associated with pain and deterioration of functional skills (Jahnsen, Villien, Stangelle and Holm 2003). In a recent study of mostly ambulant adults with CP, 20% of the sample was fatigued and a further 41% were severely fatigued (Van der Slot et al. 2012). Of concern, there was a high prevalence and relatively frequent co-occurrence of chronic pain, fatigue, and depressive symptoms (Van der Slot et al. 2012).

http://dx.doi.org/10.5772/58344

5. Physical activity

Physical activity is important for health and wellbeing across the lifespan and is an important part of healthy aging. Strong evidence has linked participation in regular moderate-intensity physical activity to a wide range of health and social benefits. Physical activity is particularly important for those living with a chronic disability such as CP, in order to maintain long term health and to prevent secondary complications of disability. Chronic conditions that cause mobility problems are known to place individuals at greater risk of inactivity (Ashe, Miller, Eng and Noreau 2009) increasing the likelihood of developing long-term negative health consequences such as cardiovascular disease or poor bone density (Carlon, Taylor, Dodd and Shields 2013). As individuals with CP age and become less active, they may also be at greater risk of lifestyle-related diseases, such as diabetes mellitus or obesity. Encouraging physical activity (PA) is important for health promotion, and may have beneficial effects on secondary conditions and on functional independence, social integration, and well being (Buffart et al. 2009).

Being active across the life span can pose additional challenges for those with physical disability. Evidence has shown that children with CP engage in significantly lower levels of habitual physical activity than their peers, and less than recommended guidelines (Carlon et al. 2013). Relatively few studies have examined physical activity of adults with CP. Most studies suggest that this group have lower levels of physical fitness and are less active than able bodied peers (Nieuwenhuijsen et al. 2011), with the exception perhaps of those with mild unilateral deficits (van der Slot et al. 2007). Data from a larger study by Jahnsen et al (Jahnsen et al. 2003) also suggests that physical activity may decline over time, with many inactive individuals reporting a reduction/cessation of physical activity over recent years. Those who were physically active were found to demonstrate a reduction in risk of mobility decline, with deterioration in gait associated with higher odds ratio for inactivity (Maltais, Dumas, Boucher and Richards 2010). Of particular interest, a recent systematic review found preliminary evidence suggesting that exercise and online support programmes can increase habitual physical activity in people with CP (Bania et al. 2011).

The determinants of regular physical activity in this group have not been well defined. To describe factors associated with PA in people with a physical disability, van der Ploeg et al. (van der Ploeg, van der Beek, van der Woude and van Mechelen 2004) proposed the Physical Activity for People with a Disability (PAD) model (Figure 3). This model explicitly considers the determinants of physical activity within the context of the environmental factors and personal factor components of the International Classification of Function model. The PAD model highlights the importance of social influence and environmental barriers and facilitators, including factors such as support and opinions of family, friends and healthcare professionals, and transport, access and assistance provision where needed. Within the personal factors component, key behavioural factors such as self-efficacy, intention and attitude are considered, in conjunction with the health condition. This framework is highly relevant to adults with CP, but as yet the determinants have not yet been explored in a comprehensive nor systematic manner.

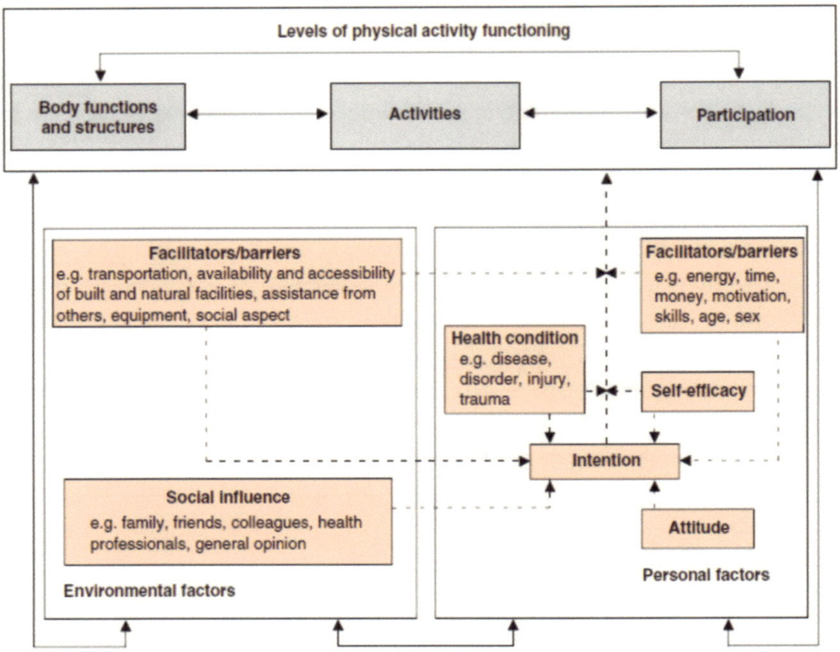

Figure 3. The Physical Activity for people with a disability (PAD) model; Van der Ploeg et al. 2004 [reproduced with permission]

Limited knowledge has identified some factors that inform factors within the PAD model. Jahnsen et al found that age and mildness of severity were significantly associated with regular physical activity (Jahnsen et al. 2003). Of particular interest, the strongest predictor of physical activity was a factor called "learnt personal responsibility for personal health". The most frequently reported motivational factor in relation to physical activity was improvement or preservation of health, and the most frequently reported reason for not being physically active was lack of initiative and motivation (Jahnsen et al. 2003). Some knowledge is also available from a wider study of young people with a range of physical disabilities including CP, suggesting that barriers to physical activity included attitude and motivation. In addition, lack of energy, existing injury or fear of new injuries, limited physical activity facilities, and lack of information appeared to be barriers. Facilitators of engagement in PA included fun and social contacts, as well as improved health and fitness (Buffart et al. 2009). Further detailed studies are needed to identify and explore the range of personal and environmental factors that influence PA in adults with CP.

6. Health care services for adults with Cerebral Palsy

Adults with chronic but changing health conditions such as CP require access to appropriate health services across the lifespan, to provide support to individuals to optimize function and health related quality of life. Health promotion is defined as 'activities directed toward increasing the level of well-being and actualizing the health potential of individuals, families, communities and societies' (Pender 1987). Health promoting behaviours, in contrast to disease management strategies, may be ongoing activities that become an integral part of one's life such as physical exercise, nutritional eating, stress management, stopping smoking. Absence of illness or disability is not a pre requisite for health; therefore individuals living with a disability can be considered 'healthy'. Of concern, health care providers may perceive that people with disabilities are 'sick', contributing to people with disabilities thinking of themselves as passive participants in their own health care, rather than as individuals responsible for, and contributing to, their well-being. In a more collaborative model of rehabilitation, a partnership paradigm is advocated where the clinician has expertise regarding disability management and the care-seeker (e.g. adult with CP) has expertise about their own life. This is consistent with components of broader approaches of models of care to support self management in those who live with chronic disease or chronic health conditions (Bodenheimer, Wagner and Grumbach 2002). This approach emphasizes self care, and promotes an active, independent and informed attitude towards lifelong rehabilitation. This direction is consistent with a shift from a focus on a 'medical' to a 'participation' model.

6.1. Lifelong access to expert health care

Adults with CP may experience both diagnosis-related and ageing-related health consequences (Svien, Berg and Stephenson 2008; Peterson, Gordon and Hurvitz 2013). As described earlier in this Chapter, many adults with CP experience new onset of symptoms such as muscle fatigue and weakness, pain, spasticity and contracture, joint dislocation or skin breakdown. Relative inactivity can result in further health related concerns such as premature sarcopenia and obesity (Peterson et al. 2013). Furthermore, adults with CP may experience psychosocial issues, as well as secondary biomedical concerns, related to their disability (Horsman, Suto, Dudgeon and Harris 2010). Medical management for co-morbidities can also result in health concerns. For example, some medications to treat epilepsy can cause osteopenia. The incidence of other diseases in adults with CP is also acknowledged to be higher than age matched comparisons; e.g. an increased incidence of cancer, chronic obstructive airways disease, pneumonia, and bowel obstruction. It has been suggested that a decreased verbal ability to convey symptoms and a reduced tendency to access regular health screening may contribute to the rise in disease incidence in this population (Svien et al. 2008). Adults with disabilities such as CP need lifelong, but not necessarily continuous, access to health and rehabilitation services to meet their changing needs and enable them to make informed choices to address any health problems that arise (Field, Scheinberg and Cruickshank 2010).

6.2. Transition from paediatric to adult health services

In the childhood of an individual with CP, there is often a supportive health facility and/or therapy organization, which is readily accessible and staffed with competent professionals. It is usual for children with CP and their families to frequently have long relationships with their therapists and medical team, who are knowledgeable and dedicated to providing services for children with CP. Although some children may become less engaged with the requirement for ongoing physiotherapy and rehabilitation services throughout adolescence, there are still a range of functional activity options available, and ready access to rehabilitation professionals as and when required.

For an adult with CP who seeks health services, it is not immediately obvious where to turn. As a result, many young adults with CP experience a 'vacuum' after leaving paediatric rehabilitation (Ng, Dinesh, Tay and Lee 2003). A lack of adequate care, together with changes in social role and in environmental expectations as they grow into adulthood, may result in unmet (health) needs of adults with physical disabilities (Ng et al. 2003). Over the last ten years or so, 'transition clinics' have been established in many countries in an attempt to bridge the gap between paediatric and adult services. Most commonly, these clinics cater for those with chronic illness or disability in the 16 to 25 age group. A variety of literature has described the optimal design of these transition services to provide adult health care for adolescents and young adults with disability, such as adequate preparation, flexible timing, care coordination, transition clinic visits, and interested adult-centred health care providers (Binks, Barden, Burke and Young 2007). The importance of an adult system that includes multidisciplinary teams that are central to the care of people with CP and other lifelong health conditions is also stressed (Bakheit et al. 2009; Field et al. 2010) with evidence that a team approach is more likely to enhance participation in society of young people with physical disabilities (Bent et al. 2002).

6.3. Challenges in accessing adult health services

Despite this knowledge, specialized health services for adults with CP are widely reported to be extremely limited (Bent et al. 2002; Ng et al. 2003; Field et al. 2010), and despite an increase in funding of transition services (Field et al. 2010), remain fragmented and challenging to navigate. The health care services that are publicly accessible to most adults with a chronic disability are frequently limited in scope, or perceived by users as inadequate and staffed by practitioners with limited knowledge and skills in disability care (Sandstrom 2007). A common experience by adults with CP is frustration with health service type and availability, facility access, staff turnover, and lack of engagement with their needs. With an attempt to increase access by driving disability services into 'community accessible models' (such as dieticians and therapists within the community health sector), the development of experience and expertise in disability by health practitioners has been limited. For example, a paediatric physiotherapist may have an exclusive caseload of children with CP with mentoring and advice readily available from colleagues within a specialist tertiary paediatric hospital, a defined career pathway within the health service, and access to professional development within the paediatric disability area. In contrast, a physiotherapist working in a community health centre may provide services to a wide variety of adults ranging from those post fracture or knee replacement, to elderly people post fall, to those recovering from stroke, and only see a few adults with CP each year. Knowledge of who are the 'experts' in management of adults with CP, and hence who to seek advice from, is frequently unclear both to adults with CP and the health practitioner community.

http://dx.doi.org/10.5772/58344

Health services for adults with developmental disability are frequently accessed and provided in an ad-hoc manner by many different organisations each with their own criteria for defining who receives health services and support and the nature of any services and support provided. Adults with CP reportedly use specialty health-care and rehabilitation services less, and emergency room care more, than their non-CP peers (Tosi et al. 2009). Anecdotally, adults with CP who present to their general medical practitioner (GP) following a fall may be referred to 'geriatric' (>65 years) services such as Falls and Balance Clinics as their GP cannot identify where alternate suitable services may be located. Tosi and colleagues (Tosi et al. 2009) reported that few medical facilities are prepared to treat adults with developmental disabilities, and adults with CP needing surgery may find themselves in paediatric environments, where personnel have not been trained in adult care. As a result, overall care is often fragmented and does not address the complex physical and psychosocial issues of adults with CP with any continuity. Horsmann and colleagues (Horsman et al. 2010) reported that decisions regarding allocation of health and support services to those living with CP are often flawed, as such decisions are viewed from an exclusively medical model rather than a participation model. For example most self-assessments by adults with disabilities identified taking part in leisure activities as a priority (participation model) whereas most social service agency assessments considered only basic health needs (medical model) when determining eligibility for hired caregivers (Horsman et al. 2010). As a result, adults with CP may be provided with unwanted or unsuitable health and support services.

6.4. An ideal model for health service delivery for rehabilitation

Effective service development for rehabilitation for adults with CP at any age requires a detailed knowledge of the likely health issues experienced by this population across the lifespan and a robust evidence base upon which to base recommendations and management. Goldstein, Chairman of the Cerebral Palsy International Research Foundation, in 2009 urged a move to a system of health services to maximise life-long functioning of people with disabilities rather than to just a new health care environment. Reddihough and colleagues (Reddihough et al. 2013) however are hopeful that the introduction of a National Disability Insurance Scheme in Australia will assist in improving the physical and social outcomes of adults with CP in this country. The scheme aims to optimise opportunities for people with disabilities to participate in the social and economic life of the community. People with disabilities will be empowered to use their own funding packages to purchase the equipment and associated therapy they require for optimal independent function and participation, while health services will be required to provide therapy that responds to health concerns and needs. However, in order to achieve a seamless rehabilitation service across the lifespan of disability, health practitioners urgently require upskilling and training to be able to deliver evidence based interventions for adults with CP experiencing functional and mobility decline at any age. Furthermore, adults with CP require information in order to make informed choices about health interventions.

Ambulant adults with CP have ongoing health needs to address age-related changes associated with their disability. Despite a growing body of evidence describing persistent unmet health needs experienced by adults with CP little appears to have changed regarding this issue. Adults with CP continue to report frustration with service type and availability, facility access, staff knowledge and skills, staff turnover, and lack of engagement with their needs. A more effective, equitable and accessible system of health care for this population is urgently needed.

6.5. Summary

Many adults with CP face challenges with declining mobility and the emergence of secondary musculoskeletal conditions as they age. Decline in walking and falls are common, potentially comprising activity, participation and health-related quality of life. Carefully constructed longitudinal studies of population-based samples are needed to evaluate and characterise the prevalence and impact of mobility decline. Evidence to guide clinical practice is currently extremely sparse. CP-specific interventions to address mobility decline and falls need to be developed and evaluated in rigorous randomised controlled trials. Similarly, the development of interventions for the secondary conditions of OA, pain, fatigue and reduced physical activity warrant similar consideration. Current healthcare services for adults with CP currently lack the evidence-based knowledge needed to develop and implement best practice clinical guidelines.

Acknowledgements

The support of the Physiotherapy Department, at The University of Melbourne, Physiotherapy Department, at Monash University and the Clinical Gait Analysis Service at Monash Health is gratefully acknowledged.

Author details

Jennifer L. McGinley[1], Dina Pogrebnoy[2] and Prue Morgan[3]

1 Department of Physiotherapy, University of Melbourne, Parkville, Australia

2 Clinical Gait Analysis Service, Monash Health, Cheltenham, Australia

3 Department of Physiotherapy, Monash University, Frankston, Australia

References

[1] Access Economics (2008). The Economic Impact of Cerebral Palsy in Australia in 2007.

[2] Allen, N. E., C. G. Canning, C. Sherrington, et al. (2010). The effects of an exercise program on fall risk factors in people with Parkinson's disease: a randomized controlled trial. Movement Disorders 25(9): 1217-25.

[3] Ashe, M. C., W. C. Miller, J. J. Eng and L. Noreau (2009). Older adults, chronic disease and leisure-time physical activity. Gerontology 55(1): 64-72.

[4] Bakheit, A., S. Easton, K. Edwards, et al. (2009). Young People with Cerebral Palsy in Transition from Paediatric to Adult Health Services - Best Practice Recommendations. Advances in Clinical Neuroscience and Rehabilitation 8: 20-21.

[5] Bania, T., K. J. Dodd and N. Taylor (2011). Habitual physical activity can be increased in people with cerebral palsy: a systematic review. Clinical Rehabilitation 25(4): 303-15.

[6] Batchelor, F. A., K. D. Hill, S. F. Mackintosh, C. M. Said and C. H. Whitehead (2012). Effects of a multifactorial falls prevention program for people with stroke returning home after rehabilitation: a randomized controlled trial. Archives of Physical Medicine and Rehabilitation 93(9): 1648-55.

[7] Bent, N., A. Tennant, T. Swift, et al. (2002). Team approach versus ad hoc health services for young people with physical disabilities: a retrospective cohort study. Lancet 360(9342): 1280-6.

[8] Berg, K., S. Wood-Dauphinee, J. Williams and D. Gayton (1989). Measuring balance in the elderly: Preliminary development of an instrument. Physiotherapy Canada 41: 304-11.

[9] Binks, J. A., W. S. Barden, T. A. Burke and N. L. Young (2007). What do we really know about the transition to adult-centered health care? A focus on cerebral palsy and spina bifida. Archives of Physical Medicine and Rehabilitation 88(8): 1064-73.

[10] Bodenheimer, T., E. H. Wagner and K. Grumbach (2002). Improving primary care for patients with chronic illness. JAMA 288(14): 1775-9.

[11] Boldingh, E. J., M. A. Jacobs-van der Bruggen, C. F. Bos, G. J. Lankhorst and L. M. Bouter (2005). Determinants of hip pain in adult patients with severe cerebral palsy. Journal of Pediatric Orthopaedics B 14(2): 120-5.

[12] Bottos, M., A. Feliciangeli, L. Sciuto, C. Gericke and A. Vianello (2001). Functional status of adults with cerebral palsy and implications for treatment of children. Developmental Medicine & Child Neurology Aug: 516-28.

[13] Buffart, L. M., T. Westendorp, R. J. van den Berg-Emons, H. J. Stam and M. E. Roebroeck (2009). Perceived barriers to and facilitators of physical activity in young adults with childhood-onset physical disabilities. Journal of Rehabilitation Medicine 41(11): 881-5.

[14] Campbell, A. J. and M. C. Robertson (2007). Rethinking individual and community fall prevention strategies: a meta-regression comparing single and multifactorial interventions. Age Ageing 36(6): 656-62.

[15] Carlon, S. L., N. F. Taylor, K. J. Dodd and N. Shields (2013). Differences in habitual physical activity levels of young people with cerebral palsy and their typically developing peers: a systematic review. Disability & Rehabilitation 35(8): 647-55.

[16] Carter, D. and B. Tse (2009). The pathogenesis of osteoarthritis in cerebral palsy. Developmental Medicine & Child Neurology 51(Suppl 4): 79-83.

[17] Cathels, B. A. and D. S. Reddihough (1993). The health care of young adults with cerebral palsy. The Medical Journal of Australia 159: 444-446.

[18] Delbaere, K., J. C. Close, A. S. Mikolaizak, et al. (2010). The Falls Efficacy Scale International (FES-I). A comprehensive longitudinal validation study. Age Ageing 39(2): 210-6.

[19] Faria, C. D., L. F. Teixeira-Salmela and S. Nadeau (2009). Effects of the direction of turning on the timed up & go test with stroke subjects. Topics in Stroke Rehabilitation 16(3): 196-206.

[20] Field, B., A. Scheinberg and A. Cruickshank (2010). Health care services for adults with cerebral palsy. Aust Fam Physician 39(3): 165-7.

[21] Gajdosik, C. and N. Cicirello (2001). Secondary conditions of the musculoskeletal system in adolescents and adults with cerebral palsy. Phys & Occ Ther Ped 21(4): 49-68.

[22] Gannotti, M. E., G. E. Gorton, 3rd, M. T. Nahorniak and P. D. Masso (2010). Walking abilities of young adults with cerebral palsy: changes after multilevel surgery and adolescence. Gait Posture 32(1): 46-52.

[23] Hanna, S. E., P. Rosembaum, D. J. Bartlett, et al. (2009). Stability and decline in gross motor function among children and youth with cerebral palsy aged 2 to 21 years. Developmental Medicine & Child Neurology 36(1): 295-302.

[24] Harada, T., S. Ebara, M. M. Anwar, et al. (1993). The lumbar spine in spastic diplegia. A radiographic study. Journal of Bone Joint Surgery (Br) 75(4): 534-7.

[25] Hilberink, S., M. Roebroeck, W. Nieuwstraten, et al. (2007). Health issues in young adults with cerebral palsy: Towards a life-span perspective. Journal of Rehabilitation Medicine 39: 605-611.

[26] Himmelmann, K. (2013). Epidemiology of cerebral palsy. Handbook of Clinical Neurology 111: 163-7.

[27] Holsbeeke, L., M. Ketelaar, M. M. Schoemaker and J. W. Gorter (2009). Capacity, capability, and performance: different constructs or three of a kind? Archives of Physical Medicine and Rehabilitation 90(5): 849-55.

http://dx.doi.org/10.5772/58344

[28] Horsman, M., M. Suto, B. Dudgeon and S. R. Harris (2010). Ageing with cerebral palsy: psychosocial issues. Age Ageing 39(3): 294-9.

[29] Horsman, M., M. Suto, B. Dudgeon and S. R. Harris (2010). Growing older with cerebral palsy: insiders' perspectives. Pediatric Physical Therapy 22(3): 296-303.

[30] Jahnsen, R., L. Villien, G. Aamodt, J. Staghelle and I. Holm (2003). Physiotherapy and physical activity - Experiences of adults with cerebral palsy, with implications for children. Advances in Physiotherapy 5: 21-32.

[31] Jahnsen, R., L. Villien, T. Egeland, J. K. Stangelle and I. Holm (2004). Locomotion skills in adults with cerebral palsy. Clinical Rehabilitation 18: 309-316.

[32] Jahnsen, R., L. Villien, J. K. Stangelle and I. Holm (2003). Fatigue in adults with cerebral palsy in Norway compared to the general population. Developmental Medicine & Child Neurology 45: 296-303.

[33] Jahnsen, R., L. Villien, J. K. Stangelle and I. Holm (2004). Musculoskeletal pain in adults with cerebral palsy compared with the general population. Journal of Rehabilitation Medicine 36: 78-84.

[34] Mackintosh, S. F., K. Hill, K. J. Dodd, P. Goldie and E. Culham (2005). Falls and injury prevention should be part of every stroke rehabilitation plan. Clinical Rehabilitation 19(4): 441-51.

[35] Maltais, D. B., F. Dumas, N. Boucher and C. L. Richards (2010). Factors related to physical activity in adults with cerebral palsy may differ for walkers and nonwalkers. American journal of Physical Medicine & Rehabilitation 89(7): 584-97.

[36] Morgan, P. and J. McGinley (2013). Falls, fear of falling and falls risk in adults with cerebral palsy: A pilot observational study. European Journal of Physiotherapy 15/2(2): 93-100.

[37] Morgan, P. and J. McGinley (2013). Gait function and decline in adults with cerebral palsy: a systematic review. Disability & Rehabilitation 36(1):1-9

[38] Morgan, P. and J. McGinley (2013). Performance of adults with cerebral palsy related to falls, balance and function: a preliminary report. Dev Neurorehabil 16(2) 113-20.

[39] Mosqueda, L. (2004). Maintaining health and function. Aging with a disability - What the clinician needs to know. B. Kemp and L. Mosqueda. Baltimore, The John Hopkins University Press.

[40] Murray, K., K. Hill, B. Phillips and B. Waterston (2005). A pilot study of falls risk and vestibular dysfunction in older fallers presenting to hospital Emergency Departments. Disability and Rehabilitation 27(9): 499-506.

[41] Nevitt, M. C., S. R. Cummings and E. S. Hudes (1991). Risk factors for injurious falls: a prospective study. Journal of Gerontology 46(5): M164-70.

[42] Ng, S. Y., S. K. Dinesh, S. K. Tay and E. H. Lee (2003). Decreased access to health care and social isolation among young adults with cerebral palsy after leaving school. J Orthop Surg (Hong Kong) 11(1): 80-9.

[43] Nieuwenhuijsen, C., W. M. van der Slot, A. J. Dallmeijer, et al. (2011). Physical fitness, everyday physical activity, and fatigue in ambulatory adults with bilateral spastic cerebral palsy. Scand J Med Sci Sports 21(4): 535-42.

[44] Nordell, E., M. Andreasson, K. Gall and K. Thorngren (2009). Evaluating the Swedish version of the Falls Efficacy Scale–International (FES-I). Advances in Physiotherapy 11: 81-87.

[45] Novacheck, T., J. Stout, J. Gage and M. Schwartz (2009). Distal femoral extension osteotomy and patellar tendon advancement to treat persistent crouch gait in cerebral palsy. Surgical technique. Journal of Bone & Joint Surgery (Am) 91(Suppl 2:): 271-86.

[46] Opheim, A., R. Jahnsen, E. Olsson and J. Staghelle (2009). Walking function, pain and fatigue in adults with cerebral palsy: A 7-year follow-up study. Developmental Medicine & Child Neurology 51(5): 381-8.

[47] Opheim, A., R. Jahnsen, E. Olsson and J. K. Stanghelle (2011). Physical and mental components of health-related quality of life and musculoskeletal pain sites over seven years in adults with spastic cerebral palsy. Journal of Rehabilitation Medicine 43(5): 382-7.

[48] Opheim, A., R. Jahnsen, E. Olsson and J. K. Stanghelle (2012). Balance in relation to walking deterioration in adults with spastic bilateral cerebral palsy. Physical Therapy 92(2): 279-88.

[49] Palisano, R., P. Rosenbaum, S. Walter, et al. (1997). Development and reliability of a system to classify gross motor function in children with cerebral palsy. Developmental Medicine and Child Neurology 39(4): 214-23.

[50] Pender, N. (1987). Health promotion in nursing practice. Nonvalk, CT, Appleton and Lange.

[51] Peterson, E. W., C. C. Cho and M. L. Finlayson (2007). Fear of falling and associated activity curtailment among middle aged and older adults with multiple sclerosis. Multiple Sclerosis 13(9): 1168-75.

[52] Peterson, M. D., P. M. Gordon and E. A. Hurvitz (2013). Chronic disease risk among adults with cerebral palsy: the role of premature sarcopoenia, obesity and sedentary behaviour. Obes Rev 14(2): 171-82.

[53] Pickering, R. M., Y. A. Grimbergen, U. Rigney, et al. (2007). A meta-analysis of six prospective studies of falling in Parkinson's disease. Movement Disorders 22(13): 1892-900.

http://dx.doi.org/10.5772/58344

[54] Pieterse, A. J., T. B. Luttikhold, K. de Laat, et al. (2006). Falls in patients with neuromuscular disorders. Journal of Neurological Sciences 251(1-2): 87-90.

[55] Podsiadlo, D. and S. Richardson (1991). The timed "Up & Go": a test of basic functional mobility for frail elderly persons. Journal of the American Geriatrics Society 39(2): 142-8.

[56] Reddihough, D. S., B. Jiang, A. Lanigan, et al. (2013). Social outcomes of young adults with cerebral palsy. Journal of Intellectual & Developmental Disability 38(3): 215-22.

[57] Root, L. (2009). Surgical treatment for hip pain in the adult cerebral palsy patient. Developmental Medicine & Child Neurology 51 Suppl 4: 84-91.

[58] Russell, M. A., K. D. Hill, I. Blackberry, L. M. Day and S. C. Dharmage (2008). The reliability and predictive accuracy of the falls risk for older people in the community assessment (FROP-Com) tool. Age Ageing 37(6): 634-9.

[59] Ruthig, J. C., J. G. Chipperfield, N. E. Newall, R. P. Perry and N. C. Hall (2007). Detrimental effects of falling on health and well-being in later life: the mediating roles of perceived control and optimism. J Health Psychol 12(2): 231-48.

[60] Sandstrom, K. (2007). The lived body - experiences from adults with cerebral palsy. Clinical Rehabilitation 21(5): 432-41.

[61] Sandstrom, K., J. Alinder and B. Oberg (2004). Descriptions of functioning and health and relations to a gross motor classification in adults with cerebral palsy. Disability &Rehabilitation 26(17): 1023-1031.

[62] Schwartz, L., J. Engel and M. Jensen (1999). Pain in persons with cerebral palsy. Archives of Physical Medicine and Rehabilitation 80: 1243-6.

[63] Shroeder, K., C. Hauck, B. Wiedenhöfer, F. Braatz and P. Aldinger (2009). Long-term results of hip arthroplasty in ambulatory patients with cerebral palsy. International Orthopaedics (SICOT) (2010) 34:(9): 335-339.

[64] Strauss, D., J. Brooks, L. Rosenbloom and R. Shavelle (2008). Life expectancy in cerebral palsy: an update. Developmental Medicine & Child Neurology 50(7): 487-93.

[65] Strauss, D., K. Ojdana, R. M. Shavelle and L. Rosenbloom (2004). Decline in function and life expectancy of older persons with cerebral palsy. NeuroRehabilitation 19: 69-78.

[66] Svien, L., P. Berg and C. Stephenson (2008). Issues in Aging with Cerebral Palsy. Topics in Geriatric Rehabilitation 24: 27-41.

[67] Tosi, L. L., N. Maher, D. W. Moore, M. Goldstein and M. L. Aisen (2009). Adults with cerebral palsy: a workshop to define the challenges of treating and preventing secondary musculoskeletal and neuromuscular complications in this rapidly growing population. Developmental Medicine & Child Neurology 51 Suppl 4: 2-11.

[68] Turk, M., C. Geremski, P. Rosenbaum and r. R. Webe (1997). The health status of women with cerebral palsy. American Journal Of Physical Medicine and Rehabilitation 78: S10-17.

[69] van der Ploeg, H., A. van der Beek, L. van der Woude and W. van Mechelen (2004). Physical activity for people with a disability: a conceptual model. Sports Medicine 34: 639-49.

[70] Van der Slot, W., C. Nieuwenhuijsen, H. J. Van den Berg-Emons, et al. (2012). Chronic pain, fatigue, and depressive symptoms in adults with spastic bilateral cerebral palsy. Developmental Medicine & Child Neurology 837-842.

[71] van der Slot, W. M. A., M. E. Roebroeck, A. P. Landkroon, et al. (2007). Everyday physical activity and community participation of adults with hemiplegic Cerebral Palsy. Disability & Rehabilitation 29(3): 179 - 189.

[72] Vogtle, L. K. (2009). Pain in adults with cerebral palsy: impact and solutions. Developmental Medicine & Child Neurology 51 Suppl 4: 113-21.

[73] Yardley, L., N. Beyer, K. Hauer, et al. (2005). Development and initial validation of the Falls Efficacy Scale-International (FES-I). Age Ageing 34(6): 614-9.

[74] Watson, W., A. Clapperton, et al. (2010). The incidence and cost of falls injury among older people in New South Wales 2006/07. Sydney, NSW Department of Health.

Chapter 4

Cerebral Palsy and Accessible Housing

Emira Švraka

Additional information is available at the end of the chapter

http://dx.doi.org/10.5772/56983

1. Introduction

Healthcare systems in transition countries must adapt to the many changes occurring in society as a whole. This is a very serious process requiring reforms that will significantly change the management and organization of healthcare at all levels. Bosnia and Herzegovina has suffered large scale destruction during the war (1992-1995), and all the medical capacities in the country sustained significant damage. The causes of demographic changes in transition countries include: the growth of urban populations, expansion of education, the modernization of society, the disintegration of the family, medical advances, increased income, decreased fertility and increased mortality (Loga S, 2011).

Community Based Rehabilitation (CBR) is strategy for rehabilitation, equal possibilities and social integration of all persons with disabilities. CBR program is implementing with joint effort of persons with disabilities, their families, community and related health, educational and social institutions. Before the 1992, medical rehabilitation in Bosnia and Herzegovina had been provided at the level of institutions, usually after the hospital or ambulant treatments. Model of Community Based Rehabilitation (CBR), which practically tested in all parts of Bosnia and Herzegovina, suggests numerous advantages when compared to the previous period, until 1992.

There is large number of health and educational institutions in the Canton of Sarajevo which are working on re/habilitation and education of children and adolescents with disabilities, but there isn't unique database about the people with disabilities, as well as with cerebral palsy. Lack of unique database indicates poor network among the institutions and Associations in the Canton of Sarajevo.

A number of issues arise from the study "Family Quality of Life: Adult School Children with Intellectual Disabilities". Four of the seemingly most important are: lack of organized community services for adults after they leave school; lack of a cantonal, state, or federal registration program that would improve coordination of health and social services and link to the European Register; necessity of conducting continuous education for the teaching staff at schools regarding effective curricula, for parents, and for health professionals; and the possibility of developing occupational and physical therapy programs for children, adolescent, and adults. The degree to which improvements such as these might affect family quality of life also needs to be examined in future study (Švraka E, Loga S, Brown I, 2011).

The goals of education and rehabilitation in Bosnia and Herzegovina, similar to most other countries of the world, are to work toward community inclusion, acceptance of diversity, optimal physical and mental health, and personal and social well-being. The focus on family quality of life is a step toward understanding how we can move closer to achieving these goals (Švraka, Loga, Brown, 2011).

The fact that concerns the most in the South-Eastern Europe is that many people with disabilities are isolated in their homes. One reason for this isolation is huge barriers that must face when they try to go out of their homes. Common premises, such as elevators, corridors and passages often are inaccessible. This only reinforces the fact that most laws on accessibility applies only to public buildings, so that investors who invest in private buildings can go unpunished for not fulfilling these regulations (Sestranetz, Adams, 2006).

1.1. Cerebral palsy

Cerebral palsy (CP) is characterized by nonprogressive abnormalities in the developing brain that create a cascade of neurologic, motor and postural deficit in the developing child. Cognitive, sensory and psychosocial deficits often compound motor impairments and subsequent functioning. Characteristically, the child with CP shows impaired ability to maintain normal posture because of a lack of muscle coactivation and the development of abnormal movement compensations. These compensatory patterns develop in certain muscle groups to maintain upright postures and move against gravity. Hyperactive responses to tactile, visual or auditory stimuli may result in fluctuations of muscle tone that often adversely affect postural control and further diminish coordinated responses in everyday activities (Rogers, Gordon, Schanzenbacher, Case-Smith, 2001).

Cerebral palsy (CP) occurs at present in about 2,2 per 1000 live born children in Sweden. Epilepsy occurs in 15% to more than 60% of children with CP, depending on the type of CP and the origin of the series, compared with 0,5% in the general population (Carlsson, Hagberg, Olsson, 2003).

According to the time of influence, *causes of cerebral palsy* can be divided to prenatal (from conception until beginning of the delivery), perinatal (beginning of the delivery until age of 28 days) and postnatal (from 29th day of age until two years of age). The majority of international studies indicates that the *prevalence of the cerebral palsy* is about 2-2,5 cases per 1000 born, although there are some reports about lower and higher prevalence rates (Nordmark, Hagglund, Lagergren, 2001).

http://dx.doi.org/10.5772/56983

Evidences indicated that 70-80 % of cerebral palsy is caused by the prenatal factors and that the birth asphyxia has a relatively minor role with the less than 10 % (Jacobsson & Hagberg, 2004).

Early diagnosis of CP symptoms followed with early intervention is crucial, as soon as possible.

With the rising incidence of CP in time, the distribution over the subtypes changed: fewer cases with diplegia and more with hemiplegic. The *motor impairments of CP*, in especially the spastic types, lead to other impairments of the musculoskeletal system; for example; among children and adolescents with quadriplegic CP, 75% have hip luxations, 73% contractures, and 72% scoliosis (Odding, Roebroeck & Stam, 2006).

About 40% of children with hemiplegic CP have normal *cognitive abilities*, while children and adolescents with tetraplegic CP are generally severely intellectually impaired (Odding, Roebroeck & Stam, 2006).

CP associated with *epilepsy* is far more frequently accompanied by intellectual disability than CP without epilepsy. Similarly, the combination of CP and intellectual disability is reported to be associated with a high risk of developing epilepsy (Carlsson, Hagberg & Olsson, 2003).

Many children with more severe spastic CP experience *communication problems* due to disturbed neuromuscular control of speech mechanism, i. e, dysarthria, that diminish the ability of the child to speak intelligible. However, substantial dysarthria are most often seen in children with severe CP and intellectual disability, while most children with mild or moderate CP and average cognitive level of functioning have normal or near-normal expressive language and articulation skills (Bottcher, 2010).

In the study of the influence of prenatal etiological factors on learning disabilities of children and adolescents with cerebral palsy in the Canton of Sarajevo, of all sample, 31 (38,75%) children with CP used nonverbal and sign communication, and 49 (61,25%) children used verbal communication (i.e. speech).

Depending of the study, the prevalence of *visuomotor and perceptual problems* among children with spastic CP varies from 39% to 100% (Stiers & Vanderkelen, 2002)

Professionals and parents need to be aware that children with cerebral palsy are at higher risk of *psychological problems* than their non-disabled peers and this may be attributable to problems in adjustment to their adverse circumstances as well as having an organic basis. Attention should be paid to the effective *management of pain*, particularly in children unable to self-report for whom a reliable instrument for assessing pain now exists. The difficulties most commonly reported here were peer problems; as these may have implications for later psychological adjustment, follow up work into adolescence and beyond will be important. It may be that for many children with cerebral palsy and their families, chronic psychological problems will have a greater impact than the physical impairments and this possibility also needs to be investigated in longitudinal studies (Parkes, White-Koning, Dickinson, Thyen, Arnaud, Beckung & all, 2008).

1.2. Occupational therapy for persons with cerebral palsy in the Canton of Sarajevo

The research was conducted through Project: „Occupational therapy for persons with cerebral palsy", in homes of participants. *The aim* was to determine accessible housing for persons with cerebral palsy.

The client was Association of persons with cerebral palsy in the Canton of Sarajevo. The Association includes 315 members. Of that number, 123 (47,13%) are children and adolescents, age 4 up to 20 years, and 138 (52,87%) are adults.

Sample was consisted of 30 members of the Association of persons with cerebral palsy of the Canton of Sarajevo, age from 4 up to 53: 8 children (4-11 years), 14 adolescents (12-20 years), and 8 adults (21-53 years); 14 male (46,67 %) and 16 (53,33 %) female.

Nine participants had private houses, and 21 were living in flats.

The principal measure used was the *Environmental Assessment – Home assessment form*. The first part should deal with accessibility of the dwelling's exterior, and the second half should be concerned with an assessment of the home's interior. During the *On-Site visit* a tape measure and home assessment form are tools (Schmitz, 1988), translated and modified by the author (Švraka, 2007).

The part about accessibility of the dwelling's exterior is made of 36 items: type of home, entrances to building or home, approach to apartment or living area (hallway, steps, door, and elevator). *"Inside home"* part consists of bedroom, bathroom, living room area, dining room, kitchen, laundry, cleaning, emergency and few other items.

The study was approved by parents of children with CP, or adults with CP, and president of the Association of persons with cerebral palsy in the Canton of Sarajevo. Before starting the data collection, the research aim and Environmental Assessment – Home assessment form were explained to parents and they agree to participate by signing consent.

Ideally, the physical and occupational therapists should accompany the patient on the home visit. They assume shared responsibility for assessing the patient's functional level at home. Depending of the specific needs of the patient and/or family, a speech therapist, social worker, or nurse also may be included on the home visit (Shmitz, 1988).

Research was conducted during 3 months period trough home visits to clients. Basic inclusion criteria were:

1. Association members with severe motor disability,

2. Lower community engagement or majority of clients are not involved in some form of institution, continuous forms of education and/or re/habilitation.

Students of Department of physiotherapy, at Faculty of Health Studies in Sarajevo, who have completed their course of studies, apply the *Environment assessment - Home assessment form* in patient's home as part of practical education, in an environment that does not have the occupational therapy program. Supervision was performed by assistant professor of Faculty of Health Studies. Based on the initial assessment of the patient in the house/ On-site assessment, individual therapeutic programs/interventions were made in order to improve occupational performance.

1. Interventions which changed requirements of occupation was bringing large gymnastic ball in the home of all 30 patients.

2. Interventions that want to affect the environment, followed after the evaluation. In co-operation with the police and local community, students were working on improvement of accessibility: free parking places in front of the building, entrance ramps, accessible elevators.

3. Interventions that want to improve the ability of the person was the education in certain exercises for the improvement and preservation of posture, balance, coordination, increase the mobility and prevention of deformities deterioration, which influenced the personal competencies, i.e. skills related to motor performance, sensor abilities, cognitive ability and general health condition.

The Association of persons with cerebral palsy in the Canton of Sarajevo is member of *Cerebral Palsy Association of Federation of Bosnia and Herzegovina* which was established at 17. October of 2011. That day was announced as *Day of persons with cerebral palsy of Federation of Bosnia and Herzegovina* (FB&H). Cerebral Palsy Association members include five Associations of persons with cerebral palsy of FB&H, from five towns/Cantons: Sarajevo, Goražde, Zenica, Široki Brijeg and Sapna.

People with CP can lead active lives and make a valuable contribution to society. Art workshop of the Association of persons with cerebral palsy in the Canton of Sarajevo consists of 9 female members, 7 with CP and 2 with paraplegia. Middle age is 37,7 years; two youngest members are 27 years old, and oldest one is 58 year old. Five members use wheelchairs (3 with CP and 2 with paraplegia), one cane, and three of them are walking independently. It is necessary to reduce the numbers of sheltered workshops, and develop supported employment and self-employment, in other to reduce segregation of persons with disabilities and give support to social inclusion.

1.3. Assistive technology

Assistive technology (AT) is an umbrella term for a wide range of products. A commonly accepted definition is "any item, piece of equipment or product system whether acquired commercially off the shelf, modified or customized that is used to increase, maintain or improve functional capabilities of individuals with disabilities" (US Statute, 1988). Therefore in terms of devices or equipment it includes from walking sticks to environmental control systems (ECS), or simple dressing aids to communication aids (Cowan & Wintergold, 2007).

Assistive devices include ortho-prosthetic devices, wheel chairs, walking aids, technical aids and adapted controls for cars. Adequate assistive devices are often financially inaccessible to many users because of their high cost despite the fact that they should be covered by social and insurance schemes. Under the current system, most assistive devices are covered only partially by the state and require user co-payments, which can be exorbitant in cost. Within the socialist system, assistive devices were generally provided for free within the public health care system. This is a crucial issue in South East Europe as one of the largest barriers to accessing assistive devices is financial. Ortho-prosthetic devices are partially subsidized by the state and in most countries, co-payments have been set up but the financial burden is still heavy, especially for mid to low-income households. For example, in Bosnia and Herzegovina, co-payments can range from 10-50%, which can range from EUR 100-1,000 depending on the device. In the UN administered province of Kosovo there is an absence of a health care financing system so patients must pay the full price for their wheelchairs or other devices (Handicap International, 2004).

Mechanical assistive technology includes equipment such as manual wheelchair, postural management equipment, equipment for active exercise, protective devices, orthoses and aids for daily living. Provision of mechanical AT for children has its own unique challenges. Children are constantly changing as they grow and their abilities change and develop. Equipment therefore needs to be chosen with these aspects in mind. Adjustable equipment enables changes to be made according to a child's needs. Adjustability within a device does tend to make equipment heavier, more complex and expensive but it will last longer and may be adjusted to fit the constantly changing needs of a child (Cowan & Wintergold, 2007).

The study of the influence of prenatal etiological factors on learning disabilities of children and adolescents with cerebral palsy in the Canton of Sarajevo was conducted with sample of 80 participants, children and adolescents with cerebral palsy in the Canton of Sarajevo, age from 6 up to 20 years; 25 children (age 6-11), and 75 adolescents (age 12-20). Mean age was 13,94 years, 47 male (58,75%) and 33 (41,25%) female. The sample was divided in two subgroups, first includes 30 participants whose mothers had problems during the pregnancy, and second includes 50 participants whose mothers didn't have problems during the pregnancy.

Cerebral palsy	**Walking ability**				**Total**
	Walks without restrictions	**Holding a hand**	**Walker**	**Wheelchair**	
	Bilateral spastic CP				
Spastic Quadriplegic CP	2	/	1	10	13
Spastic Quadriplegic CP mixta	/	/	/	3	3
Triplegia	/	/	/	1	1
Paraplegia	3	1	1	1	6
	Unilateral spastic CP				
Spastic Hemiplegic CP (right)	5	/	/	/	5
Spastic Hemiplegic CP (left)	4	/	/	/1	5
Total	14	1	2	16	33

Table 1. Structure of the sample of children with CP and epilepsy according to walking ability (Švraka, 2012)

Of 33 children with cerebral palsy and epilepsy, 14 (42,4%) were able to walk independently, 1 (3%) child needs to hold a mother's or friend's hand, 2 (6%) children walks with assistive device (walker), and 16 (48,5%) children were unable to walk, in need of wheelchair.

Of total sample of 80 participants, 34 (42,5%) were in need of wheelchair, and 46 (57,5%) were not.

Of total sample of 80 children, 42 (52,5%) were able to walk independently.

In the group of 30 participants, with illnesses during pregnancy, 13 (43,3%) were in need of wheelchair, and 17 (56,7%) were not. In the group of 50 participants, without illnesses during pregnancy, 21 (42%) were in need of wheelchair, and 29 (58%) were not.

Of 30 participants with illnesses during pregnancy, 17 (56,7%) were able to walk independently, and 13 (43,3%) were not. Of 50 participants without illnesses during pregnancy, 25 (50%), were able to walk independently, and 25 (50%) were not.

Cerebral palsy	**Mobility assistance**			**Total**
	Wheelchair	**Without assistance**	**Walker, tripod, holding**	
		Bilateral spastic CP		
Spastic Quadriplegic CP	7	2	0	9 (30%)
Spastic Quadriplegic CP mixta	5	0	0	5 (16,7%)
Triparesis	4	0	1	5 (16,7%)
Paraparesis	3	3	2	3 (26,7%)
		Unilateral spastic CP		
Spastic Hemiplegic CP	1	2	0	3 (10%)
Total	20 (66,7%)	7 (23,3%)	3 (10%)	30 (100%)

Table 2. Relation of cerebral palsy types and use of mobility assistance

Of 30 persons with cerebral palsy 20 (66.7%) use wheelchairs, 7 (23.3%) have independent mobility, without aid, and 3 (10%) persons walk with aid. Client with triparesis use a walking tripod.

Intellectual abilities	Mobility assistance			Total
	Wheelchair	Without assistance	Walker, tripod, holding	
Normal intellectual abilities	2	6	1	9 (30%)
Borderline intellectual abilities	1	0	0	1 (3,33%)
Mild intellectual disability	7	0	0	7 (23,3%)
Moderate intellectual disability	1	1	1	3 (10%)
Severe intellectual disability	7	0	0	7 (23,3%)
IQ not determined	2	0	1	3 (10%)
Total	20 (66,7%)	7 (23,3%)	3 (10%)	30 (100%)

Table 3. Relation of intellectual abilities and use of mobility assistance

Of 7 persons with independent mobility, 6 are with normal intellectual abilities, and 1 with moderate disability. All seven persons (23.3%) with severe intellectual disability use wheelchairs

The wheelchair and other wheeled seated mobility devices, such as scooters, have been and remain important technological devices in the field of rehabilitation. In North America, a substantial number of adults require wheeled seated mobility, with estimates indicating that over 179000 Canadians and over 1,5 million Americans utilize a wheelchair. With respect to the environment, wheeled seated mobility systems may increase the accessibility of the physical environment thereby increasing opportunities for interacting with the social environment (Reid, Laliberte-Rudman & Hebert, 2002).

2. Accessible home

Accessibility for all is a fundamental right, and any environmental barrier which denies access and free movement for persons with disabilities and other persons with reduced mobility is and must be recognized as discrimination (Howitt, 2003).

An accessible home is a pre-condition for independent living or self-determined living as it enables individuals to do what they need and desire to do as independently as possible within their living space. This definition is addressed to all people meeting difficulties in performing daily activities at home as a result of a disability. It means that not only people with physical disabilities, people who we automatically have in mind when talking about accessibility, but also people with sensory or intellectual disabilities or even elderly people who might have lost certain capacities and therefore meet obstacles in their homes - all need accessible housing. For some, this can be achieved with accessible features that are permanently fixed such as *wide doors, grab bars, a tub seat in the bathroom* or by using adaptable features adjustable in a short time without involving structural or material changes. For individuals with sensory disabilities, a blind person, for instance, requires *tactile markings for changes in the floor level and Braille markings on appliances*. Individuals with hearing impairments will *need visual adaptations* for things such as *telephone ringers, the doorbell and smoke alarms*. For wheelchair users, access may require *ramps at the entrances, lower counters, no thresholds, wider toilets,* a shower rather than a bathtub, and ensuring there is an *accessible lift* if the dwelling is above the ground floor (Consumer's Guide to Accessible Housing, 2007).

Accessible living space is helping to enable an independent life and to provide way that people with disabilities live in the community. With personal assistant and accessible home, people with disabilities can live independently. Inadequate housing for people with disabilities has serious consequences. In the United Kingdom one study showed that there are over 4 million of people who have difficulties to move, but only 80000 are in accessible housing. Between 1980 and 1988 the number of homeless people with disabilities has increased by 92%, not including those who live in institutions or family homes (Sestranetz, Adams, 2006).

Children with CP may have limitations in all areas of human occupation to some degree. Functional performance in self-care and independent living, school and work performance, play and recreation may all need to be addressed at some point in the child's life. Parents my require support and respite, as well as education, to care for child with CP to meet the needs of the family as a whole (Rogers, Gordon, Schanzenbacher, Case-Smith, 2001).

Another common problem people face in the region when adapting an inaccessible dwelling is that there are *no services available to provide guidance and consultation* on making the adaptations. In Calgary Canada, there is an Accessible Housing Society providing consultation services to people who wish to adapt their home. With this service, *an occupational therapist and an architect* visit individual homes to assess what needs to be adapted to suit the needs of the person and then draw up plans for modifications. They also provide information such as names of the relevant vendors and contractors, accessibility products and standards. There is no charge for the service if the client qualifies for income-tested government funding programs that include: Residential Access Modification Program, Residential Rehabilitation Assistance Program, Home Adaptations for Senior's Independence under the Alberta government housing support programs. Under these programs, applicants who qualify receive a grant to make proper adaptations. The government housing support programs contain an accessible housing registry for people seeking barrier-free dwellings. This registry refers clients to available accessible housing while documenting housing needs for future planning and construction (Disability Monitor Initiative South East Europe, 2007).

2.1. Exterior accessibility

The *entrance* should be well lighted and provide adequate cover from adverse weather conditions. If a *ramp* is to be installed, there should be adequate space. The recommended grade for wheelchair ramps is 12 inches in ramp length for every inch of threshold height. Ramps should be a minimum of 48 inches (121,9 cm) wide with a nonslip surface. *Handrails* also should be included on the ramp, 32 inches (81,3 cm) in height and extend 12 inches (30,5 cm) beyond the top and bottom of the ramp (Schmitz, 1988).

Seven studies focused on aspects of the physical environment as it relates to accessibility issues and wheelchair accident. The most wheelchair accidents occur outdoors or on ramps. There remains a need for public buildings to implement barrier-free access changes for wheelchair users. Wheelchair users voiced concern about not being included in decisions regarding the design (Reid, Laliberte-Rudman & Hebert, 2002).

For wheelchair users, the entrance should have a platform large enough to allow the patient to rest and to prepare for entry. This platform area is particularly important when a ramp is in use. The *door locks* should be accessible to the patient. The *door handle* should be turned easily by the person. The door should be open and close in a direction that is functional for the person. A cane may be hung outside the door to help the wheelchair user close the door when leaving. The *doorway width* should be measured. Generally, 32 inches (81,3 cm) to 34 inches (86,3 cm) is an acceptable doorway to accommodate most wheelchairs. (Schmitz, 1988).

Place of living	Width of the entrance door (cm)						Total
	62-75	76-82	83-92	93-102	103-112	No answer	
Flat	1	8	4	-	1	7	21 (69,93%)
Private house	2	5	0	-	-	2	9 (29,97%)
Total	3 (9,99%)	13 (43,29%)	4 (13,32%)	0	1 (3,33%)	9 (29,97%)	30 (100%)

Table 4. Relations of the place of living and entrance door width

The range of width of the entrance door was from 62cm to 112 cm. Thirteen families (43,29%) had entrance door width between 76 cm to 82cm.

Place of living	Elevator		Total
	YES	NO	
Flat	11	10	21 (69,93%)
Private house	-	9	9 (29,97%)
Total	11 (36,63%)	19 (63,27%)	30 (100%)

Table 5. Place of living and existing of elevator

Eleven families (36,63%) of persons with cerebral palsy, who live in flats have elevators. In private houses there are no elevators.

If there is raised *threshold* in the doorway, it should be removed. If removal is not possible, the threshold should be lowered to no greater than 0,5 inch (1,27 cm) in height, with beveled edges (Schmitz, 1988).

http://dx.doi.org/10.5772/56983

Threshold	Gender		Total
	Female	Male	
Raised threshold	14	9	23 (76 7%)
Without threshold	2	5	7 (23,3%)
Total	16 (53,33%)	14 (46,67%)	30 (100%)

Table 6. Entrance door thresholds

Twenty three families (76,7%) have raised entrance door thresholds made of different material: wood (16), concrete (3), metal (3), and one made of marble; 1 cm to 7 cm in height.

2.2. Interior accessibility

Inaccessible buildings and rooms crowded with furniture limit how children in wheelchairs move throughout the environment. Differences in the terrain or room surface also affect mobility. For example, a child who can run outdoors on an asphalt playground may trip and fall inside when walking on a rug. Other physical characteristics that the occupational therapist assesses relate to the type of furniture, objects, or assistive devices in the environment and whether they are usable and accessible. This includes the type of equipment, household items, clothing or toys. Sensory aspects of the physical environment often influence performance, e.g. the type of lighting, noise level, visual stimulation, and tactile or vestibular input of tasks (Shepherd, 2001).

Sufficient room should be made available for maneuvering or ambulating with an assistive device. Clear passage must be allowed from one room to the next. Unrestricted access should be provided to electrical outlets, telephones and wall switches. All *floor coverings* should be glued or tacked to the floor. This will prevent bunching or rippling under wheelchair use. Scatter rugs should be removed. Use of nonskid waxes should be encouraged. Raised *thresholds* should be removed to provide a flush, level surface. *Doorways* may need to be widened to allow clearance for a wheelchair or assistive device. *Doors* may have to be removed, reversed, or replaced with curtains or folding doors. All indoor *stairwells* should have handrails and should be well lighted. For patients with decreased visual acuity or age-related visual changes, contrasting textures on the surface of the top and bottom stair/s will alert them that the end of the stairwells is near. Circular band or tape also can be placed at the top and bottom of the handrail for the same purpose (Schmitz, 1988).

2.2.1. Bedroom

The bed should be stationary and positioned to provide ample space for transfers. Stability may be improved by placing the bed against the wall or in the corner of the room. The height of the sleeping surface must be considered to facilitate transfer activities. The *mattress* should be carefully assessed; it should provide a firm, comfortable surface. If the mattress is in relatively good condition, a bed board inserted between the mattress and box spring may suffice to improve the sleeping surface adequately. If the mattress is badly worn, a new one should be suggested. A *bed side table or cabinet* might be suggested; it will be useful to hold a lamp, a telephone, necessary medications, and a call bell if assistance is needed (Schmitz, 1988).

Participants	Height of the bed (cm)															Total
	20	21	35	36	38	40	43	44	45	46	48	50	51	55	60	
Children *						3	1			2		2				8
Adolescents	1		2	2	1	2			3			2		1		14
Adults		1		1		2		1			1		1		1	8
Total	1	1	2	3	1	7	1	1	3	2	1	4	1	1	1	30

Children (4-11 years); Adolescents (12-20 years); Adults (21-53 years)

Table 7. Height of the bed and age of the participants

Range of the height of the bed was 20 cm to 60 cm. The height of the bed of 7 participants was 40 cm.

The range of the height of the bed of children was 40 cm to 50 cm.

The range of the height of the bed of adolescents was 20 cm to 55 cm.

The range of the height of the bed of adults was 21 cm to 60 cm.

Participants	Width of the bed (cm)															Total
	55	65	68	80	90	95	100	105	110	120	125	150	162	170	220	
Children	1	1		1			2		1	1			1			8
Adolescents			1	3		1	1	2	1	1	1	2		1		14
Adults					2		2		1			2			1	8
Total	1	1	1	4	2	1	5	2	3	2	1	4	1	1	1	30

Table 8. Width of the bed and age of the participants

Range of the width of the bed was 55 cm to 220 cm. The width of the bed of 5 participants was 100 cm.

The range of the width of the bed of children was 55 cm to 162 cm.

The range of the width of the bed of adolescents was 68 cm to 170 cm.

The range of the width of the bed of adults was 90 cm to 220 cm.

Night table	Mobility assistance			Total
	Wheelchair	Without assistance	Walker, tripod, holding	
Within patient's reach from bed	6	4	0	10 (33,3%)
Without patient's reach	9	1	1	11 (36,7%)
Without night table	3	2	2	7 (23.3%)
Without answer	2	0	0	2 (6,7%)
Total	20 (66,7%)	7 (23,3%)	3 (10%)	30 (100%)

Table 9. Relations of night table and mobility assistance

Of 30 persons with cerebral palsy, for 10 (33,3%) persons night table is within patient's reach from bed, for 11 (36,7%) persons night table is not within patient's reach from bed, 7 (23,3%) persons don't have night table, and for two persons there are no answers.

Of 20 persons with wheelchair, for 6 persons night table is within patient's reach from bed, and for 9 persons night table is not within patient's reach from bed.

Of 7 persons with independent mobility, for 4 persons night table is within patient's reach from bed, for 1 person night table is not within patient's reach from bed, and 2 persons don't have night table.

2.2.2. *Bathroom*

If door frame prohibits passage of a wheelchair, the patient may transfer at the door to a chair with *casters* attached. An elevated toilet seat will facilitate transfer activities (Schmitz, 1988).

Special equipment that gives support can help the child feel safe and secure. Bath hammocks fully hold the body and enable the parent to wash the child thoroughly. A simple, inexpensive way for giving security is to use a plastic laundry basket lined with foam at its bottom. Commercially, alight, inconspicuous bath support offers good design features. The front half of the padded support ring swings open for easy entry and then locks securely, holding the child at the chest to give trunk stability. Various kinds of bath seats and shower benches are available for the older child to aid bathtub seating transfers. For the child with severe motor limitations who is lying supine in the tub in shallow water, a horseshoe-shaped inflatable bath collar serves to support the neck and keep the child's head above water level. A bath stretcher is constructed like a cot and fits inside the bathtub rim level or mid tub to minimize the caregiver's bending while transferring and bathing the child (Rogers, Gordon, Schanzenbacher, Case-Smith, 2001).

Independent toileting is an important self-maintenance milestone with wild variation among individual children. Independence in toileting includes getting on and off the toilet, managing fastener, and clothing, cleansings after toileting, and washing and drying hands efficiently without supervision. With weakness and limited range of motion, the child may be unable to manage fastenings because of hand involvement or may have problems in sitting down or getting up from the toilet seat because of hip-knee contractions or quadriceps weakness. (Shepherd, 2001).

Cerebral palsy	Use of toilet seat	Use of toilet pot	Diapers	Total
		Bilateral spastic CP		
Quadriplegic CP	6	0	3	9
Quadripl. CP mixta	1	1	3	5
Triparesis	5	0	0	5
Paraparesis	5	0	3	8
		Unilateral spastic CP		
Hemiplegic CP	2	1	0	3
Total	19 (63,3%)	2 (6,7%)	9 (30%)	30 (100%)

Table 10. Use of toilet

Of total sample, 19 (63,3%) patients use toilet seats, 2 (6,7%) use toilet pot and nine (30%) need diapers.

Cerebral palsy	Window accessibility		No window	No bathroom	No answer	Total
	YES	NO				
			Bilateral spastic CP			
Quadriplegic CP	1	5	2	0	1	9
Quadripl. CP mixta	0	5	0	0	0	5
Triparesis	1	2	2	0	0	5
Paraparesis	3	3	1	1	0	8
			Unilateral spastic CP			
Hemiplegic CP	2	1	0	0	0	3
Total	7 (23,3%)	16 (53,3%)	5 (16,7%)	1 (3,33%)	1 (3,33%)	

Table 11. Window accessibility in the bathroom

Of 30 clients window in the bathroom is accessible for 7 (23,3%), and not accessible for 16 clients. One family has no bathroom and five families have no windows in the bathroom.

Grab bars (securely fastened to a reinforced wall) will assist in both toilet and tab transfers. Grab bars should be 1,5 inches (3,8 cm) in diameter and be knurled. For use in toilet transfers, the bars should be mounted horizontally 33 inches (83,8 cm) to 36 inches (91,4cm) from the floor. The length of the grab bars should be between 24 inches (61 cm) and 36 inches (91,4 cm) on the back wall and 42 inches (106,7 cm) on the side wall. For use in tab transfers they should be mounted horizontally 24 inches (61 cm) high measured from the floor of the tub (Schmitz, 1988).

Cerebral palsy	Toilet seats grab bars		Bathing tab grab bars		Total
	YES	NO	YES	NO	
		Bilateral spastic CP			
Quadriplegic CP	1	8	1	8	9
Quadripl. CP mixta	0	5	2	3	5
Triparesis	1	4	1	4	5
Paraparesis	1	7	1	7	8
		Unilateral CP			
Hemiplegic CP	0	3	1	2	3
Total	3 (10%)	27 (90%)	6 (20%)	24 (80%)	30 (100%)

Table 12. Toilet seats and bathing tab grab bars

Of whole sample of 30 persons, 3 (10 %) patients have toilet seats equipped with grab bars, and 27 (90%) patients do not have. Six (20%) patients have bathing tab equipped with grab bars, and 24 (80%) patients don't have.

Cerebral palsy	Toilet seat height (cm)						No answer	Total
	39	40	41	42	43	45		
			Bilateral spastic cerebral palsy					
Quadriplegic CP	1	4	2	1	1	0	0	9
Quadripl. CP mixta	2	3	0	0	0	0	0	5
Triparesis	1	2	0	2	0	0	0	5
Paraparesis	3	2	2	1	0	0	0	8
			Unilateral cerebral palsy					
Hemiplegic CP	0	1	0	0	0	1	1	3
Total	7 (23,3%)	12 (40%)	4 (13,3%)	4 (13,3%)	1 (3,3%)	1 (3,3%)	1 (3,3%)	30 (100%)

Table 13. Range of toilet seat height

Range of toilet seat height was 39 cm to 45 cm. Toilet seat height for twelve clients (40%) was 40 cm, for 7 (23,3%) was 39 cm, for 4 clients (13,3 %) was 41 cm, for other 4 was 42 cm, for one 43cm and for other one was 45 cm.

2.2.3. *Kitchen*

The height of counter tops (work space) should be appropriate for the wheelchair user; the armrests should be able to fit under the working surface. The ideal height of counter surfaces should be no greater than 31 inches (79 cm) from the floor with a knee clearance of 27,5 inches (69,8 cm) to 30 inches (76,2 cm). Counter space should provide a depth of at least 24 inches (61 cm). All surfaces should be smooth to facilitate sliding of heavy items from one area to another. Slide out counter spaces are useful in providing an over-the-lap working surface. For ambulatory patients, stools (preferably with back and foot rests) may be placed strategically at the main work area/s (Schmitz, 1988).

Cerebral palsy	Fitting of wheelchair in the table		Door clearance	No answer	Without table	Total
	YES	NO				
			Bilateral spastic CP			
Spastic Quadriplegic CP	3	3	0	1	0	7
Spastic Quadripl CP mixta	4	0	0	0	1	5
Triparesis	0	2	1	1	0	4
Paraparesis	3	0	0	0	0	3
			Unilateral spastic CP			
Spastic Hemiplegic CP	0	1	0	0	0	1
Total	10	6	1	2	1	20

Table 14. Accessibility of kitchen table for patients in wheelchairs

Out of 20 persons in wheelchairs, for 10 (50%) of them kitchen table is accessible. For 6 patients (20%) with wheelchairs, kitchen table is inaccessible: wheelchairs do not fit in the table. One patient in wheelchair does not have a kitchen table, he has a dining room table (40 cm)which is inaccessible. Range of kitchen table height was from 45 up to 120 cm, 16 different heights. Four (13.33%) of patients don't have a kitchen table. Majority of patients, 5 (16.7%) have a kitchen table which is 75 cm height. Thirteen persons have a kitchen table which is from 70 to 77 cm in height.

http://dx.doi.org/10.5772/56983

CP	Opening refrigerator		Without refrigerator	No answer	total
	YES	NO			
		Bilateral spastic CP			
Quadriplegic CP	4	5	0	0	9
Quadripl. CP mixta	0	5	0	0	5
Triparesis	3	2	0	0	5
Paraparesis	5	3	0	0	8
		Unilateral spastic CP			
Hemiplegic CP	2	1	0	0	3
Total	14	16	0	0	30

Table 15. Independent opening of the refrigerator and taking food

Refrigerator was accessible for 14 (46.7%) clients, which can independently open the door and take food. Refrigerator was inaccessible for 16 (53.3%) of clients. All clients have refrigerator.

CP	Opening refrigerator		Without freezer	No answer	total
	YES	NO			
		Bilateral spastic CP			
Quadriplegic CP	3	5	0	1	9
Quadripl. CP mixta	0	4	0	1	5
Triparesis	2	2	1	0	5
Paraparesis	3	3	1	1	8
		Unilateral spastic CP			
Hemiplegic CP	2	1	0	0	3
Total	10	15	2	3	30

Table 16. Independent opening of the freezer and taking food

Freezer was accessible for 10 (33.3%) clients, which can independently open it and take food. Freezer was inaccessible for 15 (50%) of clients. Two clients do not own a freezer, and three answers are omitted.

The sink may be equipped with large blade-tape handles, and a spray-hose fixture often provides improved function. Shallow sink 5 to 6 inches (12,7 cm to 15,2 cm) in depth will improve knee clearance below. As in the bathroom, hot-water pipes under the kitchen sink should be insulated to prevent burns (Schmitz, 1988).

Cerebral palsy	Fitting of wheelchair under the sink		Door clearance	No answer	Without sink	total
	YES	NO				
		Bilateral spastic CP				
Quadriplegic CP	0	5	0	1	1	7
Quadripl. CP mixta	1	3	0	1	0	5
Triparesis	1	1	1	1	0	4
Paraparesis	1	2	0	0	0	3
		Unilateral spastic CP				
Hemiplegic CP	0	1	0	0	0	1
Total	3	12	1	3	1	20

Table 17. Sink accessibility for patients in wheelchairs

Of 20 persons in wheelchairs, kitchen sink is accessible for 3 (15%) patients, or wheelchairs fit under the sink. For 12 patients (60%) in wheelchairs, sink is inaccessible. One patient does not have a kitchen dink and for two there is no answer on this question.

Tap on the kitchen sink can open and close 15 (50%) of clients. Kitchen sink tap is inaccessible for 12 (40%) of patients. One patient does not have a kitchen dink and for one there is no answer. Kitchen sink bottom is accessible for 16 (53.3%) of patients. Kitchen sink bottom is inaccessible for 11 (36.7 %) of patients.

Cerebral palsy	Opening/closing shelves		Inaccessible kitchen	No answer	No shelves	Total
	YES	NO				
		Bilateral spastic CP				
Quadriplegic CP	3	6	0	0	0	9
Quadripl. CP mixta	2	3	0	0	0	5
Triparesis	2	2	1	0	0	5
Paraparesis	5	3	0	0	0	8
		Unilateral spastic CP				
Hemiplegic CP	2	1	0	0	0	3
Total	14	15	1	0	0	30

Table 18. Shelves and cabinets accessibility in the kitchen for all patients

http://dx.doi.org/10.5772/56983

Shelves and cabinets in the kitchen are accessible (opening and closing) for 14 patients (46.7%), and inaccessible for 15 (50%) of patients. For one patient kitchen is not accessible because is too narrow.

Equipment and food storage areas should be selected with optimum energy conservation in mind. All frequently used articles should be within easy reach, and unnecessary items should be eliminated. Additional storage space may be achieved by installation of open shelving or use of peg boards for pots and pans. If shelving is added, adjustable shelves are preferable and should be placed 16 inches (41 cm) above counter top (Schmitz, 1988).

Cerebral palsy	Transport possibility		Door clearance	No answer	Total
	YES	NO			
		Bilateral spastic CP			
Quadriplegic CP	3	6	0	0	9
Quadripl. CP mixta	2	3	0	0	5
Triparesis	1	3	1	0	5
Paraparesis	4	3	0	1	8
		Unilateral spastic CP			
Hemiplegic CP	2	1	0	0	3
Total	12	16	1	1	30

Table 19. Possibility to transport necessities (food, dishes...) around kitchen

12 (40%) of patients can carry necessities from one end of the kitchen to another, and 16 (53.3%) cannot.

For 14 patients (46.7%) inaccessible are stove switches, which means they cannot use them, and for 10 (33.3%) are not. Four patients (13.3%) do not use the kitchen, and two patients (6.7%) did not give an answer.

Eleven patients (36.7%) can operate stove doors, 12 (40%) cannot, 4 (13.3%) does not use the kitchen, and 3 (10%) of patients did not answered

3. Conclusion

Based on the results of the evaluation with *Environmental Assessment – Home assessment form*, the study *Occupation therapy for persons with cerebral palsy*, in the Canton of Sarajevo, made proposals for changes in the environment to improve the accessibility of housing.

Educating persons with cerebral palsy and members of their families, with specific exercises to improve and preserve posture, balance, and coordination, increase the volume of mobility and prevent deformities deterioration, had an impact on the personal competencies, i.e. skills related to motor performance, sensor capabilities, cognitive ability and general health.

Private homes need to be converted according to the individual needs of tenants. As for the individual adaptation, arrangement of private space so that it is accessible, it requires precise planning according to the needs of people. *Multidisciplinary team* should lead that planning, and find such design solutions that overcome the problem of architectural barriers for people with disabilities to improve their quality of life.

Ideally, the physical and occupational therapists should accompany the patient on the home visit. They assume shared responsibility for assessing the patient's functional level at home. Depending of the specific needs of the patient and/or family, a speech therapist, social worker, or nurse also may be included on the home visit.

It is necessary to open the *Services or Counseling centers* for accessible housing. As part of these services, occupational therapist and an architect should visit homes of persons with disabilities and assess what needs to be adapted to meet the needs of that person.

The significance of this research for the community is multiple: educational, scientific, humane and promotional.

The results provide a basis for further research in needs of these families and improvement of their quality of life.

4. Summary

Accessible design generally refers to houses or other dwellings that meet specific requirements for accessibility. The laws dictate standards dimensions and characteristics for such features as door widths, clear space for wheelchair mobility, audible and visual signals, grab bars switch and outlet height, and more.

The research was conducted through Project: „Occupational therapy for persons with cerebral palsy", in homes of participants. *The aim* was to determine accessible housing for persons with cerebral palsy.

Sample was consisted of 30 respondents, members of the Association of persons with cerebral palsy of the Canton of Sarajevo, age from 4 up to 53: 8 children (4-11 years), 14 adolescents (12-20 years), and 8 adults (21-53 years); 14 male (46,67 %) and 16 (53,33 %) female.

The principal measure used was the International Environmental Assessment – Home assessment form. The first part should deal with accessibility of the dwelling's exterior, and the second half should be concerned with an assessment of the home's interior. During the *On-Site visit* a tape measure and home assessment form are tools (Schmitz, 1988), translated and modified by the author (Švraka, 2007).

The Association of persons with cerebral palsy in the Canton of Sarajevo is member of *Cerebral Palsy Association of Federation of Bosnia and Herzegovina* which was established at 17. Oc-

http://dx.doi.org/10.5772/56983

tober of 2011. That day was announced as *Day of persons with cerebral palsy of Federation of Bosnia and Herzegovina* (FB&H). Cerebral Palsy Association members include five Associations of persons with cerebral palsy of FB&H, from five towns/Cantons: Sarajevo, Goražde, Zenica, Široki Brijeg and Sapna.

Assistive devices include ortho-prosthetic devices, wheel chairs, walking aids, technical aids and adapted controls for cars. Adequate assistive devices are often financially inaccessible to many users because of their high cost despite the fact that they should be covered by social and insurance schemes. Under the current system, most assistive devices are covered only partially by the state and require user co-payments, which can be exorbitant in cost. Within the socialist system, assistive devices were generally provided for free within the public health care system. This is a crucial issue in South East Europe as one of the largest barriers to accessing assistive devices is financial.

Of 30 persons with cerebral palsy 20 (66.7%) use wheelchairs, 7 (23.3%) have independent mobility, and 3 (10%) persons require the use of particular device. Client with triparetic CP use a walking tripod.

The range of width of the entrance door was from 62cm to 112 cm. Thirteen families (43,29%) had entrance door width between 76 cm to 82cm.

Eleven families (36,63%) of persons with cerebral palsy, who live in flats have elevators. In private houses there are no elevators.

Twenty three families (76,7%) have raised entrance door thresholds made of different material: wood (16), concrete (3), metal (3), and one made of marble; 1 cm to 7 cm in height.

Of 30 persons with cerebral palsy, for 10 (33,3%) persons night table is within patient's reach from bed, for 11 (36,7%) persons night table is not within patient's reach from bed. 7 (23,3%) persons don't have night table, and for two persons there are no answers.

Of total sample, 19 (63,3%) patients use toilet seats, 2 (6,7%) use toilet pot and nine (30%) need diapers.

Of 30 persons window in the bathroom is accessible for 7 (23,3%), and not accessible for 16 persons. One family has no bathroom and five families have no windows in the bathroom.

Of 30 persons, 3 (10 %) patients have toilet seats equipped with grab bars, and 27 (90%) patients do not have. Six (20%) patients have bathing tab equipped with grab bars, and 24 (80%) patients don't have.

Range of toilet seat height was 39 cm to 45 cm. Toilet seat height for 12 (40%) persons was 40 cm, for 7 (23,3%) was 39 cm, for 4 (13,3%) persons was 41 cm, for other 4 was 42 cm, for one 43 cm and for other one was 45 cm.

Three persons can enter the bathroom with wheelchairs. From whole sample, 3 (10 %) persons have toilet seats equipped with grab bars, and 27 (90 %) persons do not have.

Out of 20 persons in wheelchairs, for 10 (50%) of them kitchen table is accessible. For 6 patients (20%) with wheelchairs, kitchen table is inaccessible: wheelchairs do not fit in the table. One patient in wheelchair does not have a kitchen table, he has a dining room table (40 cm)which is inaccessible. Range of kitchen table height was from 45 up to 120 cm, 16 different heights. Four (13.33%) of patients don't have a kitchen table. Majority of patients, 5 (16.7%) have a kitchen table which is 75 cm height. Thirteen persons have a kitchen table which is from 70 to 77 cm in height.

Refrigerator was accessible for 14 (46.7%) clients, which can independently open the door and take food. Refrigerator was inaccessible for 16 (53.3%) of clients. All clients have refrigerator.

Freezer was accessible for 10 (33.3%) clients, which can independently open it and take food. Freezer was inaccessible for 15 (50%) of clients. Two clients do not own a freezer, and three answers are omitted.

Of 20 persons in wheelchairs, kitchen sink is accessible for 3 (15%) patients, or wheelchairs fit under the sink. For 12 patients (60%) in wheelchairs, sink is inaccessible. One patient does not have a kitchen dink and for two there is no answer on this question.

Tap on the kitchen sink can open and close 15 (50%) of clients. Kitchen sink tap is inaccessible for 12 (40%) of patients. One patient does not have a kitchen dink and for one there is no answer. Kitchen sink bottom is accessible for 16 (53.3%) of patients. Kitchen sink bottom is inaccessible for 11 (36.7 %) of patients.

Shelves and cabinets in the kitchen are accessible (opening and closing) for 14 patients (46.7%), and inaccessible for 15 (50%) of patients. For one patient kitchen is not accessible because is too narrow.

12 (40%) of patients can carry necessities from one end of the kitchen to another, and 16 (53.3%) cannot.

For 14 patients (46.7%) inaccessible are stove switches, which means they cannot use them, and for 10 (33.3%) are not. Four patients (13.3%) do not use the kitchen, and two patients (6.7%) did not give an answer.

Eleven patients (36.7%) can operate stove doors, 12 (40%) cannot, 4 (13.3%) does not use the kitchen, and 3 (10%) of patients did not answered

Ideally, the physical and occupational therapists should accompany the patient on the home visit. They assume shared responsibility for assessing the patient's functional level at home. Depending of the specific needs of the patient and/or family, a speech therapist, social worker, or nurse also may be included on the home visit.

It is necessary to open the *Services or Counseling centers* for accessible housing. As part of these services, occupational therapist and an architect should visit homes of persons with disabilities and assess what needs to be adapted to meet the needs of that person.

The results provide a basis for further research in needs of these families and improvement of their quality of life.

http://dx.doi.org/10.5772/56983

Acknowledgements

I thank the children and families, members of the Association of persons with cerebral palsy of the Canton of Sarajevo, Bosnia and Herzegovina, who participated in this Project.

Author details

Emira Švraka

University of Sarajevo, Faculty of Health Studies in Sarajevo, Physiotherapy Department, Bosnia and Herzegovina

References

[1] Bottcher, L. (2010). Children with spastic cerebral palsy, their cognitive functioning, and social participation: a review. *Child Neuropsychology*, 16: 209-228.

[2] Carlsson, M.; Hagberg, G. and Olsson, I. (2003). Clinical aetiological aspects of epilepsy in children with cerebral palsy. *Developmental Medicine & Child Neurology* 2003, 45: 371-376

[3] "Consumer's Guide to Accessible Housing" available at: http://www.abledata.com/abledata_docs/icg-hous.htm In: *Disability Monitor Initiative South East Europe*. (2007) Free movement of people with disabilities in South East Europe.

[4] Cowan, D. and Wintergold, A. (2007). Assistive technology. In: *Physiotherapy for Children*. pp 139-160 Butterworth Heinemann Elsevier ISBN-13; 978 0 750 68886 4

[5] Disability Monitor Initiative South East Europe (2007). Free movement of people with disabilities in South East Europe. An Inaccessible Right?

[6] Handicap International "Beyond De-Institutionalisation: The Unsteady Transition to an Enabling System in South East Europe", Disability Monitor Initiative (Belgrade: Handicap International: 2004): 58 In: *Disability Monitor Initiative South East Europe*. (2007) Free movement of people with disabilities in South East Europe.

[7] Howitt, R. (2003). Member of European Parliament, President of the Disability Intergroup of the European Parliament. In: *Disability Monitor Initiative South East Europe*. (2007) Free movement of people with disabilities in South East Europe.

[8] Jacobsson, B. and Hagberg, G. (2004). Antenatal riscs factors for cerebral palsy. *Best Pract Clinic Obstetric Gynaecol*, 18 (3), 425-436

[9] Loga, S. (2011) Transition of Bosnian-Herzegovinian society and its impact on health protection. University of Sarajevo. International symposium. *Proceedings: Bosnia and Herzegovina – 15 years of Dayton peace agreement.* pp 225-241

[10] Nordmark, E.; Hagglund, G. & Lagergren, J. (2001). Cerebral Palsy in south Sweden. Prevalence and clinical features. *Acta Pediatrica* 90: 1271-1276

[11] Odding, E.; Roebroeck, M. E. & Stam, H. J. (2006). The epidemiology of cerebral palsy: Incidence, impairments and risk factors. *Disability and Rehabilitation,* 28(4): 183-191.

[12] Parkes, J.; White-Koning, M.; O Dickinson, H.; Thyen, U.; Arnaud, C.; Beckung, E. & all. (2008). Psychological problems in children with cerebral palsy: a cross-sectional European study. *The Journal of Child Psychology and Psychiatry* 49: 4, p. 405-413.

[13] Reid, D.; Laliberte-Rudman, D. & Hebert, D. (2002) Impact of wheeled seated mobility devices on adult users' and their caregivers' occupational performance: A critical literature review. *Canadian Journal of Occupational Therapy.* Volume 69, Number 5, pp 261-281 ISSN-0008-4174

[14] Rogers, S. L.; Gordon, C. Y.; Schanzenbacher, K. E.; Case-Smith, J. (2001) *In: Case-Smith J. Occupational Therapy for Children, fourth edition.* Mosby. An Affiliate of Elsevier Science. St Louis, London, Philadelphia, Sydney, Toronto. ISBN 0-323-00764-3

[15] Schmitz, T. J. (1988) Chapter 13: Environmental assessment. In: *Physical rehabilitation: Assessment and treatment.* Second edition. pp. 237- 251 ISBN 0-8036-6698-5

[16] Sestranetz, R. & Adams, L. (2006) In: *Disability Monitor Initiative South East Europe.* (2007) Free movement of people with disabilities in South East Europe.

[17] Shepherd, J. (2001) Self-Care and Adaptations for Independent Living. In: Case-Smith J. *Occupational Therapy for Children, fourth edition.* Mosby. An Affiliate of Elsevier Science. St Louis, London, Philadelphia, Sydney, Toronto. ISBN 0-323-00764-3

[18] Stiers, P. & Vanderkelen, R. (2002). Visual-perceptual impairment in a random sample of children with cerebral palsy. *Developmental Medicine & Child Neurology,* 44: 370-382.

[19] Švraka, E. (2007) *Another side of life – Learning difficulties of children with cerebral palsy.* Second enlarged edition. TDP d.o.o. Sarajevo ISBN 978-9958-9214-7-6

[20] Švraka, E.; Loga, S.; Brown I. (2011) Family quality of life: adult school children with intellectual disabilities. *Journal of Intellectual Disability Research* 55, pp 1115-1122.

[21] Švraka, E. (2012) Chapter: Children with cerebral palsy and epilepsy. *In: Epilepsy-Histological, Electroencephalographic and Psychological Aspects.* Edited by Dejan Stevanovic p. 251-276 INTECH Open Access Publisher of Scientific Books and Journals ISBN 978-953-51-0082-9 Printed in Croatia www.intechopen.com

Section 2

Improving the Quality of Life

Neuroprotection in Perinatal Hypoxic-Ischemic Encephalopathy — Pharmacologic Combination Therapy

Mª Carmen Carrascosa-Romero and
Carlos de Cabo-de la Vega

Additional information is available at the end of the chapter

http://dx.doi.org/10.5772/57459

1. Introduction

Hypoxic ischemic encephalopathy (HIE) currently constitutes one of the non-excluding causes of child cerebral palsy (CP) and, together with prematurity, is potentially preventable. For this reason there is an increasing interest in prevention policies as well as in research on neuroprotection therapies that minimize cerebral lesion and concomitant disabilities.

In the last few decades there has been an explosion of studies employing either animal models of global or focal hypoxia or cell cultures investigating the preventing effect of many chemicals on neuronal lesion. Recent clinical research has shown that certain pharmaceuticals have neuroprotective effects, suggesting that their use could be generalized for clinical practice in a near future. However, the use of some of these chemicals, such as nicardipine (calcium blocker) or magnesium (blocking NMDA-receptors), has been investigated in clinical trials showing no beneficial effects while causing severe hemodynamic adverse events. Therefore there is no generally accepted standard of care in the brain-oriented pharmacologic therapy for full-term neonates sustaining cerebral hypoxia–ischemia (H-I). In fact, neuroprotective treatment for HIE in the clinical practice is limited to the application of hypothermia in the newborn which is accepted now as a meaningful therapy, since no pharmaceutical has shown any benefit when administered by itself yet.

Future advances in the understanding of preconditioning may lead to the administration of neuroprotective agents earlier before childbirth. Although most of these neuroprotective strategies have not yet entered clinical practice, there is a significant hope that further developments will allow to incorporate them besides hypothermic neuroprotection. More specifically, maternal administration of allopurinol (xanthine oxidase inhibitor/ anti-oxidant) has been proposed as prebirth treatment when there is suspicion of an adverse event eliciting perinatal asphyxia.

Since it is conceivable that hypothermia postpones secondary energy failure, application of hypothermia immediately after the hypoxic event could prolong the window for pharmacotherapeutic intervention; furthermore, there is accumulating preclinical evidence that adjunctive therapies can enhance hypothermic neuroprotection. The question that still remains is whether a combination of therapeutic agents would be more efficient in reducing brain damage due to hypoxia-ischemia than applying just one pharmaceutical. The hypothesis is that combinations of therapies intervening at different levels in the cascade might lead to more prominent reduction of brain injury.

In this chapter we review the mechanisms of action of chemicals that have shown potential neuroprotection effect, with special regard to those already approved for use in the newborn and show no side effects. Finally, we propose a model of off-label combined neuroprotective therapy using a staggered design according to the severity of the asphyxia /encephalopathy.

2. Neonatal encephalopathy

The incidence of birth asphyxia is 9.4/1000 live term births whereas the incidence of neonatal encephalopathy secondary to intrapartum hypoxia-ischemia (H-I) is very low, estimated between 0.27 per 1000 (Palsdottir et al, 2007) and 1.5 per 1000 live full-term births; and about 15% to 20% of affected newborns die in the postnatal period, and an additional 25% of the survivors exhibiting permanent neuropsychological deficits (Kurinczuk et al, 2010). Birth asphyxia often appears to be a secondary symptom of an otherwise sick baby; thus, it is not the primary cause of CP in the majority of cases. One study of children with CP found that in only about 8% (15/183) of all the children with spastic CP was intrapartum asphyxia the possible cause of their brain damage; and the contribution of intrapartum events and obstetric mismanagement to overall CP rates is probably less than was previously throught (Blair & Stanley, 1998). In fact, due to this type of research, the term birth asphyxia has been replaced with the term neonatal encephalopathy because this later term does not imply a causal relationship (Fehlings et al, 2007).

Perinatal asphyxia can be defined as the injury caused to the fetus or newborn as a result of both reduced oxygen (O2) supply to the brain and the sustained reduction in blood flow due to an inadequate cerebral perfusion. The term "asphyxia" is not synonymous with **HIE** despite been closely related: "asphyxia" is the *cause* whereas HIE is the *effect*. Besides, asphyxia does not always produce brain damage. **HIE** is defined as the neurological syndrome occurring in the newborn after a hypoxia / ischemia episode that affects consciousness in different degrees, with decrease of spontaneous movements, tone and reflexes, as well as the appearance of convulsions in the most severe cases. Hypoxic-ischemic events may cause multisystemic failure with impairment of pulmonary, cardiovascular, digestive, renal, hematological and metabolic functions, constituting the post-asphyctic syndrome. However, it is important to

http://dx.doi.org/10.5772/57459

emphasize that the clinical features of hypoxic ischemic encephalopathy are nonspecific, and a diagnosis of perinatal asphyxia should be made with caution and only after careful consideration of all data collected in the clinical history.

The essential criteria required to define an acute intrapartum hypoxic event as sufficient to cause CP were established by by both the American College of Obstetricians and Gynecologists and the American Academy of Pediatrics, and the International Cerebral Palsy Task (ACOG, 2003), are listed as follows:

Essential criteria (must meet all four)

- Evidence of a metabolic acidosis in fetal umbilical cord arterial blood obtained at delivery (pH < 7 and base deficit =12 mmol/L).
- Early onset of severe or moderate neonatal encephalopathy in infants born at 34 or more weeks of gestation.
- Cerebral palsy of the spastic quadriplegic or dyskinetic type.
- Exclusion of other identifiable etiologies such as trauma, coagulation disorders, infectious conditions, or genetic disorders.

Criteria that collectively suggest an intrapartum timing (within close proximity to labor and delivery, eg, 0-48 hours) but are nonspecific to asphyxial insults

- A sentinel (signal) hypoxic event occurring immediately before or during labor
- A sudden and sustained fetal bradycardia or the absence of fetal heart rate variability in the presence of persistent, late, or variable decelerations, usually after a hypoxic sentinel event when the pattern was previously normal
- Apgar scores of 0-3 beyond 5 minutes
- Onset of multisystem involvement within 72 hours of birth
- Early imaging study showing evidence of acute nonfocal cerebral abnormality

Neonatal encephalopathy is a clinically defined syndrome of disturbed neurological function in the infant at or near term during the first week after birth, manifested by difficulty with initiating and maintaining respiration, depression of tone and reflexes, altered level of consciousness, and often seizures. Sarnat and Sarnat (Sarnat & Sarnat, 1976) were the first to define this syndrome as neonatal encephalopathy following foetal distress (TABLE 1). The clinical features and severity of encephalopathy have been well defined. They distinguished three stages of encephalopathy: stage 1, or mild encephalopathy associated with hyperalertness, sympathetic overdrive, and a normal EEG; stage 2, or moderate encephalopathy marked by obtundation, hypotonia, multifocal seizures, and an EEG showing periodic or continuous delta activity; and stage 3, or severe encephalopathy in which infants were stuporous and flaccid with an isoelectric or periodic EEG.Infants who did not enter stage 3 and who had signs of stage 2 for fewer than 5 days were normal on follow-up, but persistence of stage 2 for a week or failure of the EEG to normalise predicted later neurological impairment or death.

	State 1 Mild	Stage 2 Moderate	Stage 3Severe
Level of Consciousness	Hyperalert	Lethargic or obtunded	Stuporous
Neuromuscular Control			
Muscle tone	Normal	Mild hypotonia	Flaccid
Posture	Mild distal flexion	Strong distal flexion	Intermittent decerebration
Stretch reflexes	Overactive	Overactive	Decreased or absent
Segmental myoclonus	Present	Present	Absent
Complex Reflexes			
Suck	Weak	Weak or absent	Absent
Moro	Strong; low threshold	Weak; incomplete; high threshold	Absent
Oculovestibular	Normal	Overactive	Weak or absent
Tonic neck	Slight	Strong	Absent
Autonomic Function	Generalized sympathetic	Generalized parasympathetic	Both systems depressed
Pupils	Mydriasis	Miosis	Variable; often unequal; poor light reflex
Heart Rate	Tachycardia	Bradycardia	Variable
Bronchial and Salivary Secretions	Sparse	Profuse	Variable
Gastrointestinal Motility	Normal or decreased	Increased; diarrhea	Variable
Seizures	None	Common; focal or multifocal	Uncommon (excluding decerebration)
Electroencephalogram Findings	Normal (awake)	Early: low-voltage continuous delta and thetaLater: periodic pattern (awake) Seizures: focal 1-to 1-Hz spike-and-wave	Early: periodic pattern with Isopotential phasesLater: totally isopotential
Duration	<24 h	2-14 h	Hours to weeks

Table 1. Sarnat and Sarnat's [236] Clinical Stages of Perinatal Hypoxic Ischemic Brain Injury

The diagnosis HIE has often not been proven and has been assumed from a variety of clinical markers that do not accurately reflect hypoxia and ischaemia of either acute or chronic origin. Over 75% of cases of neonatal encephalopathy have no clinical signs of intrapartum hypoxia. Dammann et al (Dammann et al, 2011), under the title "Neonatal Encephalopathy or Hypoxic-

Ischemic Encephalopathy? Appropriate Terminology Matters" propose that by simply calling "neonatal encephalopathy" what is now called "HIE," we might not only help reduce the number of unjustified convictions of obstetricians, midwifes, nurses, and hospitals but also increase the amount of much needed research in perinatal brain injury not necessarily related to the hypoxia-ischemia paradigm.

Although in many newborns who have cerebral dysfunction, the brain injury may have occurred before the onset of labor and delivery, there exists a relatively small but clearly identifiable group of term newborns who sustain significant intrapartum H-I cerebral insult and who subsequently develop an acute encephalopathy during the first week of life. The Sarnat evaluation of staging of encephalopathy by clinical examination correlates well with subsequent neurodevelopmental impairment in infancy and childhood (Robertson, 2003; Badawi et al, 2005; Ambalavanan et al, 2006). The presence of abnormal neurologic examination results in the first few days of life highly predicts a brain insult in the perinatal period. Neonates with mild encephalopathy usually do not have an increased risk of motor or cognitive deficits. Neonates with severe encephalopathy have a high risk of death (up to 85%) and an increased risk of CP and intellectual disability among survivors. Neonates with moderate encephalopathy have significant motor deficits, fine motor disability, memory impairment, visual or visuomotor dysfunction, increased hyperactivity and delayed school readiness (Shankaran et al, 1991. Robertson, 2003; Marlow et al, 2005; Gonzalez& Miller, 2006. De Vries & Jongmans, 2010).

3. Neuropathologic aspects of hypoxic–ischemic brain damage

3.1. Gestational/perinatal age

H-I insults during critical cellular and tissue differentiation processes have a serious impact on brain maturation. Thus, the gestational/perinatal age of the infant is one of the main variables in determining the neuropathological picture of H-I brain injury. For this reason alone, the developmental stage at which a H-I insult occurs is of great importance. However, the mechanism(s) underlying the neuronal damage and death triggered by H-I in the developing brain leading to various forms of neurological disabilities remains inadequately understood. Due to selective ischemic vulnerability, the damage affects the gray matter in term newborns and white matter (WM) in preterm newborns with the typical neuropathological aspects of laminar cortical necrosis in the former and periventricular leukomalacia (PVL) in the latter (Volpe, 2003).

Reactive oxygen species (ROS) play a pivotal role in the development of PVL. The major type of injury involves cerebral white matter and the principal cellular target is the developing oligodendrocyte. The specific phase of the oligodendroglial lineage affected has been defined from the study of both human brain and experimental models. This premyelinating cell (pre-OL) is vulnerable because of a series of maturation-dependent events. The pathogenesis of pre-OL injury involves two upstream mechanisms, H-I and systemic infection/inflammation, both of which are common occurrences in premature infants. This differential susceptibility to injury can also be related to the development of interneuronal connections and excitatory glutamate receptors that create the possibility of excessive neurotransmitter release and receptor overstimulation. Moreover, most forms of H-I in neonates cause injury to selected areas of the brain rather than the entire brain (Johnston, 2001; Johnston et al, 2001). Most importantly, elucidation of these factors has led to delineation of a series of potential therapeutic interventions, which in experimental models show marked protective properties. The critical next step, i.e., clinical trials in the living infant, is now on the horizon (Volpe et al, 2011).

3.2. Developing human cortex

The subplate zone (SPZ) is a transient cytoarchitectonic compartment of the fetal telencephalic wall, situated between the fetal WM (i.e. intermediate zone) and the cortical plate, and is the crucial laminar compartment for the development of the human cerebral cortex. The subplate contains numerous neurons of various morphological types and molecular phenotypes, including differentiated projection (glutamatergic) neurons and local (GABA and peptidergic) interneurons. The developing human cortex goes through three major early stages of functional development: (1) between 13 and 15 postconceptional weeks (PCW): initial-transient fetal circuitry, centered at the SPZ, which is endogeneously (spontaneously) driven; (2) 15 and 30 PCW: perinatal dual circuitry (co-existence of endogeneously driven subplate-centered transient circuitry with developing cortical plate-centered permanent circuitry), that slowly disappears towards the end of gestation and during the early postnatal period; and (3) postnatally established permanent (externally driven) cortical circuitry, centered at the cortical plate (that is, developing cortical layers I-VI). While the SPZ disappears during the perinatal and early postnatal period, numerous subplate neurons survive and remain embedded in the superficial (gyral) WM of adolescent and adult brain as so-called *interstitial neurons* (Judaš et al, 2010). The growth of the axonal pathways preterm explains their vulnerability and plasticity. In neonates the vulnerability is related to the intracortical circuitry. The neuronal elements in transient fetal zones form a developmental potential for plasticity after perinatal cerebral lesions (Kostovic & Judas, 2006).

3.3. Gender differences

This new information about gender differences in neuronal death pathways in experimental models is probably directly relevant to gender differences reported in the response of infants and children to brain injuries. Quantitative imaging showed that male premature infants are more vulnerable than girls to white matter injury from intraventricular hemorrhage (IVH), but girls are more vulnerable to gray matter injury. It follows directly from this information that an infant's gender could influence the efficacy of neuroprotective agents and the cell types most at risk. A striking example of this effect was reported from the prospective "Indomethacin Intraventricular Hemorrhage Prevention Trial" (Ment et al, 2004). In this study, indomethacin eliminated parenchymal hemorrhage and improved verbal scores in boys at ages 3 to 8 years, but had no effect on girls. "It is becoming increasingly clear that gender differences are not simply a result of hormonal influence but are profound properties of individual cells" (Fatemi et al, 2009).

http://dx.doi.org/10.5772/57459

4. Physiopathological and biochemical processes of H-I cerebral injury

The principal biochemical mechanisms of cellular death in hypoxemia, ischemia and asphyxia are very similar and derive from O2 deprivation. Besides the main role of perinatal asphyxia, a key factor in the genesis of HIE is the loss of "cerebral blood flow (CBF) autoregulation", a protective mechanism that maintains stable CBF velocity (CBFV) in normal infants regardless of variations of systemic arterial pressure. At the cellular level, the reduction in CBF and oxygen delivery initiates a cascade of deleterious biochemical events. Clinical and experimental observations demonstrate that HIE is not a single "event" but is rather an "evolving process", and reflect the evolution of a delayed cascade of molecular events triggered by the initial insult (Fatemi et al, 2009).

4.1. Physiopathological mechanisms of cerebral injury

H-I injuries develop in two phases: the ischemic phase, dominated by necrotic processes, and the reperfusion phase, dominated by apoptotic processes extending beyond ischemic areas. This second phase takes place two - six hours after H-I insult, such latency constituting a useful window in which therapeutic measures can be able to stop the evolution of cerebral damage (Hammerman & Kaplan, 2000; Inder &Volpe, 2000).

4.1.1. Acute injury: Primary energy failure (associated with anaerobic metabolism)

There is an initial immediate phase (FIGURE 1) characterized by alterations in glucose metabolism; the anaerobic metabolism is an energy-inefficient state, since anaerobic glycolysis produces lactate and only 2 ATP, whereas aerobic glycolysis produces 38 ATP. HI rapidly lowers the neonatal brain's stores of glucose and high energy phosphates (ATP and phosphocreatine), resulting in energy depletion and accumulation of lactate and inorganic phosphate; and a metabolic acidosis develops due to accumulation of lactic acid, with local and systemic consequences such as impaired vascular tone and cardiac contractility. In this phase, the crucial event triggering a cascade of chain reactions is represented by ATP depletion secondary to anaerobic glycolysis and metabolic acidosis induced by hypoxia. Reduced ATP availability determines the dysfunction of ATPase systems, in particular Na+, K+-ATPase and glial-ATPase. Na+, K+-ATPase dysfunction leads to membrane depolarization in neurons mainly causing intracellular calcium accumulation, and sodium and water accumulation with cytotoxic edema and/or cell lysis followed by inflammatory reaction with cytokines release. Disturbances of intracellular calcium homeostasis result in activation of calpains (*calcium-activated proteases*). At the same time, neuronal depolarization induces glutamate release which tends to accumulate in the intersynaptic and intercellular spaces because of the dysfunction of glial-ATPase (an astrocytic enzyme normally in charge of glutamate reuptake). Thus extracellular glutamate increses due to both enhanced release and lower recapture. Glutamate, a neuroexcitatory aminoacid, stimulates specific NMDA and AMPA neuro-glial receptors and hence determines a massive intracellular entry of calcium that activates some endocellular enzymes including proteases, endonucleases and phospholipases. Proteases degrade neurofilaments causing cytoskeleton rupture and disintegration of the cellular body. Furthermore, excessive amount of glutamate can cause excitotoxicity leading to celldeath of neurons and glial cells. These events involve major mechanisms of fast neuronal death (edema, inflammation, necrosis) due to either direct or indirect neurotoxicity [mediated by free-radicals (FR) and nitric oxide (NO) generated during the first minutes / hours of anoxic insult]. In asphyxiated human neonates, the extent of depletion of high-energy phosphates, and the extent of accumulation of lactate, measured by magnetic resonance spectroscopy, correlate with the severity of eventual neurologic impairment (Hanrahan et al, 1999).

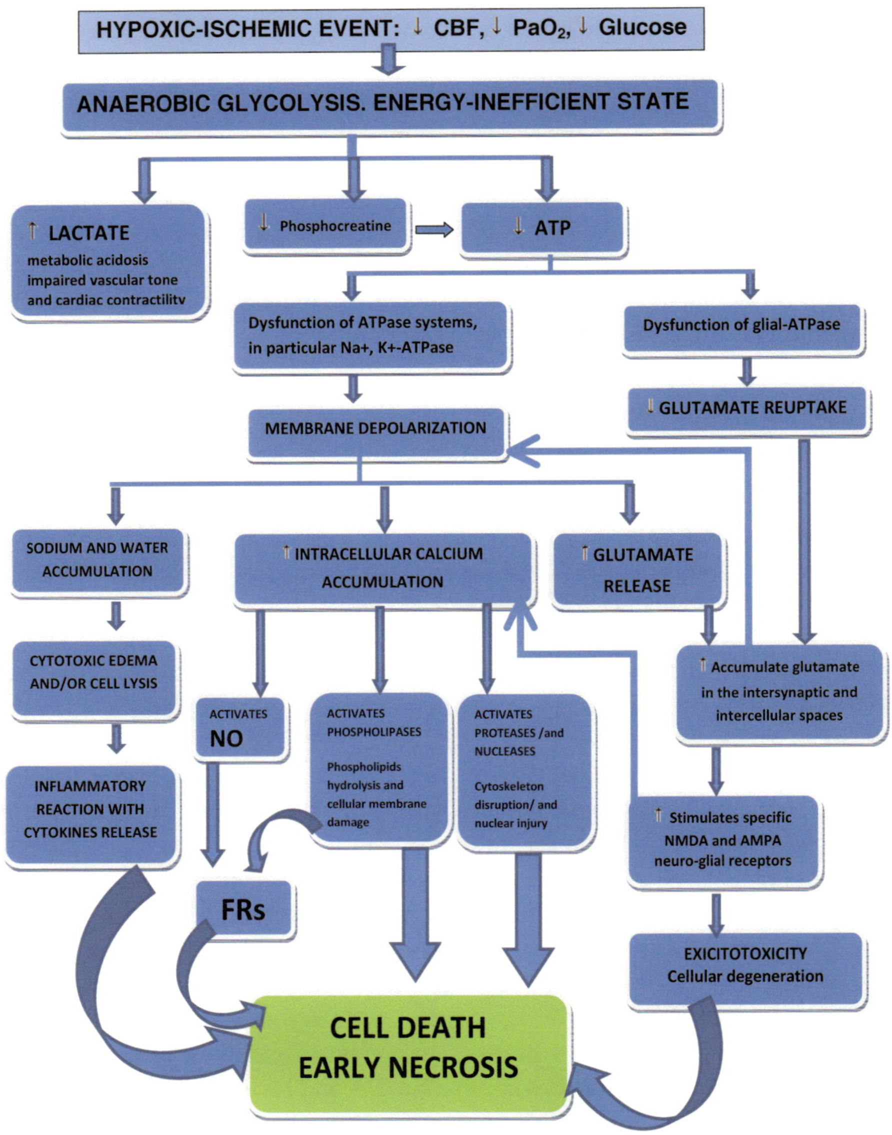

Figure 1. ACUTE INJURY: Primary energy failure (associated with anaerobic metabolism).

http://dx.doi.org/10.5772/57459

4.1.2. Delayed brain damage: Secondary energy failure (associated with reperfusion)

Secondly, a complex cascade of pathogenic mechanisms associated with reperfusion (or recovery of the blood flow) is triggered and this response is proportional to the severity of the first response. The secondary cerebral energy failure, also known as "delayed injury" (FIGURE 2), occurs from 6 to 48 hours after the primary event and may continue for days or weeks. Extended reactions from primary insults (eg, calcium influx, excitatory neurotoxicity, oxygen free radicals, or NO formation) secondary involve mitochondrial dysfunction. This mitochondrial impairment may lead directly to caspase-dependent and -independent apoptosis. This second phase also has an undoubtable prognostic value since it is associated with delayed neuronal death (apoptosis) during hours or days after the initiation of injury, and represents a window for therapeutic intervention. The exact mechanisms of secondary energy failure remain unclear but appear to be related to oxidative stress, excitotoxicity, and inflammation. In this phase, recovery of ischemia increases O2 availability and hence activates xantine-oxidase (via metabolizing the hypoxanthine formed during the initial period of HIE to uric acid) and cyclooxygenase enzymes and generates reactive oxygen species (ROS) responsible for oxidative cellular damage. Due to the increased intracellular calcium from the previous period, the reperfusion phase shows a sustained activation of phospholipase A2 that hydrolyze phospholipids and can damage cellular membrane and induce the consequent release of free fatty acids (FFA), especially arachidonic acid (AA). Reperfusion also activates cyclooxygenase that catalyses the formation of prostaglandins and generate -among others- superoxide free radicals. The production of vasodilator prostaglandins that give rise to reperfusion of ischemia and build-up of platelet-activating factor (PLF) in brain tissues. Collectively, these processes lead to a surge of the superoxide free radical, which plays a central role in further production of free radicals and other toxic compounds (Lorek et al, 1994).

4.2. Biochemical mechanisms of injury

Key players in the pathophysiology of neonatal cerebral injury are accumulation of cytosolic calcium, oxidative stress, excitotoxicity, and inflammation leading to apoptotic and necrotic neuronal death (Grow & Barks, 2002; Hossain, 2005).

4.2.1. Oxidative stress: Formation of free radicals (FRs)

Definitions of Oxidative stress: The imbalance between free radical (FR) generation and free radical scavenging that leads to cell injury.

All biological systems that consume oxygen generate FRs, which are molecules with one or more unpaired electrons in their outer orbit. They readily accept electrons from iron and other metals to form more reactive radicals, which attack other biomolecules, especially lipids, proteins, and nucleic acids, generating more radicals that damage the developing brain. In aerobic cells, oxygen free radicals (reactive oxygen species –ROS-) are produced within the cytoplasm and mitochondria. The three most common ROS are superoxide ($O2^-$), hydroxyl radical (OH^-), and hydrogen peroxide (H2O2). Two important sources of ROS are the byproducts of xanthine (derived from the breakdown of ATP) and prostaglandin synthesis (derived

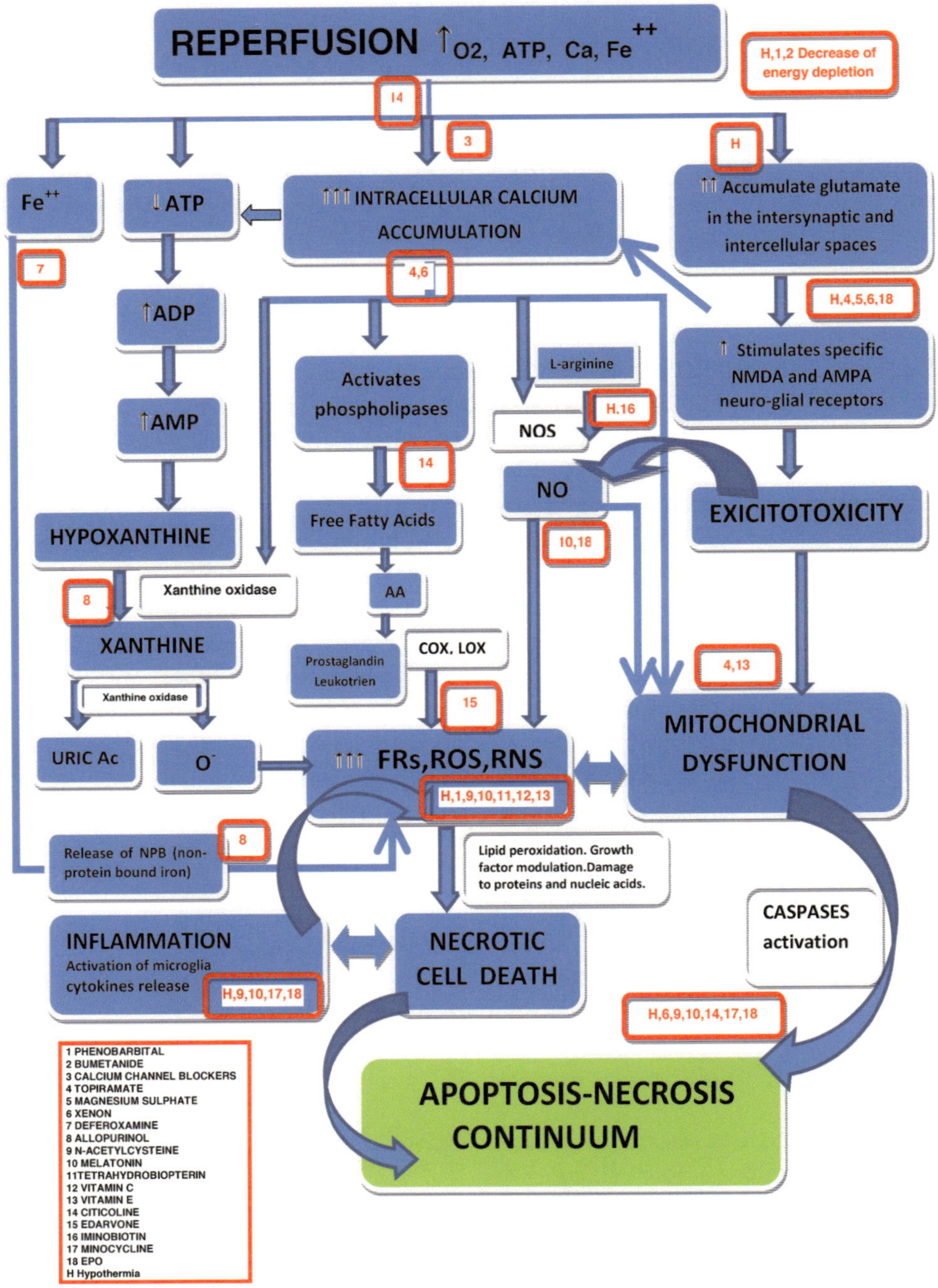

Figure 2. DELAYED INJURY: Secondary energy failure (associated with reperfusion). Neuroprotective interventions drugs according to predominants effects.

from the breakdown of free fatty acids). Low levels of ROS are indispensable in many biochemical processes, including intracellular messaging in the cell differentiation and cell progression or the arrest of growth, apoptosis, immunity, and the defense against micro-organisms. Interestingly, more recent data strongly suggest that low levels of NO and ROS are involved in normal events such as gene transduction control (Kroncke, 2003). In contrast, high doses and/or inadequate removal of ROS result in oxidative stress, which may cause severe metabolic malfunctions and damage to biological macromolecules. Increased production of ROS contributes to the pathogenesis of neonatal H-I brain injury. Acute restoration of blood flow after ischemia leads to the production of ROS, which are directly toxic to neurons and glia, and which may exacerbate leukocyte accumulation, microvascular thrombosis, and NO-mediated injury (Matés et al 1999).

Under normal physiologic conditions, low concentrations of $O2^-$ and H2O2 are produced as a byproduct of mitochondrial electron transport, a balance is maintained between the production of ROS and the capacity of the antioxidant enzyme system; however, if this balance breaks down, ROS can exert toxic effects. The oxygen FR are destroyed rapidly by the endogenous antioxidants: they are scavenged enzymatically by superoxide dismutase (SOD), catalase (CAT), and glutathione peroxidase (GPX), and nonenzymatically by reaction with antioxidant molecules, such as alpha-tocopherol (vitamin E), and ascorbic acid (vitamin C), glutathione (GSH), b-carotene, and vitamin A (Grow & Barks, 2002). Treatment with the iron chelator deferoxamine, N-acetylcysteine (NAC), or with allopurinol (a xanthine oxidase inhibitor that also acts as a FR scavenger), attenuates H-I damage in the immature rat (Palmer et al, 1993; Palmer et al, 1994). A recently discovered antioxidant enzyme family, peroxiredoxin (Prdx), is also an important scavenger of FR: Prdx1 expression is induced at birth, whereas Prdx2 is constitutively expressed, and Prdx6 expression is consistent with the classical antioxidant enzymes (SOD, CAT, and GPX) (Shim & Kim, 2013).

It is recognized that during intra- and peripartum asphyxia the generation of ROS is increased exceeding the capabilties of the protective mechanisms and causing oxidative stress. FRs produced from xanthine by products and prostaglandin synthesis attack polyunsaturated fatty acids (PUFAs) of the plasma membrane, increasing membrane permeability, endothelial cell death compromises the blood-brain barrier (BBB), resulting in vasogenic edema and hemorrhage. The mitochondrial respiratory chain is also a major source of ROS, and mitochondrial dysfunction contributes to cellular necrosis as well. FRs are another example of the inability of the immature brain to handle reoxygenation. Although FR concentrations rise during hypoxia, a significant secondary elevation occurs during resuscitation.

The relationships between FR generation and perinatal brain damage are complex there are a number of potential mechanisms of FR generation (Mishra & Delivoria-Papadopoulos, 1999; Robertson & Finer, 1993):

1. **Accumulation of intracellular Ca^{2+} and subsequent activation of:**

   - **Phospholipase a2** leading to increased generation of oxygen FRs from cyclooxygenase (COX) and lipoxygenase (LOX) pathways, **and phospholipase C** leading to IP3 formation resulting in release of Ca2+ from intracellular stores. Neonates also have high concentrations of PUFAs that break down to form more oxygen FR (Shalak & Perlman,

2004). ROS contribute to tissue injury by attacking the PUFAs component of the cellular membrane, resulting in membrane fragmentation and cell death.

- **Nitric oxide synthase (NOS)** leading to peroxinitrite formation and generation of FRs. NO is a weak free radical formed during the conversion of L-arginine to L-citrulline by NO synthase (NOS). NOS is strongly activated during H-I and reperfusion, producing a large amount of NO for extended periods. When combined with superoxide, NO generates a potent radical, peroxynitrite, which activates lipid peroxidation. In addition, NO enhances glutamate release (Palmer, 1995). The generation of ROS which interact with endothelial cell-derived NO leading to the formation of reactive nitrogen species (RNS). ROS and RNS target endothelial cells and neuronal cells, and both the oxidative stress and the nitrosative stress have been implicated in animal models of perinatal brain damage (Grow & Barks, 2002; McQuillen & Ferriero, 2004).
- **Proteases** leading to conversion of **xanthine oxidase** resulting in increased FR generation.

2. **Reduction of electron transport chain components including ubiquinone** (a component that undergoes autooxidation to produce FRs) (Turrens et al, 1985).
3. **Release of iron from ferritin under the condition of depleted cellular high energy compounds.** Physiologically, iron is maintained in a nontoxic ferric state and is bound to proteins (ferritin, transferrin). During H-I, free ferric iron is released from these proteins and reacts with peroxides to generate potent hydroxyl radicals. In addition, free ferric iron is reduced to a ferrous form, which further contributes to FR injury. Iron is a major mediator of cell damage. Differentiating OLs are particularly susceptible to FR damage because they are rich in iron, which is required for differentiation.

and

4. **Increased degradation of ATP during hypoxia and increased substrate for the xanthine oxidase reaction** Hypoxanthine accumulates as a product of ATP degradation and cannot be reconverted to ATP by the salvage pathway in anaerobic conditions. Xanthine dehydrogenase is converted to xanthine oxidase through the activation of a specific protease by calcium and this reaction produces ROS.

4.2.2. Excitotoxicity (Excitatory neurotransmitter release)

Definitions of Excitotoxicity: Excessive levels of extracellular neurotransmitters (*NT's*) causing excitatory overstimulation and neuronal damage.

The fundamental process responsible for H-I damage of neurons is called excitotoxicity, a term popularized in 1970s by John Olney that refers to cell death caused by excessive stimulation of extracellular excitatory amino acid (EAAs) receptors (Choi &, Rothman, 1990). Both hypoxia and ischemia result in the failure of energy-dependent ion pumps and causes cellular depolarization, increasing the release of EAAs -glutamate and aspartate- into the extracellular space in the cerebral cortex and basal ganglia.

http://dx.doi.org/10.5772/57459

The amino acid glutamate is the major excitatory NTs in the central nervous system (CNS). During neurotransmission, glutamate is released from pre-synpatic neurons by means of depolarisation of the pre-synaptic neuronal endplate, and then diffuses across the synaptic cleft to activate post-synaptic glutamatergic receptors. Several studies indicate that glutamate receptor-mediated excitotoxicity is a key player in neuronal cell death and that it is more critically involved in the developing brain than in the adult brain (Johnston, 2001). The neuronal injury occurring with cerebral H-I has been attributed to overstimulation of the Nmethyl-D-aspartate (NMDA) and alfa-amino-3-hydroxy- 5-methyl-4-isoxazole-propionic acid (AMPA) subtypes of excitatory amino acid glutamate receptors. Acute energy deprivation leads to the excessive release of extracellular glutamate and uncontrolled activation of ionotrophic glutamate receptors NMDA, AMPA, and kainate, impedes energy-dependent reuptake of glutamate and causes the rise in intracellular Ca2+ concentration. The massively increased levels of intracellular second messenger Ca2+ trigger activation of toxic intracellular pathways involving kinases, phosphatases, proteases, endonucleases, FRs, mitochondrial dysfunction, inflammation, DNA damage, and, ultimately, irreversible neuronal injury and death (Choi & Rothman, 1990; McDonald, 1998). Another important component of glutamate synaptic dysfunction caused by H-I is postsynaptic neuronal membrane depolarization with secondary opening of voltage-sensitive channels. Membrane depolarization due to high levels of synaptic glutamate produces maximum channel opening and flooding of calcium and sodium into neurons. Although high levels of glutamate can produce some degree of membrane depolarization under normal mitochondrial function, maximal excitotoxicity probably occurs when there is synergism between high synaptic glutamate levels due to disruption of glucose delivery and oxidative stress. The prominent role of the NMDA receptor–channel complex in perinatal H-I is related in part to its special transient role in brain development. The NMDA receptor subunits in the developing brain create populations of receptors that input more calcium, open more easily and block less frequently than mature forms, allowing them to serve these special developmental roles. However, this makes the immature brain more susceptible to excitotoxic injury if critical levels of energy failure are exceeded. For instance, neurotoxicity mediated by NMDA is more enhanced in the neonatal brain than the adult brain. Besides, NMDA receptors play an important role in activity-dependent neuronal plasticity during development Therefore, development-dependent changes in the expression of NMDA receptor subunits and their composition are, at least in part, responsible for the fact that immature brains are far more excitable and epileptogenic than the adult brain.

A second important mechanism for the destruction of ion pumps is the lipid peroxidation of cell membranes, where enzyme systems, such as the Na^+/K^+-ATPase, are located. This leads to water influx and cell swelling, causing cell death. EAAs also increase the local release of NO, which may exacerbate neuronal damage, although its mechanisms are unclear. It is quite possible that EAAs disrupt factors that normally control apoptosis, increasing the pace and extent of programmed cell death. The regional differences in injury severity may be explained by the fact that EAAs particularly affect the hippocampus, the developing oligodendroglia, and the subplate neurons along the borders of the periventricular region in the developing brain. This may be the basis for the disruption of long-term learning and memory faculties in infants with HIE (Barks & Silverstein, 1992).

4.2.3. Inflammatory mediators (Microglial and astrocyte activation)

Neuroinflammation, caused by activated microglia and astrocytes, plays a key role in the pathogenesis of CP. Maternal intrauterine infection and inflammation are risk factors for the development of PVL (characterized by focal necrosis around the ventricles, and diffuse microglial and astrocyte activation in the immature WM) and CP in the neonate (Haynes et al, 2003; Leviton et al, 2010). The microglia -immune cells in the brain-, play an important role in remodeling and growth during the fetal and postnatal periods, and they are proposed to be involved in the pathophysiological mechanism for the development of CP in humans. Activation of these cells can result in an exaggerated inflammatory response with formation of FR, excitotoxic metabolites, and pro-inflammatory cytokines, leading to brain injury (Dommergues et al, 2003; Li et al, 2008). In severe inflammation, astrocytes that normally participate in the protection of neurons and in preventing oxidative injury, are unable to maintain their neuroprotective role (Maragakis & Rothstein, 2006). H-I injury in the neonate is progressive, producing lesions of variable severity including focal necrotic cell death, diffuse WM injury, cystic or cavitary infarction and the resulting neuropathies linked to the activation of neuroinflammatory processes (inflammatory cytokines, chemokines, and matrix metallo-proteinase (MMP) activity) that occur in response to the initial wave of cell death.

Cytokines may be classified as interleukins (IL), interferons, tumor necrosis factors, chemokines (proteins that stimulate leukocyte motility), and growth factors. Systemically, cytokines may be ''proinflammatory'' (eg, IL-1b, TNFa, IL-6, IL-8) or ''antiinflammatory'' (eg, IL-4, IL-10, TGFa). However, this subdivision may not apply to the CNS. Cytokines that are produced peripherally can send signals to the brain, but more importantly, they are produced locally within the CNS in response to acute insults, including H-I. Cytokines may act on neurons, astrocytes, microglia, and endothelium, and may influence CNS injury indirectly by way of systemic parameters, such as blood flow and temperature. Cytokines can also play trophic roles during development, and their effects may differ depending on cell type (Grow & Barks, 2002). In experimental models, there is strong evidence that increased IL-1b gene expression and bioactivity after HI contributes to the pathogenesis of H-I injury in the immature and adult brain, (Hagberg et al, 1996). Clinical data also suggest that inflammatory mediators play a role in the pathogenesis of H-I brain injury: In human infants, CSF IL-6 and IL-8 concentrations are increased after birth asphyxia in comparison with controls; the magnitude of the increases correlates with the severity of the encephalopathy (Savman et al, 1998).

Chemokines are cytokines that regulate leukocyte migration and activation. H-I induces increased expression of two potent monocyte chemokines (MCP-1 and MIP-1a). This chemokine response precedes, and may mediate, the recruitment and activation of monocytes and microglia. There is no direct evidence that these chemokines contribute to neonatal HI brain injury. A large body of evidence implicates neutrophils and leukocyte adhesion molecules in the pathogenesis of focal cerebral ischemia in the adult brain, but there is only limited information to substantiate a role for these mediators in the pathogenesis of neonatal cerebral H-I (Hudome et al, 1997).

Neural injury after H-I is exacerbated due to neuroinflammatory signaling from activated microglia and peripheral infiltration of macrophage, cell death via necrotic and apoptotic mechanisms, and astrogliosis. Activation of microglia and macrophage infiltration is associated with the necessary phagocytosis of cellular debris, but it also results in a burden of increased production of neurotoxic substances, such as RNS and ROS, pro-inflammatory cytokines, and the NMDA agonist quinolinic acid. Thus, prolonged activation of these monocyte-derived cells for at least a week may elicit further deleterious changes in the brain (Nakajima & Kohsaka, 2004).

4.3. Mechanisms of neuronal cell death after hypoxia-ischemia

Excitotoxic cell death occurs through necrosis and/or apoptosis (also known as programmed cell death), the balance between modes of death may be influenced by the severity of the insult, cell phenotype and location, and maturational stage. One important characteristic feature of ischemia is selective neuronal necrosis (SNN), which is characterized predominantly by neuronal death whereas the astroglial cells are spared, at least initially. Necrosis predominates in more severely affected areas; apoptosis is seen more in penumbral areas of the injury. Several important features distinguish apoptotic from necrotic death (Northington et al, 2001):

1. Apoptosis is a physiologic process. Is the mechanism for refining cell connections and pathways during development by removing many neurons that will not be needed in adulthood, it is an essential part of normal brain development. This fact may underlie the observation that the propensity for experimental H-I to induce apoptosis peaks in the early, postnatal period.

2. Apoptosis is an active process, dependent on activation of a family of proteases called caspases that are modulated by other proteins (eg, those of the Bcl-2 family). The biochemical cascade of apoptosis can be blocked pharmacologically at several points.

3. Necrosis is characterized by cell swelling and eventual loss of cell membrane integrity followed by cell lysis. Resultant inflammation is a significant part of necrotic death. In contrast, apoptosis is an energy-dependent process and is characterized by a shrinking of the cytoplasm, condensation of the nucleus and eventual fragmentation of the cell body into smaller bodies. Markers of apoptosis such as cytoplasmic and nuclear condensation, as well as nuclear DNA fragmentation, appear in neurones, particularly in the infarct and penumbra of cerebral injury.

4. The time course of apoptotic death in H-I is slower than that of necrotic death. This feature may provide a more prolonged window of opportunity for therapeutic intervention than for necrotic death.

Early H-I-induced neuronal death occurs through necrosis (primary damage). Delayed neuronal death (secondary damage) occurs hours or days later through a series of complex and highly regulated biochemical and molecular events leading to apoptosis. Accumulating data suggest that apoptosis plays a prominent role in the evolution of H-I injury in the neonatal brain and may be more important than necrosis after injury. During neonatal brain injury, excitotoxicity, oxidative stress, and inflammation all contribute to accelerated cell death by means of either apoptosis or necrosis, depending on the region of the brain affected and the severity of the insult (Blaschke, 1996). This apoptosis–necrosis morphological continuum of neuronal death after H-I is similar to that observed in neonatal rats after excitotoxic activation of NMDA and non-NMDA glutamate receptors, suggesting that H-I neuronal injury is triggered by the excitotoxic cascade (Portera-Cailliau, 1997; Martin, 1998; Bittigau, 1999).

Recent explosive progress toward dissecting the molecular basis of apoptosis revealed the existence of family acting proteases, known as caspases. These protein-splitting enzymes act in a cascade form when signals are transmitted from the various receptors. Caspases sequentially activate each other and several proapoptotic protein kinases and disable DNA repair mechanisms (e.g., by cleaving poly (ADP-ribose) polymerase, PARP). Apoptosis can be elicited through two pathways:

- Intrinsic pathway: in which translocation of cytochrome c from the mitochondria to the cytoplasm is an early step; when translocated cytochrome c combines with ATP, apoptosis protease activating factor-1 (Apaf-1) and procaspase-9, caspase-9 is cleaved to its active state.
- Extrinsic pathway initiated by cell membrane death receptor activation (e.g. TNF receptor). Death receptor-ligand interactions lead to caspase-8 activation.

Activation of either caspase-8 or -9 (initiator caspases) leads to cleavage and activation of caspase-3 (effector caspase). Caspase activation is modulated by the BCL-2 family of proteins, which includes proapoptotic (eg, Bax) and antiapoptotic (eg, Bcl-XL) members. ROS can induce apoptosis in neurons. This effect is mediated by mitochondrial cytochrome c release (Green, 2000). Current data implicate the caspase-9 pathway as predominant in acute neuronal injury. Ca2+ is an important inducing agent in the mitochondria-dependent apoptotic pathway. Increased free cytosolic Ca2+ may lead to uncoupling of mitochondrial oxidative phosphorylation, inducing the mitochondrial the release of cytochrome c. Once released from mitochondria, cytochrome c specifically activates caspase 3, which triggers a biochemical cascade involving activation of many other caspases and other substrates. Known substrates of activated caspase-3 include PARP (a nuclear enzyme that participates in DNA repair), and DNA-dependent kinase, endonucleases, and cytoskeletal proteins (eg, fodrins, actin, lamin). Caspase-3 is strongly activated in animal models of H-I, resulting in increased activity of PARP in neuronal nuclei, indicating activation of the DNA repair pathway (Mehmet, 1994). On the other hand, pancaspase inhibitors are strongly neuroprotective when given several hours after the insult (McDonald, 1997).

Severe cerebral ischemia also results in a major increase in intracellular Ca2+, which is closely related to NO. Both play a fundamental role in signal transduction –controlling cell processes such as proliferation (Ashkenazi, 1998). NO has several important physiologic functions in the CNS. These include control of central and peripheral functions, modulation of synaptic plasticity, perception of pain and neuronal damage, and protection. High NO levels may be neurotoxic and induce apoptosis or necrosis. Changes in intracellular Ca2+ or NO may alternatively lead to blockade or activation of the cell cycle, and the decision as to whether the cell lives or dies is presumably a well-regulated phenomenon in which the duration and intensity of the Ca2+ and NO signals may play a fundamental role. Identification of enzymes and substrates for these phosphorylation pathways would shed light on hypoxia-induced processes leading to cell recovery or death (Esplugues. 2002).

NMDA channel overactivity, calcium entry into neurons, production of NO, and mitochondrial dysfunction are intimately linked in a potentially lethal cycle. There is abundant evidence that excitotoxic injury to neuronal mitochondria is linked to activation of neuronal death, and mitochondria appear to play a central role in determining expression of injury as necrosis and/or apoptosis, determining whether neurons live or die following hypoxic–ischemic insult. The death receptor and mitochondrial apoptosis pathways converge at the caspases that cleave multiple cellular substrates, ending in DNA fragmentation and cell death. It is noteworthy that activation of apoptosis-executing caspases is much greater in the immature brain than in the adult brain (Hu, 2000).

From the above it follows, that one of the main difficulties for developing a neuroprotective pharmacotherapy is the existence of multiple cellular death mechanisms, occurring in different cells or even in the same cell, which may all need to be inhibited. In fact, the widely accepted apoptosis-necrosis dichotomy is being replaced by a more complex view involving a third type of cell death, named autophagic cell death, characterized by the presence of intense autophagy (Clarke, 2008). Whereas the roles of apoptosis and necrosis in such conditions have been studied intensively, the implication of autophagic cell death has only recently been considered. Autophagy is an essential pathway for the degradation and recycling of intracellular macromolecules. The most important autophagic mechanism, macroautophagy, consists in the sequestration of long-lived proteins and damaged organelles in multimembrane vesicles, named autophagosomes, which then fuse with lysosomes to degrade their contents. Whereas basal macroautophagy (called hereafter autophagy) plays a central physiological function in maintaining cellular homeostasis, induced autophagy may have both survival and deleterious roles. In neurons, autophagy has been demonstrated to be induced during development, starvation, neurodegeneration, and also after different excitotoxic stimuli. More recently, an involvement of enhanced autophagy in neuronal death following cerebral ischemia has been proposed (Koike, 2008).

4.4. Neurogenetic and gliogenetic processes after H-I injury

Recent studies in brains damaged by H-I insult have demonstrated that some cellular mechanisms can be activated in order to repair cerebral injuries. These mechanisms involve the neural stem/progenitor cells (NSPs) normally resident in the subventricular zone (SVZ) of the mammalian brain that can be stimulated by H-I to proliferate and differentiate into both neurons and OLs. These new cells are likely to play an important role in repairing neuronal and glial losses related to HIE. Perinatal H-I results in brain injury, whereas mild hypoxic episodes result in preconditioning, which can significantly reduce the vulnerability of the brain to subsequent severe H-I. "In vivo", hypoxic-preconditioning has been shown to enhance cell survival and differentiation of progenitor cells in the CNS, stimulating SVZ proliferation and neurogenesis. This phenomenon may be a positive adaptation for an efficient repair and plasticity in the event of a H-I insult (Ara, 2013). Therefore, the NSPs in the SVZ could be a valuable target for therapeutic strategies to enhance recovery after cerebral H-I injury (Scafidi, 2008).

5. Therapeutic window

Brain injury following H-I insult is a complex process evolving over hours to days, which provides a unique window of opportunity for neuroprotective treatment interventions. The seminal concept emerging from both experimental and clinical studies is that brain cell death does not necessarily occur during H-I (the 'primary' phase of injury), but rather that the injurious event may precipitate a cascade of biochemical processes leading to delayed cell death, hours or even days afterwards (the 'secondary' phase). Experimental studies in piglets, immature rats, and fetal sheep have demonstrated the existence of both a primary phase of energy failure during H-I, a 'latent' phase during which oxidative metabolism normalizes, followed by secondary failure of oxidative metabolism. Consistent with these studies, although with exceptions (some newborn infants exposed to profound asphyxia show no initial recovery of oxidative metabolism after birth and typically have very severe brain injury and high mortality), in many cases infants show initial, transient recovery of cerebral oxidative metabolism followed by a secondary deterioration, with cerebral energy failure from 6 to 15 hours after birth. The severity of secondary energy failure correlates closely with the severity of neurodevelopmental outcome at 1 and 4 years of age. Critically, for understanding labor insults, experimental studies show that a single 'sub-threshold' insult that causes either minor or no neural injury can lead to a phase of increased vulnerability to further insults in a similar window of around 6 or more hours (Gunn, 2009). Advances in neuroimaging, brain monitoring techniques, and tissue biomarkers have improved the ability to diagnose, monitor, and care for newborn infants with neonatal encephalopathy, as well as to predict their outcome. The importance given to the role of oxidative stress in newborn morbidity with respect to the higher risk of FR damage in these babies is growing. However, challenges remain in early identification of infants at risk for neonatal encephalopathy, determination of timing and extent of H-I brain injury, as well as optimal management and treatment duration (Buonocore, 2012).

The literature review reinforces the notion that the spectrum of H-I encephalopathy outcomes represents a continuum, which has important implications for the prediction of outcome and the indications for intervention. Perlman & Shah (Perlman & Shah, 2011) summarize predictive clinical criteria at 3 time points: the first 6 hours of life, 6-72 hours of life, and at hospital discharge. They highlight the predictions at pivotal decision making times: at 0-6 hours (the earlier the better), to initiate or not to initiate neuroprotection, and at 6-72 hours, to identify, implement, and achieve the goal of withdrawing lifesustaining therapy.

The **"therapeutic window"** is this interval after reperfusion ("*reperfusion* window": *narrow therapeutic time-window within 6 hours of insult)*, during which an intervention might be efficacious in reducing the severity of the ultimate brain damage. As we previously mentioned, the cascade of biochemical and histopathological events initiated by H-I can extend from days to weeks after the insult is triggered, which may provide a "therapeutic window of opportu-

http://dx.doi.org/10.5772/57459

nity", probably more extensive (ranging from 6 to 72 hours, called *"cytoprotection window")*, for intervening in the pathogenesis of the developing brain. But such interventions are currently limited by insufficient knowledge of the timing and duration of the so-called therapeutic window in newborns (Barks, 2008; Levene, 2010). Recent data suggests that interventions for perinatal HIE may be combined to enhance the protective and reparative processes, and thought must be given to the best time to administer these interventions. As we described before, injury evolves over long periods of time with different mechanistic phases, adding to the already slow process of neuronal necrosis and apoptosis that can extend for several hours to a day or more. Therefore, therapies will also need to be administered over long periods of time, with different combination of drugs aimed at these temporally evolving targets.

6. Neuroprotective therapies

Asphyxia thus constitutes one of the leading avoidable causes of morbi-mortality in the newborn, and, therefore neuroprotection strategies oriented to prevent neuronal lesion (or at least minimize its consequences) are currently one of the most important lines of research in perinatal medicine. Before the introduction of neuroprotective therapies, the management of neonates with HIE in Neonatal Intensive Care Unit (NICU), was limited to supportive intensive care, including resuscitation in the delivery room followed by stabilization of hemodynamic and pulmonary disturbances (hypotension, metabolic acidosis, and hypoventilation), correction of metabolic disturbances, (glucose, calcium, magnesium and electrolytes), monitoring for multiorgan dysfunction and treatment of seizures.

There has been significant research progress in HIE over the last 2 decades, and many new molecular mechanisms have been identified. Despite all these advances, therapeutic interventions are still limited. At present, magnesium sulfate in preterm- and hypothermia in term newborns are currently the only treatments recognized as effective in neonatal HIE, and they are the focus of completed or continuing clinical trials.

6.1. Hypothermia

Hypothermia has been shown to be neuroprotective at critical cellular and vascular sites of cerebral injury in fetal/neonatal models of HI by inhibiting many steps in the excitotoxic-oxidative cascade. The mechanism of action of hypothermia includes: 1) decrease in energy use, decrease in cerebral oxygen consumption and amelioration of secondary energy failure; 2) inhibition of the increase of lactic acid in the brain; 3) reduction/suppression of extracellular EAA –glutamate- accumulation; 4) inhibition of NOS activity and NO concentrations; 5) suppression of FR activity and lipid peroxidation; 6) decrease of inflammation: inhibits platelet activating factor (PAF); decreases the level of interleukin-1beta and the release of toxic cytokines themicroglial/glial cells and inhibits protease activation, 7) attenuation of secondary energy failure; 8) inhibition of mitochondrial failure; 9) inhibition of necrosis and decrease of caspase-3 activation and morphologic evidence of apoptosis. Hypothermia has been shown to reduce cerebral metabolism, prevent edema and loss of membrane potential, decreas brain energy use, prolong the latent phase, reduce infarct size, decrease neuronal cell loss, and the extent of brain injury and epileptic activity, relieves the permeability of BBB and intracranial pressure, helps to retains sensory motor function, and preserves hippocampal structures (Shankaran, 2012).

Factors that influence the effects of hypothermia include the target body temperature, mode of hypothermia induction (selective head cooling vs total body cooling), duration of hypothermia, and rate of rewarming. Hypothermia after neonatal HIE, to a target temperature of 33–34°C for 72 hours has currently been shown to reduce death or disability at 18 months or increase the rate of survivors without disabilities. The safety and effectiveness of hypothermia as neuroprotection agent has been reported in in many randomized controlled trials to date, enrolling infants born at greater than or equal to 36 weeks or greater than or equal to 35 weeks gestation, within the therapeutic window of 6 hours (Gluckman, 2005; Shankaran et al., 2005, Shankaran et al., 2008; Azzopardi, 2009; Edwards, 2010; Rutherford, 2010; Zhou, 2010; Simbruner, 2010; Jacobs, 2011; Tagin,2012).

These trials have all included moderate/severe stage of encephalopathy (Sarnat & Sarnat, 1976) as eligibility criteria in random assignment studies at < 6 hours of age. Three trials included the additional criteria of abnormal amplitude-integrated electroencephalography (aEEG).(Gluckman, 2005; Azzopardi 2009; Simbruner 2010).

The Cool Cap trial (Gluckman, 2005) involved 234 term infants with moderate or severe encephalopathy and abnormal aEEG. Death or disability occurred in 66% conventional care and 55% cooled group (adjusted odds ratio [OR] [95% CI] 0.61 [0.34–1.09], P 5.10). Predefined subgroup analysis suggested that head cooling had no effect in infants with the most severe aEEG changes but was beneficial in infants with less severe aEEG changes.

The National Institute of Child Health and Human Development (NICHD) Neonatal Research Network (NRN) randomized controlled trial (RCT) of whole-body hypothermia for neonates with HIE: 208 term infants with moderate or severe encephalopathy were randomly assigned to whole-body cooling to an esophageal temperature of 33.5º C for 72 hours or usual care. Death or severe disability occurred in 62% of the usual care group and 44% of the hypothermia group (relative risk [RR] [95% CI] 0.72 [0.54–0.95], P 5.01) (Shankaran et al., 2005).

The Total Body Hypothermia for Neonatal Encephalopathy (TOBY) trial (Azzopardi, 2009) enrolled 325 infants. Death or severe disability occurred in 53% of the standard care group and 45% of the hypothermia group (RR 0.86 [0.68–1.07], P 5.17). The rate of CP was lower, and improved mental and psychomotor indices were noted in the hypothermia group as compared with the usual care group (all P<.05). (Rutherford, 2010).

Hypothermia is currently the only treatment recognized as effective in neonatal HIE at present, and the only recommends therapeutic as a standard practice (Biban, 2011). Hypothermia is currently being recommended since 2010 by health care policy makers:

- The international Liaison Committee on Resuscitation (ILCOR). The most recent International Consensus Conference was held in Dallas in February 2010 and the published

http://dx.doi.org/10.5772/57459

conclusions and recommendations from this process form the basis of these 2010 ERC Guidelines (guideline of ILCOR CoSTR 2010) (Hazinski, 2010).

- American Heart Association Guidelines for Cardiopulmonary Resuscitation and Emergency Cardiovascular Care (Kattwinkel, 2010).
- The International Consensus on Cardiopulmonary Resuscitation and Emergency Cardiovascular Care Science with Treatment Recommendations (Perlman, 2010).
- The 2010 guidelines released by the European Resuscitation Council (ERC) (Nolana, 2010).
- and the American Academy of Pediatrics (AAP) (Shelov, 2011).

The guides recommend therapeutic hypothermia as a standard practice for term or near term infants with moderate to severe HIE, stating that during the postresuscitation period in greater than or equal to 36-week gestation neonates with evolving moderate or severe encephalopathy, hypothermia should be offered in the context of clearly defined protocols similar to published trials. Further trials to determine the appropriate techniques of cooling, including refinement of patient selection, duration of cooling and method of providing therapeutic hypothermia, will refine our understanding of this intervention (Cochrane Database; Jacobs, 2013).

6.2. Neuroprotective drugs

Although therapeutic hypothermia is a significant advance in the developed world and improves outcome, the rate of death or disability following hypothermia for neonatal HIE is unacceptably high and ranges from 31% to 55% in the published trials, therefore it is becoming evident that the association of moderate hypothermia with another neuroprotective interventions drugs may enhance the outcome.

Therefore, there still is an urgent need for other treatment options. Further, there are currently no clinically established interventions that can be given antenatally to ameliorate brain injury after fetal distress. One of the major limitations to progress is what may be called "the curse of choice" (Robertson, 2012). A large number of possible neuroprotective therapies have shown promise in pre-clinical studies (Painter, 1995; Kelen & Robertson, 2010). The lines of research currently include: 1) antenatal therapy for fetuses with a diagnosis of antenatal fetal distress at term; 2) and postnatal therapy of infants with moderate to severe neonatal encephalopathy. To date, there is no consensus on which drugs have a higher chance of success in preventing the continued neuronal loss for either antenatal or postnatal treatment in human neonates who have suffered from HIE (reviewed in Degos, 2008; Gonzalez & Ferriero, 2008; Kelen & Robertson, 2010; Rees et al, 2011; Buonocore et al, 2012; Robertson et al, 2012).

The majority of the drugs used as neuroprotective therapy in animal models does not penetrate in good condition the BBB. Thus, drug delivery across the BBB to target cells to treat diffuse brain injury is the hardest challenge. The development of new biocompatible dendrimers has become an important objective for the companies involved in Biotechnology. Among the potential applications of these *nanopolymer biopharmaceuticals* it would be improving the drug transport to increase both the bioavailability and the active fraction, as well as the controlling the release of pharmaceuticals in order to custom and /or prolong the drug delivery over time. In this way, the use of nanoparticles opens a new door to cerebral HIE treatment.

Bellow, we review the drugs that have shown both higher neuroprotection and clinical safety and, therefore, could be applied to neonates. Since many of them have several mechanisms of action, they have been classified according to their predominant effect (see FIGURE 2).

6.2.1. Anticonvulsant drugs

The developing brain has both a higher incidence of seizures in human and animal models, and experiences seizures that can produce long-lasting consequences that are also stage-dependent (Ben-Ari, 2006). Perinatal H-I brain injury is one of the most important causes of epilepsy (Carrascosa et al, 1996), which occur in the majority (15–60%) of children with CP. Epilepsy is a disorder in which the balance between cerebral excitability and inhibition is tipped towards uncontrolled excitability. Selected neuronal circuits as well as certain populations of glial cells die from the excitotoxicity triggered by HI. The presence of seizure, practically occurring within the first hours, predicates a poor outcome of HIE.

The treatment of seizures is an essential component of the HIE management. Seizures are generally self-limited to the first days of life but may significantly compromise other body functions, such as maintenance of ventilation, oxygenation, and blood pressure. Additionally, seizures should be treated early and be well controlled, since even asymptomatic seizures (seen only on the EEG) may continue to injure the brain. The energy metabolism can be compromised by the hyperactive neurons, and both acute energy deprivation after HI insult and seizures are implicated in excitotoxicity. Thus, the therapeutic value of antiepileptic drugs (AEDs) may include not only the control of seizure activity but also the potentiall benefit for the compromised cellular energy metabolism (Aicardi, 2008).

Basic and clinical studies indicate that seizures in neonates have long-term neurodevelopmental and psychiatric consequences, highlighting the need for novel pharmacotherapeutics. The two most common classes of AEDs are GABAA receptor agonists and NMDA receptor antagonists. Currently, the first-line medical treatment for neonatal seizures is composed of drugs that increase GABA receptor channel chloride currents, like barbiturate and benzodiazepines (Calabresi et al, 2003).

- **Phenobarbital (FDA-approved)**

Phenobarbital (PB) increases GABA subtype A (GABAA)-receptor channel chloride currents. It is also important to acknowledge that PB has multiple potential modes of action in addition to anticonvulsant effects that could contribute to neuroprotection in this setting, including reduced cerebral metabolic demand, antioxidant effects and decreased cerebral edema.

PB controls seizures in less than half of newborns (Painter et al, 1999). This reduced efficacy of GABA-enhancing AEDs has been linked to neuronal chloride transport in the developing brain. In the adult nervous system, due to low intracellular levels of Cl-, GABA inhibits most neurons by activation of GABAARs, causing Cl- influx, membrane hyperpolarization, and inhibition. In immature neurons, the Cl-exporting activity of KCC2 is lower than in mature neurons and in the context of NKCC1 expression, neuronal $[Cl^-]_i$ is higher and $GABA_A$ reversal potential ($E_{GABA)}$ is more depolarized. High levels of expression of the Na+-K+-2Cl- (NKCC1) cotransporter in immature neurons cause the accumulation of intracellular chloride and, therefore, a depolarized Cl- equilibrium potential. This results in the outward flux of Cl- through GABA(A) channels, the opposite direction compared with mature neurons, in which GABA(A) receptor activation is inhibitory because Cl- flows into the cell (**for review see De Cabo-de la Vega et al 2006**). This outward flow of Cl- in neonatal neurons is excitatory and contributes to a greater seizure propensity and poor electroencephalographic response to GABAergic anticonvulsants such as PB and benzodiazepines (Khanna et al, 2013). It is also intriguing to consider that recent insights regarding the impact of maturational changes in neuronal chloride transporter expression on GABA receptor function may provide strategies that could improve the neuroprotective efficacy of PB in the neonate. Specifically, blocking the neonatal neuronal chloride transporter with bumetanide can augment the inhibitory activity of GABA agonists such as PB (Costa et al, 2006).

On the other hand, recent research have found that administration of PB is potentially harmful drug, associated with widespread apoptotic neurodegeneration throughout the brain when administered to immature rodents during the period of the brain growth spurt (Bittigau et al, 2003). Compounds that may cause neuronal apoptosis in the developing brain include antagonists of NMDA receptors (ketamine), agonists of GABAA receptors (barbiturates and benzodiazepines), and sodium-channel blockers (phenytoin, valproate) (Olney et al, 2002). Nevertheless, PB remains the preferred drug for the treatment of seizures in neonates with HIE (Volpe, 2008; Slaughter et al 2013). It is still a controversial issue whether PB treatment should be administered before the seizure attacks. In a small randomized trial, treatment of infants with HIE with PB within 6 hours of birth resulted in a decrease in death or disability at three years of age (Hall et al, 1998). At the present time, anticonvulsant therapy to term infants in the immediate period following perinatal asphyxia cannot be recommended for routine clinical practice, other than in the treatment of prolonged or frequent clinical seizures (Evans et al, 2007).

In a neonatal rodent model the early post- H-I administration of PB may augment the neuroprotective efficacy of therapeutic hypothermia (Barks et al, 2012). Sarkar et al. (Sarkar et al, 2012) found that the PB treatment before cooling did not improve the composite outcome of neonatal death or the presence of an abnormal post-hypothermia brain MRI. Whether this combination treatment could also result in improved neuroprotective efficacy in conjunction with cooling is an interesting question for future research.

- **Bumetanide (FDA-approved)**

Low concentrations of the diuretic bumetanide have been shown to alter the ion gradient that underlies the excitatory effects of GABA. Blocking the NKCC1 transporter with bumetanide prevents outward Cl- flux and causes a more negative GABA equilibrium potential in immature neurons. While several studies have reported anti-convulsant effects of bumetanide (Kahle et al, 2009), others have found no significant anti-convulsant effect. The alteration of Cl- transport by bumetanide reduces electrographic seizures, and the combination of bumetanide and PB is significantly more effective than PB alone on seizure occurrence, frequency, and duration (Dzhala et al, 2008). Currently there are two clinical trials evaluating bumetanide as a treatment for neonatal seizures. A phase I trial (NCT00830531) is actively enrolling patients in a randomized, double-blind, controlled dose-escalation study of bumetanide as an add-on therapy to treat refractory seizures caused by HIE. A second trial (NCT01434225) is being performed by a large, multi-center European group in an "open-label," dose escalation fashion to assess the effect of bumetanide in addition to PB for the treatment of neonatal seizures caused by HIE. Data from these pilot studies will be utilized to guide the design of larger Phase III trials that will determine the efficacy of bumetanide in the treatment of neonatal seizures (Khanna et al, 2013).

6.2.2. Anti-calcium drugs: Calcium channel blockers

Calcium channel blockers such as **Nimodipine, Flunarizine** and **Nicardipine**.

Calcium antagonist pretreatment attenuates HI damage in immature animals models. However, there is no clinical evidence, that such agents are neuroprotective in asphyxiated newborn infants. This may be attributable to lack of selectivity of the available drugs for neuronal calcium channels, poor BBB penetration, or the inability of these agents to affect intracellular calcium stores. Besides, the use of calcium channel blockers in severely asphyxiated newborn infants has been associated with clinically important hypotension and fall in cerebral blood-flow velocity, and if there is no cerebral autoregulation, may cause further cerebral hypoperfusion (Levene et al, 1990).

6.2.3. Antiexcitatory drugs: Glutamate antagonists

Glutamate antagonists are the most studied neuroprotective agents.

A. Inhibition of glutamate release:

- **Adenosine A2A receptor antagonist including clonidine and dexmedetomidine (FDA-approved):** have been show to have neuroprotective potential in animal models of perinatal H-I. Adenosine acts as a NTs in the brain through the activation of four specific G-protein-coupled receptors (the A1, A2A, A2B, and A3 receptors). The A1 receptor has long been known to mediate neuroprotection, mostly by blockade of Ca2+ influx, which results in inhibition of glutamate release and reduction of its excitatory effects at a postsynaptic level. However, the development of selective A1 and A2 receptors agonists as anti-ischemic agents has been hampered by their major cardiovascular side effects (Abbracchio & Cattabeni, 1999).
- **Riluzole (FDA-approved):** a 2-aminobenzothiazole, is a drug inhibiting glutamate reléase, neuronal excitability, and interferes with the effects of proteins activated upon NMDA-receptor stimulation. After discovery of its neuroprotective effects in 1994, riluzole was approved by the FDA for amyotrophic lateral sclerosis (ALS). However, animal experimentation showed that modest neuronal losses in a H-I model evoked by excitotoxicity have a severe impact on locomotor network function, and that they cannot be satisfactorily blocked

by strong neurodepression with riluzole, suggesting the need for more effective pharmacological approaches (Sámano et al, 2012).

B. The blockers of glutamate receptors:

The blockers of glutamate receptors [non-nmda (particularly AMPA receptors) and NMDA receptors], have conflicting results. Antagonists of the NMDA receptors (NMDARs) for glutamate are potent neuroprotective agents in several animal models of perinatal brain lesions. Administration of pharmacologic antagonists of the NMDA, AMPA/kainate or metabotropic receptors attenuates H-I-induced neuronal injury. Administration of an AMPA/kainate antagonist (but not NMDA antagonists) attenuates H-I oligodendroglial injury (Follett et al, 2000). Conversely, increasing the expression of glutamate transporters has been shown to be neuroprotective in "in vitro" models of ischemic damage and an "in vivo" model of amyotrophic lateral sclerosis. Collectively, these data indicate a neuroprotective role of astrocytes by glutamate uptake via their glutamate transporters. Combination therapy with non-NMDA and NMDA antagonists could achieve a dual beneficial.

- **AMPA/kainate antagonist:**

Topiramate (FDA-approved)

Topiramate (TPM) is a novel anticonvulsant agent currently used in adults and children older than 2-yr-of-age, characterized by good absorption, high bioavailability, and good tolerability (Elterman et al, 1999). TPM has multiple mechanisms of action, including: blocking sodium channels, high voltage-activated calcium currents, enhancing GABAinduced influx of chloride, and inhibiting kainate/ AMPA glutamate receptors; but also blockade of carbonic anhydrase isoenzymes, and mitochondrial permeability transition pore). TPM blockade of AMPA and kainate receptors (glutamate receptors non-NMDA) shows little neurotoxicity, although their effects on other stages of brain development such as synaptogenesis have not been evaluated, prevention of excitotoxicity with TPM appears to be a particularly promising approach, since the agents shown to be effective experimentally are likely to be clinically safe, at least in developing animals (Glier et al, 2004). TPM protected pre-OL against excitotoxic or H-I death (Follett et al, 2004), a key event in the pathophysiology of WM lesions in preterm infants, and protected the PVWM against damage induced by an AMPA-kainate agonist in newborn mice (Sfaello et al, 2005). Subsequent studies that demonstrated that TPM treatment alone confers neuroprotection on H-I brain injury in neonatal rats, with more prolonged treatment (4 doses over 48h). It is more likely that greater neuroprotective efficacy is attributable to intrinsic AMPA-antagonist properties (Noh et al, 2006). TPM exert neuroprotective effects against PVL (Follett et al, 2004).

In clinicals models no adverse effects attributable to TPM were detected (Filippi et al, 2010). Filippi et al. (Filippi et al, 2012) are doing a project (three-centre phase II pilot study entitled "Safety and Efficacy of Topiramate in Neonates With Hypoxic Ischemic Encephalopathy Treated With Hypothermia (NeoNATI)") to evaluate whether the efficacy of moderate hypothermia can be increased by concomitant topiramate treatment, at 10 mg/kg once a day for the first 3 days of life. Any favourable results from this research might open new perspectives about the reduction of cerebral damage in asphyxiated newborns.

- **Blockade of NMDA receptors:**

NMDA receptors play key roles in successive steps of brain development, including the proliferation, migration, survival, and differentiation of neurons (Lujan et al, 2005). Therefore, blocking NMDA receptors at specific neurodevelopmental stages might adversely affect brain development. Thus, in rats studied during the postnatal growth spurt, transient NMDA-receptor blockade with the potent noncompetitive antagonist **MK-801** led to massive cell death by apoptosis (Ikonomidou et al, 2002). These findings constitute a strong argument against the prolonged use of potent NMDA receptor antagonists during brain development.

Although glutamate receptor antagonists have shown excellent neuroprotective effects in animal studies (Lippman-Bell et al, 2013), these effects have not been validated in clinical studies. These agents all cause a similar spectrum of neuropsychological symptoms, and several have important cardiovascular effects. As a result, studies of several NMDA antagonists- **Selfotel** (CGS19755), **Gavestinel**, **Eliprodil** and **Aptiganel** (Cerestat, CNS 1102) - have been halted; **Dextrorphan** is not efficacious and may be harmful at higher doses (Labiche & Grotta, 2004; Jaffer et al, 2011). Other drugs that may interfere with glutamatergic neurotransmission include **Amantadine** and **Memantine**, two NMDA receptor antagonists devoid of the psychotomimetic and neurotoxic effects of phencyclidine (PCP) or dizocilpine (MK-801) when administered to adults. These drugs are neuroprotective in adults with conditions closely related to excitotoxicity. Unlike MK-801 and other NMDA blockers, **memantine** also appears to be relatively safe in the developing rat brain (Manning et al, 2010); their efficacy and safety in newborns need to be determined.

C. Other modulatory sites on the NMDA receptor complex:

- **Magnesium sulphate (FDA-approved)**

Magnesium is a nonspecific competitive blocker of calcium channel and plays many important roles in maintaining body homeostasis. Magnesium is involved in multiple physiological processes that may be relevant to cerebral ischaemia, including antagonism of glutamate release, NMDA receptor blockade, calcium channel antagonism, maintenance of CBF, cell membrane permeability, mitochondrial functions, the ionic membrane current in conducting cells. Prehypoxic treatment of magnesium sulfate ameliorates the severity of brain damage, but posthypoxic treatment deteriorates it. This deleterious effect may be attributable to hypotension caused by high-dose magnesium sulfate, which further worsens cerebral perfusion (Sameshima et al, 1999).

Use of antenatal magnesium sulphate **(MgSO4) for fetal** neuroprotection **of the preterm infant:** MgSO4 has been used for decades in pregnant women for different indications (to obtain tocolysis and to treat eclampsia), with no reported adverse effects in the neonates. MgSO4 was neuroprotective in a model of neonatal WM damage. In a retrospective case control study, preterm infants exposed to antenatal magnesium sulfate were found to have a reduced risk of developing cystic PVL (FineSmith et al, 1997). It has been shown to be effective for neuroprophalxis to decrease the risk of moderate to severe CP. The first multicenter controlled clinical trial, where mothers at risk of delivering before 30 weeks of gestation were given magnesium, was completed in 2003 by the Australasian Collaborative Trial of Magnesium Sulphate (ACTOMg SO4) Collaborative Group (Crowther et al, 2003). Significant perinatal side effects occurred, and neurodevelopmental benefits were noted in survivors examined at 2-yr-of-age: substantial gross motor dysfunction (3.4% vs 6.6%; RR, 0.51; 95% CI, 0.29-0.91) and combined death or substantial gross motor dysfunction (17.0% vs 22.7%; RR, 0.75; 95% CI, 0.59-0.96) were significantly reduced in the magnesium group. No serious harmful effects were seen. Cochrane Systematic Review has confirmed this effect (Doyle et al, 2009), the review concludes that antenatal magnesium sulphate therapy given to women at risk of preterm birth substantially reduced the risk of CP in their children. Clinical practice guidelines on Magnesium Sulphate prior to preterm birth for neuroprotection have been developed in Australia (Clinical Practice Guidelines on Magnesium, 2011) and Canada (Magee et al, 2011) by the Societies of Obstetricians and Gynaecologists; and lately, by the American College of Obstetricians and Gynecologists and the Society for Maternal-Fetal Medicine (Committee opinion, 2013).

Postnatal therapy: In a randomised controlled trial performed by the "Intravenous Magnesium Efficacy in Stroke (IMAGES) Study" investigators (Muir et al, 2004), a mortality slightly higher in the magnesium-treated group than in the placebo group was shown, and did not reduce the chances of death or disability significantly. Magnesium also has a number of vascular effects, being that it inhibits NOS in a non-selective form, which in the case of eNOS would be harmful; therefore its use in postnatal therapy of infants with neonatal encephalopathy would be contraindicated.

- **Xenon (Not FDA Approved)**

Medical gases are pharmaceutical molecules which offer solutions to a wide array of medical needs. More specifically however, gases such as oxygen, helium, xenon, and hydrogen have recently come under increased exploration for their potential theraputic use with various brain disease states including H-I, cerebral hemorrhages, and traumatic brain injuries. A colorless, heavy, odorless noble gas, xenon has been of particular interest to researchers because of its possible neuroprotective properties (Dingley et al, 2006). Since the discovery of xenon as an NMDA receptor antagonist, there has been growing interest in its potential use as a neuroprotectant. Some of the features of xenon that specifically interest scientists and researchers include its rapid introduction into the brain, favorable hemodynamic profile with little or no toxicity, as well as its inability to be be metabolized (Liu et al, 2011). Xenon induces anesthesia and exerts its analgesic actions by inhibiting the NMDA receptor signaling pathway. Additional studies have demonstrated that xenon may act on the secondary messenger signaling pathway via increases in cyclic guanosine monophosphate. Xenon's role in antiapoptotic mechanisms also demonstrates its neuroprotective qualities. Furthermore, Xenon plays an important role in the anti-inflammatory process. Xenon can influence mechanisms regulating the Ca2+ release channel on plasma membranes (inhibits plasma membrane calcium ATPase pump activity), resulting in an increase in neuronal Ca2+ concentration and an altered excitability in these cells.

Researchers have also explored the possibility of using Xenon in combination with other therapeutic strategies to evaluate its possible synergistic neuroprotective capabilities; Xenon may offer haemodynamic benefits in clinical neuroprotection studies (Chakkarapani et al, 2012).

6.2.4. Antioxidative drugs

The neonatal brain has a high rate of oxygen consumption and low concentration of antioxidants, making it susceptible to damage. In humans, mature OL carry increased antioxidant enzymes compared with the pre-OL present in the immature brain, which may partially explain the susceptibility of premature infants to WM damage (Haynes et al, 2005). In an effort to reduce oxidative damage to the neonate, a number of protective interventions have been used, including **(1)** FR reducers, **(2)** ROS scavengers: antioxidant enzymes, and FRs nonenzymatic scavengers, **(3)** lipid peroxidation inhibitors, and **(4)** NOS inhibitors. Antioxidant strategies have been used successfully to diminish ischemic cerebral tissue damage in animals, but the utility of a pharmacological agent as a clinically relevant therapeutic strategy may depend, in part, on its ability to cross the BBB, since although ischemic injury disrupts the integrity of the BBB, this disruption is by no means complete.

A. FR reducers:

- **Deferoxamine (DFO)** - FDA-approved-. The free iron induces the formation of ROS, and exogenous iron significantly exacerbates excitotoxic and aggravates cystic PLV in newborn mice (Dommergues et al, 1998). DFO is an iron chelator that decreases FR production by binding with iron and decreasing the production of OH^{-}. DFO is protective during exposure to H2O2 or excitotoxicity "in vitro", and in animal models of H-I (Sarco et al, 2000).

- **Polyphenols**

Resveratrol (Res) -(FDA)-approved- could be a prophylactic factor in the prevention of ischemia/reperfusion (I/R) injury, they attenuates I/R injury in cardiomyocytes by preventing cell apoptosis, decreasing LDH release and increasing ATPase activity. NO, cGMP, PKC and K (ATP) may play an important role in the protective role of Res. Moreover, Res enhances the capacity of anti-oxygen FR and alleviates intracellular calcium overload in cardiomyocytes (Shen et al, 2012).

- **Allopurinol (FDA Approved)**

The xanthine-oxidase inhibitor allopurinol (ALLO) reduces FR formation, thereby limiting the amount of I/R damage. Hypoxanthine accumulates in the ischemic brain, and with reperfusion is oxidized to uric acid and superoxide. Elevated uric acid concentrations in the first postnatal day identify a subset of premature infants who are at high risk for having subsequent hemorrhagic or ischemic injury (Perlman et al, 1998). Furthermore, ALLO also has a non-protein bound iron (pro-radical) chelating and direct FR (hydroxyl) scavenging effect. Animal research in asphyxiated pigs demonstrated beneficial effects of postnatally administrated ALLO on cerebral energy status and cytotoxic oedema (Peeters-Scholte et al, 2003). Treatment with ALLO reduces FR production following ischaemia and it reduces tissue injury in "in vitro", and high doses (50-200 mglkg) of ALLO effects cerebral protection in animal experiments and also exerts benefits on reduction of cerebral edema and neuropathological damage after neonatal HIE (Palmer et al, 1993). In humans, the first work on the neuroprotective effect of allopurinol was carried out by Russell and Cooke (Russell & Cooke, 1995), in a randomized controlled trial of allopurinol prophylaxis in very preterm infants (between 24 and 32 weeks of gestation). In this trial of prophylactic ALLO for the prevention of PVL in preterm babies, no protective effect was apparent. A prospective randomized study in human neonates, examining the effects of ALLO in term asphyxiated neonates, showed an improvement of electrocortical brain activity and a reduction in FR formation after neonatal ALLO administration (Van Bel et al, 1998). A more recent paper by Gunes et al (Gunes et al, 2007) reports an improved neurological outcome after postnatal ALLO administration (40 mg/kg/day, 3 days, within 2 hours after birth) compared to a placebo in term asphyxiated neonates. Benders et al (Benders et al, 2006), however demonstrated that ALLO was not effective if administrated 3 to 4 hours after the hypoxic incident to severely asphyxiated neonates. However, when the most severely asphyxiated children were excluded from the study, a beneficial effect of ALLO was seen on neurological development. Apparently, no advantage of neonatal treatment is seen anymore when the interval to the initiation of treatment is too long or when the brain damage is too severe. This has probably been the major disadvantage of late postneonatal treatment with ALLO on the NICU. ALLO administrated at the NICU is likely to be given too late to provide adequate neuroprotection during the early period of reoxygenation in which the vast amount of FR is being produced. Apparently, when the asphyxia has been too severe, the inflicted brain damage can no longer be reversed. It is conceivable that earlier ALLO treatment, i.e. the use of ALLO during labour in case of suspected foetal hypoxia, provides the opportunity to start earlier with the treatment, thereby limiting the amount of I/R injury and improving neurological outcome. Animal and human studies suggest that administration of ALLO immediately prior to delivery in case of suspected foetal asphyxia might reduce HIE. In a study in the chronically instrumented foetal sheep, they were able to show evidence of cardio-and neuroprotection after antenatal ALLO administration to the pregnant ewe during repeated periods of ischaemia (Derkset al, 2006). Maternal administration of ALLO has been proposed as prebirth treatment when there is suspicion of an adverse event eliciting perinatal asphyxia. A prospective randomized placebo controlled pilot study, in which they administered ALLO to the pregnant woman when foetal asphyxia was imminent, showed an inverse correlation between the levels of ALLO and the amount of S100B, a biomarker for brain tissue damage, in cord blood (Torrance et al, 2009). A clinical trial of antenatal allopurinol is in progress (Kaandorp et al, 2010).

B. Antioxidant enzymes (Endogenous antioxidants):

SOD, GPX, and CAT, are considered the classical antioxidant enzymes. One therapeutic approach for the destruction of oxygen FRs generated during and after H-I is the administration of specific enzymes known to degrade highly reactive FR to a nonreactive component. SOD and CAT are enzymes poorly soluble and with short half life, so that should be conjugated to polyethylene glycol, which prolongs their circulatory half-life and facilitates penetration of the BBB. Notwithstanding that, the latency time to initiate therapeutic effects is unacceptably long, making them useful only as preventive. Because of their large molecular sizes, they are restricted to the vascular space. In newborn animals, neuroprotection has only been shown when these agents have been administered several hours before the HI insult (Shimizu et al, 2003).

C. Free radical nonenzymatic scavengers:

- **N-Acetylcysteine (NAC) (FDA Approved)**

NAC is a precursor of glutathione and can therefore act as an anti-oxidant and is also a scavenger of ROS. NAC is used clinically for mucolysis and as an antidote for paracetamol intoxication in high doses. It also reduced oxidative stress, inflammation, and minimized H-I-induced brain injury in various acute models. In addition to reducing total tissue loss, NAC reduced WM injury, prevented endotoxin-induced degeneration of OL progenitors and hypomyelination in developing rat brain (Paintlia et al, 2004). The mechanism of NAC neuroprotection appears to be related to reduced oxidative stress, preservation of the scavengers GSH and Trx2, attenuated activation of apoptotic proteases (caspase-3, calpain), and reduced inflammation. The protective effect of NAC was much more pronounced than that produced by another FR scavenger, melatonin, when administrated before and after lipopolysaccharide-sensitized H-I. NAC was also effective when administered directly after H-I (three days after) (Wang et al, 2007).

NAC is transported across the placenta and it is considered safe during pregnancy. Therefore, although transport across the BBB is believed to be poor, it has been assumed that NAC has potential therapeutic value in humans, performing its neuroprotective action at the level of the vascular bed (Schaper et al, 2002). NAC is associated with adverse reactions ranging from nausea to death (most of the latter due to incorrect dosing) which limits its use in humans. (Sandilands & Bateman, 2009; Knudsen et al, 2005).

NAC is the most effective therapy for acetaminophen (APAP) toxicity and is currently available for children for oral and intravenous (IV) administration; both routes are equally effective and safe (Green et al, 2013). Currently, there are three protocols that are used in acetaminomiphen ingestion:

72-hour oral-NAC: 140 mg/kg of oral NAC followed by 70 mg/kg every 4 hours for an additional 17 doses.

20- hours IV-NAC: consists of a continuous IV infusion - 300 mg/kg- of NAC, patiens are given a loading dose of 150 mg/kg over 15 minutes, followed by 50 mg/kg over 4 hours and then 100 mg/kg over the next 16 hours.

48-hours IV-NAC: consisted of a loading dose of 140 mg/kg followed by 12 doses of 70 mg/kg every 4 hours.

The most frequently used protocol is the Simplified N-acetylcysteine dosing regimen -standard preparation of IV-NAC 30 g in 1 L of 5% dextrose in water, with a 150-mg/kg loading dose administered over 1 hour followed by an infusion of 14 mg/kg/h for 20 hours – this single intravenous bag protocol is effective and well tolerated, and there is infrequent interruption of therapy by dosing errors (Johnson et al, 2011). Both the IV and oral NAC have generally mild adverse drug reactions. Nausea and vomiting have been the most common reported adverse events and were more common with oral treatment (9%). Anaphylactoid reactions were more common with IV administration (2%). The changing between administration routes, introducing deviations from "standard" treatment, may decrease these side effects (Bebarta et al, 2010). In a medication error in prescribing paracetamol for closing a patent ductus arteriosus in a preterm infant, NAC was indicated without showing adverse drug reactions (Brener et al, 2013).

There is no consensus on a neuroprotective dose, and a wide range of concentrations have been used in experimental studies. After neonatal I/R in piglets, NAC at doses of 150-mg/kg bolus and 20 mg/kg/h, IV for 24 hours, reduced cerebral oxidative stress with improved cerebral oxygen delivery and reduced caspase-3 and lipid hydroperoxide concentrations in cortex (Liu et al, 2010). With smaller dosages (30 mg/kg bolus then 20 mg/kg/h infusion) in newborn piglets with I/R, postresuscitation administration of NAC lowered cerebral lactate levels, reduced cerebral oxidative stress (significantly attenuated the increase in cortical H2O2, but not NO, concentration) and improved cerebral perfusion (Lee et al, 2008). Furthermore, the anti-inflammatory effect reduced lung edema and neutrophil influx into the lung and partly reversed surfactant dysfunction in the meconium aspiration syndrome model (Mokra et al, 2013). Combination therapy of NAC and systemic hypothermia induced immediately after neonatal H-I improves infarct volume, and reduced both white and grey matter damage after focal HI injury injury in neonatal rats (Jatana et al, 2006). By contrast, Olsson et al (Olsson et al, 1998), report that free NAC at 100 mg/kg showed some efficacy in attenuating inflammation and oxidative injury in the brain, but the improvement did not translate into myelination, neuronal counts or motor function in CP. The only randomized clinical trial in preterm newborns demonstrated that continuous infusion of NAC for 6 days after birth did not improve the incidence of chronic lung disease but did appear to reduce the incidence of PVL (Ahola et al, 2003).

Dendrimer-based N-acetyl-l-cysteine (D-NAC): Dendrimers are a nanopolymer biopharmaceutical emerging as potential intracellular drug delivery vehicles. The efficacy of the activity "in vitro" of anionic polyamidoamine (PAMAM-COOH) dendrimer-N-acetyl cysteine (DNAC) was significant, even at the lowest dose, and its activity compared to free NAC increase 16x higher dosage (Wang et al, 2009). The bioavailability "in vivo" of free NAC is poor (the terminal half-life was 5.58 h after IV administration and 6.25 h after oral administration. Oral bioavailability of total NAC was 9.1%). However, IV administration of a single 10 mg/kg dose of D-NAC resulted in a significant improvement in neuronal injury and motor function in CP kits. Moreover, the improvements seen with NAC-100 were similar to that seen with D-NAC at 1% of the dose improved uptake and efficacy of D-NAC when compared to free NAC in activated microglia, as shown previously "in vitro", delivery of a higher drug-payload to the target cells (activated microglia and astrocytes) by the dendrimer "in vivo", and decreased toxicity of the drug for neurons when conjugated with the dendrimer. The effectiveness of the D-NAC treatment, administered in the postnatal period for a prenatal insult, suggests a window of opportunity for treatment of CP in humans after birth (Kannan et al, 2012). The nanoparticles D-NAC open a new door to cerebral palsy treatment (Crunkhorn, 2012; Andón et al, 2012).

- **Melatonin (FDA Approved)**

Melatonin is an indoleamine that is formed in higher quantities in the adult. It is produced mainly by the pineal gland and is a naturally occurring hormone that binds to specific receptors and allows the entrainment of circadian rhythms in several biological functions. But it can also function as neuroprotective antioxidant (as a direct scavenger of ROS and NO), and has anti-apoptotic effects. Because of its lipophilic properties, melatonin easily crosses most biological cell membranes, including the placenta and the BBB. Several animal studies have shown neuroprotective benefits from melatonin treatment, both when given before and after birth. It has been found to provide long-lasting neuroprotection in experimental H-I and focal cerebral ischemic injury. Melatonin may exert some of its protection on developing WM in H-I sheep model via an anti-microglial effect (Welin et al, 2007). Newborns treated with melatonin were also found to have decreased proinflammatory cytokines (Gitto et al, 2004 and 2005).

Clinically, melatonin has been used safely in children with sleep abnormalities related to neurological disease (Jan & O'Donnell, 1996) and in septic newborns (Gitto et al, 2001) without serious adverse effects. Melatonin appears to have beneficial effects when given to asphyxiated newborns. It was shown to significantly reduce plasma levels of malondialdehyde and nitrate/nitrite, two robust indicators of oxidative stress (Fulia et al, 2005). In a small clinical trial, melatonin was given orally to newborn babies who had suffered birth asphyxia. In terms of mortality, 3 out of 10 asphyxiated babies died in the vehicle treated group, whereas there were no deaths in the post-asphyxia babies treated with melatonin. Importantly, this study did not report any adverse effects arising from the melatonin treatment, and clinical use of melatonin in the neonatal period has now been proposed (Gitto et al, 2009).

The optimal neuroprotective dose still needs to be determined, although Robertson et al (Robertson et al, 2013), in a piglet model of perinatal asphyxia, demonstrate that the therapeutic hypothermia plus IV melatonin (5 mg/kg/h) significantly reduced the H-I-induced. The safety and improved neuroprotection of a potential treatment with a combination of melatonin and cooling support the initiation of phase II clinical trials in infants with moderate and severe neonatal encephalopathy.

- **Tetrahydrobiopterin (FDA Approved)**

BH4 is an important co-factor for a number of enzymes, such as aromatic amino acid hydroxylases, which converts phenylalanine into tyrosine (phenylketonuria), tyrosine into L-dopa, and tryptophan into 5-hydroxytryptophan and NOS. BH4 may function as a FR, but it has also been reported that inhibition of biopterin synthesis reduces ischemic brain damage (Kidd et al, 2005). BH4 is a developmental factor determining the vulnerability of fetal brain to H-I. There is evidence that BH4 deficiency can exacerbate oxidative injury (Madsen et al, 2003) and that neonatal H-I can cause relative BH4 deficiency (Fabian et al, 2010). Fujioka H et al (Fujioka et al, 2008) found neuronal iNOS expression and increase of NO production in the acute phase of H-I in a newborn-piglet model, but brain biopterin did not increase despite plasma biopterin five-fold elevation These findings suggest that the capacity of biopterin production in the CNS is inferior to that in other organs in the acute phase of H-I. The BBB is thought to prevent the transport of biopterin from the blood to the brain, therefore the initial shortage of biopterin in a H-I brain may affect the severity of brain damage. Maternal treatment with BH4 increased fetal levels in basal ganglia and significantly ameliorated motor deficits and decreased stillbirths (Vasquez-Vivar et al, 2009).

- **Vitamin C (FDA Approved)**

Ascorbic acid (AA, vitamin C) is an important enzyme cofactor and water-soluble reducing agent that is highly concentrated in the adrenal gland and CNS. AA concentrations are lowest in plasma (0.01–0.1 mM), intermediate in cerebrospinal and extracellular fluid (0.05–0.5 mM) and highest in neuropil of the brain (1–3 mM). AA is a potent antioxidant, which scavenges various types of ROS, and its neuroprotective effect has not been established yet. Because vitamin C does not penetrate the BBB, therapeutic, nonenzymatic scavenging of FRs can be accomplished by AA only at very high physiological concentrations (Jackson et al, 1998). Some studies have shown that AA has a neuroprotective effect; however, the issue is still controversial. Since higher dose of AA may cause side effects such as oxaluria and kidney stone (Massey et al, 2005), erythrocyte damage leading to hemolytic anemia and hyperbilirubinemia in infant (Ballin et al, 1988), it might also have harmful effects on brain. AA can act as pro-oxidant and cause neurotoxicity "in vitro" by reducing transition metal ions under certain conditions (Buettneret al, 1996). Only one prospective, randomized double-blinded controlled clinical study in term infants with perinatal asphyxia has been performed, which found that the combination of AA with ibuprofen had no effect on outcome at 6 months of age (Aly et al, 2009).

Although the antioxidant vitamin C does not penetrate the BBB, its oxidized form, **dehydroascorbic acid (DHA)**, enters the brain by means of facilitated transport. AA can be oxidized to DHA in the stomach. Several studies demonstrated that AA and DHA have neuroprotective effects in adult animal models of HI. IV DHA would improve outcome after stroke because of its ability to cross the BBB and increase brain antioxidant levels. A dose of 250 mg/kg or 500 mg/kg DHA administered at 3 hours post-ischemia reduced infarct volume by 6- to 9-fold, to only 5% with the highest DHA dose ($P < .05$). Teratological and adverse effects have not been documented (Huang et al, 2001). AA may be given to the mothers if the safety of its use is established, since AA has been reported to cross the placental barrier (Rybakowski et al, 1995).

- **Vitamin E (FDA Approved)**

Vitamin E, a clinically safe agent, has been shown to be effective in cultured pre-OL models (Back et al, 1998). Clinical trials in very preterm infants have been confined to vitamin E for the prevention of IVH (Sinha et al, 1987). In rat pups, there was no benefit from postnatal treatment after H-I with Mito vitamin E, a mitochondrial antioxidant (Covey et al, 2006). However, in a study in asphyxiated rat pups, the combination of methylprednisolone with vitamin E therapy reduced H-I brain damage significantly (Daneyemez et al, 1999). Nakai et al. (Nakai et al, 2002) reported that the combined administration of AA and alpha-tocopherol (vitamin E) to pregnant rats before transient intrauterine ischemia was effective against secondary mitochondrial dysfunction in the neonatal rat brain.

D. Bioactive lipid mediators (Membrane "stabilizers"):

Considerable experimental evidence supports a pathogenetic role for lipid mediators in perinatal H-I brain injury (Grow et al, 2002). Bioactive lipid mediators that may play a role in cell signaling include arachidonic acid (AA), prostaglandins, leukotrienes, thromboxanes, and PAF. Membrane phospholipids are hydrolyzed by phospholipase A2 (PLA2) to release FFA, including AA, and lysophospholipid. Many studies have associated ROS-mediated damage with a disturbance of cell membrane integrity and subsequent increase in intracellular $Ca^{2+,}$ which, in turn, activate a number of Ca^{2+} dependent enzymes including PKC and PLA, damaging membranes directly and initiating the production of lipid mediators, including AA and PAF. H2O2 has been shown to cause AA release in numerous cell systems, including cells in the CNS (Samanta et al, 1998). ROS produced during post I/R react with membrane phospholipids to form oxidized lipids, some of which have PAF-like activity. High concentrations of PAF or PAF-like oxidized lipids may contribute to neuronal injury by increasing intracellular calcium concentrations, by stimulating production and release of pro-inflammatory mediators from neurons or microglia, or by upregulating cyclo-oxygenase-2 (COX-2). Additionally, the involvements of different PLA2s, including cPLA2, iPLA2, and sPLA2, have been implicated in the oxidative-mediated AA release process (Martinez et al, 2001). XU et al (XU et al, 2003) showed that the response of astrocytes to oxidant compounds such as H2O2, which stimulated signaling pathways leading to the activation of cPLA2 and iPLA2 (Ca^{2+}independent) and the increase in AA reléase.

- **Citicoline (FDA Approved)**

Cytidine-5-diphosphocholine (CDP-choline as an endogenous compound that can also be administered exogenously as *citicoline)* is an endogenous nucleoside that is an essential intermediate in the synthesis of phosphatidylcholine, a major neuronal membrane lipid. Citicoline and its hydrolysis products (cytidine and choline) play important roles in the generation of phospholipids which are involved in membrane formation and repair, and it is known to have neuroprotective effects. The mechanisms that may explain the **neuroprotective actions** of citicoline include prevention of FFA release, stimulation of phosphatidylcholine synthesis, preservation of cardiolipin and sphingomyelin levels, increase of glutathione synthesis and glutathione reductase activity, restoration of Na^{+}/K^{+}-ATPase activity, and antiglutamatergic effects. The neuroprotective properties of CDP-choline seem to be related on glutamate-mediated cell death (Adibhatla et al, 2002; Hurtado et al, 2005). Citicoline might decrease the extracellular level of glutamate by inhibition of neuronal glutamate efflux and increased astrocytic glutamate uptake. It has been suggested that the neuroprotective effect of this compound is related to inhibition of the glutamate induced apoptotic pathway of cell injury (Mir et al, 2003). The protective effect of CDP-choline might also be associated with its actions on cell membrane stability because citicoline has distinctive membrane-modulating properties (Secades et al, 2006). Citicoline decreases phospholipase A2 stimulation and hydroxyl radical generation in cerebral ischemia (Adibhatla et al, 2003). Lately, it has been suggested a possible contribution of the gene *PNPLA4,* and codes for calcium-independent PLA2, as one of the mecanisms through which the citicoline may act (Carrascosa-Romero et al, 2012); and regulative effect on the expression of intercellular adhesion molecule-1 (ICAM-1) mRNA in neonatal brain with H-I damage (Miao et al, 2005). The upregulation of the inflammatory genes and their products precedes leukocytes' adhesion to endothelial cells and their migration into the ischemic tissue, suggesting that these upregulated adhesion molecules on brain capillary endothelium play an important role in leukocyte migration into ischemic brain tissue (Wang et al, 1995). "in vitro", Matyja et al (Matyja et al, 2008) demonstrated that CDP-choline exerts neuroprotection in progressive motor neurons injury in a model of chronic excitotoxicity. It inhibited mainly neuronal apoptotic changes, whereas necrotic and autophagocytic abnormalities were not reduced. This confirms the suggestion that citicoline might protect neurons against the glutamate-induced apoptotic pathway probably via a negative effect on activation of the caspase cell death pathway (Mir et al, 2003); markedly reduced caspase-3 activation and Hsp70 expression 24 h after the insult, and dose-dependently attenuated brain damage in a rat model of birth asphyxia (Fiedorowicz et al, 2008). Diederich et al (Diederich et al, 2012) demonstrated that citicoline (100 mg/kg) for 10 consecutive days starting 24 hours after ischemia induction, have an extended therapeutic window.

In addition citicoline has convincingly been shown to also have **neuroregenerative effects** although the underlying mechanisms are unknow. As a first mechanism contributing to this more favorable neurological outcome, they could identify increased neurogenesis in the SVZ and migration of neural progenitors to the lesion with increased neurogenesis also within the peri-infarct area. A second component of the regeneration-enhancing effect of citicoline was a shift toward excitation in the perilesional cortex (Diederich et al, 2012).

Citicoline has been shown to have neuroprotective effects in a variety of CNS injury models, including focal and global cerebral ischemia (Trovarelli et al, 1981). In clinical trials, citicoline administered after acute ischemic stroke in adults, improved neurological outcome with mild adverse effects (Davalos et al, 2002; Labiche & Grotta, 2004; Saver, 2008). A recent drug surveillance study on acute ischemic stroke in 4,191 patients (Cho and Kim, 2009) showed that oral citicoline (500-4000 mg/day) administered within less than 24 h after acute ischemic stroke improved neurological, functional and global outcomes without significant safety concerns. By contrast, in a randomised, placebo-controlled study (ICTUS trial), citicoline, (1000 mg every 12 h IV during the first 3 days and orally thereafter for a total of 6 weeks) in patients with moderate-to-severe acute ischaemic stroke, was not efficacious in the treatment of moderate-to-severe acute ischaemic stroke (Dávalos et al, 2012). These differences could be due to dosage, interaction with other drugs when they are administered simultaneously or to the fact that necrosis phenomena are more marked than those of apoptosis. However, the apoptosis in the newborn plays a prominent role in the development of H-I injury i and may be more important than necrosis after injury. Additionally, the neurogenesis in the SVZ and migration of neural progenitors are more marked in the newborn than in adults. CDP-Choline administration to newborn infants (100 mg/kg/day, IV) was well tolerated without side effects (Valls et al, 1988; Wang et al, 1997). In the context of the well-known excellent safety profile of citicoline, these data suggest that success in the clinical evaluation of the efficacy of this drug in human neonatal asphyxia may be warranted.

- **Edaravone (Not FDA Approved)**

Edaravone, 3-methyl-1-phenyl-2-pyrazolin-5-one, (MCI-186) significantly decreased lipid peroxidation (thiobarbituric acid reactive substance levels) of the damaged brain hemisphere (Keda et al, 2002). Edaravone is a FR scavenger that improves the outcome after cerebral ischemia in humans and is used for treatment after acute stroke (Group EAIS, 2003) and traumatic brain injury (Itoh et al, 2009). Since edaravone has been approved in Japan for use in patients with cerebral infarction, it could be a promising candidate for the treatment of neonatal HIE.

E. nNOS inhibitors (Not FDA approved):

NO, a water-soluble, diffusible gas, has many physiological roles including regulation of gastrointestinal motility, vasorelaxation, and furthermore performs an important role in synaptic neurotransmission (intercellular messenger and signaling molecule), and is important for neuronal survival, differentiation, and precursor proliferation. NO can also contribute to tissue injury, and has been shown to play a dichotomous regulatory role in the brain; neuronal destruction and protection (Chen et al, 2004). NO prevents apoptosis of neuronal cells via two mechanisms; first, inhibition of caspase-3 activity through S-nitrosylation of cysteine residues in the protease, and second, cGMP-dependent (Lipton, 1995). NOS catalyzes the synthesis of NO from the conversion of arginine to citrulline. susceptibility to HI damage (Ferriero et al, 1996). Cerebral ischemia stimulates production of NO by neurons and microglia.

NO is generated by three distinct nitric oxide synthases: neuronal (nNOS), endothelial (eNOS), and inducible synthases (iNOS). Constitutively expressed nNOS and eNOS are activated by increased intracellular calcium, by way of the NMDA receptor stimulates; nNOS plays a physiologic role in excitatory neurotransmission; eNOS produces vascular smooth muscle relaxation; iNOS is upregulated by hypoxia, cytokines, or endotoxin in monocyte /macrophage/ microglia and is calcium independent; its induction can result in the production of large quantities of NO (Moncada et al, 1991). Selective inhibition of nNOS or iNOS has shown potential as a neuroprotective strategy, but nonspecific blockade of nNOS and eNOS is not protective (Marks et al, 1996). During I/R, nNOS plays a role in NO production, but iNOS only contributes to NO production during reperfusion.

Selective nNOS inhibitors:

The traditionally most employed nNOS inhibitor has been **7-nitroindazole (7-NI)** (Muramatsu et al, 2000), in adult stroke, protection by 7- NI has been inconclusive, this may reflect non-specific inhibition of eNOS, causing a decrease in CBF (Willmot et al, 2005). At present, the new inhibitors **HJ619 and JI-8** are at least few hundred fold more specific than 7-NI; the new compounds tested in a perinatal model of H-I, inhibited fetal brain NOS activity "in vivo", reduced NO concentration, and dramatically ameliorated the number of deaths and CP in a rabbit model (Ji et al, 2009). These compounds are water-soluble and can be given IV. The starting dose is unknown. Adverse effects are also unknown (Robertson et al, 2012).

- **Iminobiotin (Not FDA approved)**

2-iminobiotin, a dual inhibitor with combined inhibition of nNOS and iNOS. Iminobiotin exhibits neuroprotective effects in rats following H-I (van den Tweel et al, 2005), showed protection only in female rat pups after H-I, but the protection was independent of the NO pathway (Nijboer et al, 2007). Orphan designation (EU/3/09/701) was granted by the European Commission to Neurophyxia for 2-iminobiotin for treatment of perinatal asphyxia on the basis of potential activity. At the time of submission of the application for orphan designation, no clinical trials with the designated product in patients with perinatal asphysia had been started and 2-iminobiotin was not authorised anywhere in the EU for perinatal asphyxia.

6.2.5. Anti-inflammatory drugs

The CNS has its own resident immune system, in which glial cells (microglia, astrocytes, and OL) not only serve supportive and nutritive roles for neurons but also engage from time in several "inflammatory" processes that defend the CNS from pathogens and help it to recover from stress and injury. Cytokines and activated microglia/macrophages may extend neuronal injury and/or sensitize the developing brain to a second insult. Therefore, interference with their effects would be expected to reduce subsequent neurological deficits. However, cytokines such as IL-1 β or IL-6 were found to exert trophic effects on neurons, at least in cell cultures (Otten et al, 2000). Similarly, activated microglia/macrophages, in addition to exerting toxic effects, can display protective properties, such as scavenging of excess glutamate through increased expression of glutamate transporters (Vallat-Decouvelaere et al, 2003).

Accumulating evidence suggests that targeting delayed neuroinflammatory mechanisms may be a promising avenue for therapeutic intervention (Leonardo & Pennypacker, 2012). Anti-inflammatory interventions have shown promise in experimental models of combined gray and white matter injury: **cytokine antagonists (IL-1receptor antagonist)** (Hagberg et al, 1996), **platelet activating factor (PAF) antagonist** (Liu et al, 1996; Zhang et al, 1994) and **induced neutropenia** (Hudome et al, 1997). **Corticosteroids** theoretically may interrupt the inflammatory cascade that occurs during H-I; experimental and epidemiological studies support a protective role for antenatal steroids against PWMD (Whitelaw & Thoresen, 2000; O'Shea & Doyle, 2001), but this beneficial effect must be weighed against the adverse effects during a critical period of brain development of postnatal high-dose steroids used to prevent or to treat chronic lung disease in premature infants (Baud et al, 2004). Early synthetic glucocorticoid dexamethasone (DEX) exposure may lead the neonatal brain to be more vulnerable, exacerbated HI-induced injury on P7 by a glucocorticoid receptor-mediated mechanism. The aggravating effect of neonatal DEX treatment on HI-induced brain injury was correlated with decreased glutamate transporter-1 (GLT-1)-mediated glutamate reuptake.

- **Minocycline (FDA Approved)**

Minocycline is a semisynthetic second-generation tetracycline and a potential neuroprotective intervention following brain injury. However, despite the recognized beneficial effects of minocycline in a multitude of adult disease states, the clinical application of minocycline in neonates is contentious. Tetracyclines are broadspectrum antibiotics that have anti-inflammatory effects independent from their antimicrobial activity, but, as a class, are not usually administered to neonates. Minocycline inhibited microglial activation, reduces inflammation and protected neurons against ischemia in adult and developing rats in several studies (Yrjanheikki et al, 1999; Fan et al, 2006, Buller et al, 2009; Lechpammer et al, 2008). Minocycline treatment prevents the formation of activated caspase-3, a known effector of apoptosis, as well as the appearance of a calpain cleaved substrate, a marker of excitotoxic/necrotic cell death (Arvin etal, 2002). Although another study found that minocycline exacerbated H-I cortical injury in neonatal mice (Tsuji etal, 2004). Nevertheless, minocycline is not without clinical hazard, and further study of this agent and related analogs is needed (Volpe et al, 2011).

6.2.6. Anti-apoptotic drugs

One of the hottest topics in neurobiology is apoptosis, the highly orchestrated and possibly "controllable form of cell death" in which cells enter into a programmed suicide by chopping themselves into membrane-packaged bits. In stroke, neurons in the penumbra zone, deprived of oxygen and glucose tissue, gradually die because ischemic injury triggers their suicide programmes (Miller & Marx, 1998; Barinaga, 1998). Control of apoptosis involves a balance between expression of numerous apoptotic and anti-apoptotic proteins after injury, providing many potential approaches to modifying outcome (blockade of downstream effects).

- **Erythropoietin (EPO) (FDA Approved)**

EPO is a 165 amino acid glycoprotein produced mainly by peritubular cells in the adult kidneys and by hepatocytes in the fetus. EPO acts on the later stages of erythroid progenitor cells development, allowing maturation of erythroid precursors by inhibiting apoptosis and thus regulating red cell production. Recombinant human EPO (rhEPO) is currently effective and widely used to treat anemia of prematurity. Epo, the major haemopoietic growth factor, is now considered to have beneficial effects in various nervous system disorders based on the effects of prevention of metabolic compromise, neuronal and vascular degeneration, and inflammatory cell activation (Maiese et al, 2008). EPO is required for normal brain development in mammals (Juul, 2002; Yu *et al,* 2002). Exogenously administered EPO exhibits neuroprotective effects in numerous animal models, through the activation of anti-apoptotic, anti-oxidant and anti-inflammatory pathways as well as through the stimulation of angiogenic and neurogenic events. (for a review see Kumral A et al, 2011; Juul,2012; Subirós et al, 2012). Moreover, EPO reduced the excitotoxic effect of glutamate and AMPA upon cortical neuron cultures (Sinor & Greenberg, 2000). EPO also prevented apoptosis induced by NMDA or by NO in neurons of cerebrocortical cultures (Digicaylioglu & Lipton, 2001).

Recombinant human EPO (rhEPO)- induced neurogenesis has been studied in *"in vivo"* and *"in vitro"* experiments, it has been shown that EPO regulates neurogenesis in the adult mouse brain (Shingo *et al,* 2001). Epo is not an appropriate antenatal therapy because it hardly crosses the human placenta (Widness et al, 1995). The capability of EPO to cross the BBB after systemic administration and its effective therapeutic window are advantages for H-I therapy. It requires small volumes, so it does not impose a fluid burden. The most commonly used treatment regimens used to stimulate erythropoiesis in neonates is 400 U/kg 3 times a week given subcutaneously or 200 U/kg daily given IV. The optimal dose, number of doses, or dosing interval for Epo neuroprotection in humans have not yet been determined. Neonatal Epo treatment has been studied in randomized controlled trials of erythropoiesis, with few reported adverse effects, and the medication is thought to be safe at doses ranging as high as 2100 units/kg/week (Fauchere et al, 2008; Juul et al, 2008). Complications seen in adults (eg, hypertension, clotting, seizures, polycythemia, and death) (Ehrenreich *et al.* 2009) have not been observed in infants. Angiogenesis may be an important adverse effect in preterm infants at risk for retinopathy of prematurity. The first trial with EPO in full term neonates with moderate to severe H-I demonstrated that it reduced death and disability at 18 months from 44% (controls) to 25% (EPO-treated) with no adverse effect (Zhu et al, 2009). Human trials are just beginning, but they show promise (Elmahdy et al, 2010; Lakic et al, 2010).

Elmahdy et al (Elmahdy et al, 2010) in a prospective case-control study with 45 neonates, of which 15 were infants with mild/moderate HIE, received human recombinant EPO (2500 IU/kg, subcutaneously, daily for 5 days), demonstrated the feasibility of early administration of EPO to neonates with HIE for protection against encephalopathy.

- **EPO-Mimetic Peptides (Not FDA Approved)**

The possibility of developing Epo-mimetic that have specific subsets of Epo characteristics has been of great interest, because these molecules might circumvent unwanted clinical effects or provide improved permeability with the ability to cross the placenta or BBB. The tissue protective functions of Epo can be separated from its stimulatory action on hematopoiesis, and novel Epo derivatives and mimetics, such as **asialo-Epo and carbamylated- Epo**, have been developed. No studies have been done to assess safety or efficacy of these compounds as perinatal treatments (see Robertson et al, 2012).

Neuro-EPO is a variant with a low-sialic acid content and a short half-life. Drug transport from the nasal cavity directly to the brain has been shown to be feasible, even for challenging drugs such as small polar molecules, peptides and proteins, in animals and humans. Intranasally administered Neuro-EPO exhibits neuroprotective effects in gerbil models of brain ischemia. The use of the nasal route as a new delivery pathway to the brain is aimed to achieve quick delivery of neuroprotective concentrations to the nervous tissue using small drug doses (see Subirós, 2012).

- **Neurotrophic factors (Neurotrophins)**

Trophic factors are emerging as potential cytoprotective agents, although their role may be more important in the recovery phase (Labiche & Grotta, 2004). Neurotrophins are important cues for the migration and differentiation of neural stem cells (SCs). Neurotrophins are a family of growth factors that act through tyrosine kinase receptors and regulate the development and maintenance of brain cells by affecting growth, differentiation, maturation, maintenance and neuronal survival, as well as synaptogenesis and brain plasticity. They also exhibit neuroprotective activity in multiple neuronal populations after injury. The first neurotrophin discovered was neuronal growth factor (NGF). Further work identified other members of the family such as Glial Derived Neurotrophic Factor (GDNF), Brain Derived Neurotrophic Factor (BDNF), and Neurotrophin-3 (NT-3).

The neurotrophins play a vital role in development, but also in the maintenance of the neuronal systems throughout life (Sizonenko et al, 2007). During the neonatal period, neurotrophins and their receptors are essential for brain development. After brain insult, neurotrophins levels increase, suggesting that they have an endogenous protective mechanism that limits neuronal cell death. Since discovery of the potent survival-promoting effects of neurotrophic factors (possibly through angiogenic mechanisms), they have been proposed as potential tools to be tested for the treatment of diseases of the CNS **(e.g.: basic fibroblast growth factor-bFGF)** (Binder & Scharfman, 2004). The protection of neurons, and perhaps most cells, from excitotoxicity and H-I injury may require extracellular ligand–receptor interactions and the activation of specific intracellular signaling cascades. In the CNS, these signals are provided, at least in part, by neurotrophic growth factors. Several neurotrophic factors (such as **platelet-derived growth factor, insulin-derived growth factor, and glial cell line-derived neurotrophic factor**) that have been reported to protect against excitotoxicity and H-I injury in immature animal models may act by inhibiting apoptosis (Nozaki et al, 1993; Hossain et al, 1998; Wang et al, 2013), these have not been studied in humans. Neurotrophins could guide migration and differentiation of stem cell transplants after brain injury, and once at the site of injury, enhance neuronal differentiation (Douglas-Escobar et al, 2012).

- **Neuropeptide-inhibitor**

Neuropeptides modulate neuronal activity and may therefore modulate glutamate-induced neuronal cell death. Neuropeptides are inactivated by enzymatic proteolysis, indicating that proteolysis inhibition may hold therapeutic potential. Among the peptidases identified, neural endopeptidase (NEP or neprilysin) is involved in the regulation and metabolism of a variety of biologically active peptides including tachykinins/ neurokinins (see Degos, 2008). Interestingly, the **NEP inhibitor Racecadotril** (Tiorfan®) is used in clinical practice to treat diarrhea, with a remarkably good safety profile. Racecadotril is rapidly and entirely metabolized to its active metabolite thiorphan. A recent study showed that systemic administration of thiorphan was neuroprotective against excitotoxic neuronal cell death in newborn mice (Schwartz, 2000). This neuroprotective effect was long-lasting and was still observed when thiorphan was administered 12 h after the insult, indicating a wide window for therapeutic intervention (Medja et al, 2006).

6.3. Neurorestorative therapies

There is increasing evidence from "in vitro" and "in vivo" preclinical studies that stem/ progenitor cells may have multiple beneficial effects on the outcome after H-I injury. Stem cell (SCs) treatment can be administered by stimulating endogenous SCs (neurotrophic factors/ growth factors such as EPO, insulin-like growth factor and brain-derived neurotrophic factor) or by transplanting exogenous SCs. **Stem/progenitor cells** (umbilical cord SCs, mesenchymal stromal cells, and bone marrow mesenchymal SCs) have been used for the experimental treatment of neonatal HIE models, and have shown great promise in animal studies in decreasing neurological impairment. **Neural stem/progenitor cells** (NSCs) have also been transplanted in animal models of HIE, migrating long distances to ischemic brain areas and differentiating into neurons. SCs therapies are one of the promising options for the treatment of neonatal neurological diseases in the future. However, the mechanisms of action of the SCs, and the optimal type, dose, and method of administration remain surprisingly unclear, and some studies have found no benefit. Although cell-based interventions after completion of the majority of secondary phase cell-death appear to have potential to improve functional outcome for neonates after HI, further rigorous testing in translational animal models is required before randomized controlled trials should be considered. (Review articles: Pimentel-Coelho & Mendez-Otero, 2010; Bennet et al, 2012; Pabon et al, 2013). Additionally, their clinical use is strongly limited by the existence of the BBB that makes the human brain refractory to targeting of cell-sized agents delivered through the peripheral system. Intracerebral transplantation to bypass the BBB is a very invasive delivery method that cannot be proposed for human newborns.

- **Cord blood:** Umbilical cord blood cells (UCBCs), which are readily available at birth, have been shown to reduce sensorimotor and/or cognitive impairments in several models of brain damage, representing a promising option for the treatment of neurological diseases. The possible cell types and mechanisms involved in the therapeutic effect of UCBC transplantation, including neuroprotection, immunomodulation and stimulation of neural plasticity and regeneration have been recently reviewed by Pimentel-Coelho et al. (Pimentel-Coelho et al., 2012).

- **Mesenchymal stem cells (MSC):** The beneficial effect of MSC transplantation to treat neonatal brain injury might be explained by the great plasticity of the neonatal brain. The neonatal brain is still in a developmentally active phase, leading to a better efficiency of MSC transplantation than that observed in experiments using adult models of stroke. Enhanced neurogenesis and axonal remodeling likely underlie the improved functional outcome following MSC treatment after neonatal H-I brain injury. With respect to the mechanism of repair by MSCs, MSCs do not survive long term and replace damaged tissue themselves. MSC treatment after H-I reduced contralesional rewiring taking place after HI and increased the connection between the impaired forepaw and the ipsilesional motor cortex. These intrinsic adaptive properties of MSCs make them excellent candidates for a novel therapy to treat the devastating effects of HIE in the human neonate (van Velthoven et al, 2012). Intranasal MSC treatment may become a promising non-invasive therapeutic tool to effectively reduce neonatal encephalopathy (Donega et al, 2013).

- **Neural stem/progenitor cells (NSCS):** NSCs have also been transplanted in animal models of HIE, migrating long distances to ischemic brain areas and differentiating into neurons (Llado et al, 2004). The survival of transplanted NSCs was limited in these experiments for several potential reasons (Sato et al, 2008).

- **Adipose stromal cells (ASC):** In a rat middle cerebral artery occlusion model of ischemic brain injury, intracerebral transplantation of human ASC was followed by migration of these cells to areas of ischemic damage and by expression of neuronal specific markers in conjunction with functional benefit. Therefore, the use of ASC could have potential to develop treatments to reverse or prevent the effects of H-I injury. (Kang et al, 2003)

7. Combined therapy

Following HIE, a complicated cascade of pathophysiologic processes is unleashed including excitotoxicity, oxidative stress, inflammation, and cell death via necrosis and apoptosis. These processes can lead to long-term neurologic injury. The post-injury time course can be divided into a latent (0–6 hours), secondary (6–72 hours) and tertiary phase (>72 hours) (Perlman, 2011). Studies in laboratory animals have shown that the immature brain responds differently to treatment than does the mature brain, which leads us to believe that an optimal treatment for a neonate would differ from that for a toddler and probably none of them would be the best option for an adult. Specific vulnerabilities that distinguish the response of the immature brain from that of the mature brain include:

Primary/latent phase (0–6 hours)

- Greater neuronal metabolism: The neonatal brain has a high rate of oxygen consumption. Such energetic costs seem also to exert a selective pressure towards metabolically efficient neural morphology, leading to metabolically efficient patterning of dendritic arborizations, neural codes and brain wiring patterns (Holliday, 1986).
- Antioxidant system insufficiency.
- The NMDA receptor subunits in the developing brain open more easily and block less frequently than mature forms, responsible for the fact that immature brains are far more excitable and epileptogenic than the adult brain.

Secondary phase (6–72 hours)

Increased susceptibility to excitotoxicity and FR injury, due to the production of large amounts of FR, high concentrations of PUFAs, and antioxidant system insufficiency (Grow & Barks, 2002).

Tertiary phase (>72 hours)

- Greater tendency to apoptotic death: activation of apoptosis-executing caspases is much greater in the immature brain than in the adult brain. During the tertiary phase, neurons and glial cells are lost due to chronic loss of trophic factors, loss of synaptic input from neighboring cells, and loss of or failure of recruitment of new progenitor neural stem cells and glial progenitor cells (heightened vulnerability of immature OL). Cell death involving a cell-autonomous active contribution of catabolic enzymes (or *apoptosis*) plays a prominent role in the evolution of H-I injury in the neonatal brain and it is at least as important for the loss of neurons as unregulated cell death (or *necrosis)* (Zhu et al, 2007). The prominence of autophagic neuronal death in the ischemic penumbra and the neuroprotective efficacy of postischemic autophagy inhibition indicate that autophagy should be a primary target in the treatment of neonatal cerebral ischemia (Puyal etal, 2009). Therapy designed to ameliorate brain injury in adults may worsen outcomes in neonates, possibly by accentuating the apoptotic cell death cascade.

Neurogenetic and gliogenetic processes after H-I injury: neuroblast proliferation and migration from the neurogenetic niches take place after the lesion. Some cells differentiate into neurons (neuronal differentiation) and migrate to the injured area. Some other cells from the SPZ give rise to radial glia that contribute to neural progenitor expansion and support neuroblast migration. Thus, the neuronal elements in the transient fetal zones represent a potential for plasticity after perinatal cerebral lesions and neuronal migration could play a central role in brain repair (Kostovic & Judas M, 2006, Cayre et al, 2009, Distefano and Praticò, 2010). It is possible that the same chemical mediators that have deleterious effects during the initial stages of ischemia may also be involved in the ulterior process of neurorestoration. Therefore, the lack of effect found for some neuroprotective drugs could be due to either untimely or too prolonged period of administration, thus interfering with a given metabolic route at the time that is involved in the endogenous mechanisms of repair (Lo, 2008).

With the advent of hypothermia as therapy for term HIE, there is hope for repair and protection of the brain after a profound neonatal insult. However, hypothermia alone would not be sufficient to provide the required protection or stimulate the repair to ensure a normal neurodevelopment. Based on the theory of "secondary energy failure", hypothermia may provide the possibility to "buy time", in order to successfully use other (pharmacologic) interventions, by preserving energy metabolism elongating the therapeutic time window (Gunn & Gunn, 1997). At present, no individual neuroprotective agent has been proven safe and effective for the protection of neonates from neurological sequels after H-I insults as monotherapy. The persistent clinical failures might be due to many factors, including: heterogeneity in the causes of neural death in humans, associated toxicity at the doses required for drug-efficacy, the lack of adequate CNS penetration across the BBB and or the limited time window available to start the treatment (in *real-life* clinical practice, delays in the initiation of therapy are difficult to avoid).

Since there are many mechanisms involved in H-I process, it is reasonable to assume that the combination of several drugs or the use of molecules that combine two or more neuroprotective actions can exert synergistic effects by blocking diverse metabolic pathways. Traditionally, the above mentioned simultaneous drug co-administration has been discouraged, since it involves potential risks of interference, side effects and error in the sequence of administration. However, the use of hypothermia plus adjuvant therapies has been extensively reviewed (Cilio & Ferrriero, 2010; Robertson et al, 2012; Buonocore et al, 2012; Shankaran, 2012). According to the existing literature, the combined therapy of hypothermia and other neuroprotective strategies would be expected to increase the therapeutic time window, enhance neural repair and improve the neurological outcomes of HIE. Although both clinical observations and animal experimentation suggest that the cascade of damaging events in the developing brain may last for several days -thus extending the window of opportunity for intervention- the most successful outcome is likely to result from the earliest possible delivery of therapy (Rees et al, 2011). Few studies have examined possible interactions of medications with hypothermia and whether combination therapies augment neuroprotection. Based on the preclinical studies, ongoing trials in neonates include: inhaled xenon and cooling (NCT01545271 and NCT00934700), safety of erythropoietin (NCT00719407), darbepoetin and hypothermia (NCT0147105), and topiramate plus hypothermia (NCT01241019).

Additional neuroprotective strategies to combat perinatal brain injury are urgently needed that may be used as possible synergies for therapy. Presumably the best outcome will be achieved by a multi-modal therapeutic approach such as a combination of hypothermia with anti-oxidants and glutamate receptor antagonists, using drugs with multiple effects (affecting multiple injury cascades and with neuroregenerative potential) without toxicity, no apparent interaction and previously used in children. Furthermore, the timing of the administration of medication may be critical to optimize the benefits and avoid neurotoxicity (e.g., early acute treatments targeted at amelioration of the neurotoxic cascade compared with subacute treatment that may promote regeneration and repair) (Kelen & Robertson, 2010).

8. Staggered design for a "off-label* combined therapy"

Given the urgency to find better therapies for HIE, and in the absence of ongoing clinical trials, we propose a model of **"off-label therapy"** based on hypothermia /antiepileptic drugs in combination with antioxidants, phospholipase A2 inhibitors, glutamate receptor antagonists and EPO using a staggered design in function of the intensity of the perinatal asphyxia and severity of the encephalopathy. However, we believe that a multicenter interventional randomized controlled pilot phase II clinical trial would be necessary.

Premises

- All drugs have been approved by the FDA.
- All the drugs have been currently available in infants without serious side effects.
- All drugs have demonstrated neuroprotection
- All drugs have and synergy with hypothermia, increasing the neuroprotective efficacy of therapeutic hypothermia in perinatal H-I brain injury – animals/clinicals- models.
- All drugs have different mechanisms and multiple potential modes of action.

Timing of the administration of medications:

At the time of initiation of hypothermia, stage 1 drugs would also be administered for at least 72 hours. If moderate / grave encephalopathy (according to clinical and encephalographic evaluation) persists, administration of the drugs assigned to the next stages would successively proceed. It should be noted that the use of hypothermia delays the pathophysiological response to H-I (postpones secondary energy failure) and therefore "widens" the therapeutic window. This phenomenon allows staggering the application of the adjuvant drugs, thus providing the physician with the necessary time to assess the clinical situation of the patient and decide whether to proceed to the next therapeutic phase.

8.1. Primary/latent phase (0–6 hours)

- **Anticonvulsants for neonatal seizures: PHENOBARBITAL:** PB remains the preferred drug for the treatment of seizures in neonates with HIE. **Ways of action:** anticonvulsant effects by increase GABA subtype A (GABAA)-receptor channel chloride currents, reduced cerebral metabolic demand, antioxidant effects and decreased cerebral edema. **Doses:** loading dose 20 mg/kg IV, followed by 3- 5 mg/kg/day, every 12 hours
- **Antioxidative drugs: Free radical nonenzymatic scavengers (N-acetylcysteine and Vitamin E)**
- **N-acetylcysteine (NAC). Ways of action:** act as an anti-oxidant and is also a scavenger of oxygen FRs, reduce oxidative stress and inflammation; prevents endotoxin-induced degeneration of OL progenitors and hypomyelination; attenuated activation of apoptotic proteases.

Continuous IV infusion, by simplified N-acetylcysteine dosing regimen – Standard preparation of IV-

Prepare standard solution IV: 50ml NAC 20% + 200ml SG5% (5% dextrose in water) = Solution NAC 40mg/ml. Doses: 150mg/kg (3,75ml x kg) IV loading dose administered over 1 hour, followed by an infusion of 12 mg/kg/h (7.2ml x Kg) for 24 hours. Continue with infusion of 150 mg/kg/24 h (3.75 ml x Kg) for 24 hours, to complete 72 hours

Or 72-hour oral-NAC: 140 mg/kg of oral NAC followed by 70 mg/kg every 4 h. for an additional 17 doses.

- **Vitamin E. Ways of action:** protects OL during this special vulnerability maturation period from the oxidative stress-induced death caused by glutathione depletion. It also ameliorates secondary mitochondrial failure. **Doses:** 50 U.I. orally, followed by 1 UI/kg/24 hours. Equivalency: 1 U.I. Vitamin E = 1 mg tocopherol acetate.

8.2. Secondary phase (6–72 hours)

- **Antiexcitatory drugs: Topiramate. Ways of action:** blocking sodium channels, high voltage-activated calcium currents, enhancing GABAinduced influx of chloride, and inhibiting kainite/ AMPA glutamate receptors. It also to blocks carbonic anhydrase isoenzymes and the mitochondrial permeability transition pore. **Doses:** 10 mg/kg/24 hous, divided into 2 doses, orally administered by orogastric tube.
- **Bioactive lipid mediators: Citicoline. Ways of action: Neuroprotective effects -** decrease of phospholipase A2 stimulation and hydroxyl radical generation; increase of glutathione synthesis and glutathione reductase activity, restoration of Na^+/K^+-ATPase activity, and antiglutamatergic effects. Roles in the generation of phospholipids involved in membrane formation and repair; prevention of fatty acid release, stimulation of phosphatidylcholine synthesis, preservation of cardiolipin and sphingomyelin levels. **Neuroregenerative effects -** related to inhibition of the glutamate induced apoptotic pathway, markedly reduces caspase-3 activation; increases neurogenesis in the SVZ and migration of neural progenitors

to the lesion area with increased neurogenesis also within the peri-infarct area. **Dosis:** 100 mg/kg/24 hours, every 12 h, intravenously during the first 3 days, and orally administered by orogastric tube thereafter for a total of 6 weeks.

8.3. Tertiary phase (>72 hours)

- **Anti-apoptotic drugs: Erythropoietin (EPO). Ways of action: Neuroprotective effects -** activation of anti-apoptotic, anti-oxidant and anti-inflammatory pathways as well as through the stimulation of angiogenic and neurogenic events; reduces the excitotoxic effect of glutamate and a glutamate receptor agonist (AMPA) on cortical neuron cultures. **Neuroregenerative effects** - prevents apoptosis induced by NMDA or by NO in neurons from cerebrocortical cultures and regulates neurogenesis. **Dosis:** Recombinant human EPO (rhEPO) given subcutaneously 400 U/kg daily for 5 days, thereafter 3 times a week.

9. Conclusion

Since hypoxic ischemic encephalopathy (HIE) is a potentially preventable cause of cerebral palsy (CP), much interest has been focused on prevention as well as research on neuroprotection therapies. Neuroprotective treatment for HIE in the clinical practice has been limited to the application of hypothermia in the newborn which is now accepted as a significant therapy, since so far no drug has shown any benefit when administered on its own. However, hypothermia alone may not provide complete protection or stimulate the repair that is necessary for a normal neurodevelopmental outcome. As we have described in this chapter, many mechanisms can be involved in the H-I process. It is therefore a reasonable assumption that the combination of several drugs involving two or more neuroprotective actions may exert synergistic effects by tackling several metabolic pathways at one time. We propose a model of **"off-label combined therapy"** based on hypothermia /antiepileptic drugs in combination with antioxidants, phospholipase A2 inhibitors, glutamate receptor antagonists or EPO using a staggered design in function of the intensity of the perinatal asphyxia and severity of the encephalopathy.

Note

*"Off- label" use is the use of already authorized pharmaceutical drugs for an unapproved indication or in an unapproved age group, unapproved dosage, or unapproved form of administration.

The term "*compassionate use*" (also known as *compassionate exemption* or *expanded access*) is used to define treatment options that allow the use of an unauthorised medicine. It may be applied to patients who cannot be treated satisfactorily by an authorised medicinal product or cannot enter a clinical trial. Although sometimes "off label use" has been considered a type of "compassionate use", those two concepts should not be mistaken since they have different legal requirements. For reference and discussion see:

- Randall S. Stafford. "Regulating Off-Label Drug Use — Rethinking the Role of the FDA". N Engl J Med 2008; 358: 1427–1429.
- Bombillar-Sáenz F.M. "The "Compassionate Exemption" In Spain: Not Asking For Compassion, Op. J., Vol. 2/2010, Paper n. 1, pp. 1 - 25, http://lider-lab.sssup.it/opinio, online publication July 2010.
- Guideline on compassionate use of medicinal products, pursuant to article 83 of regulation (EC) No 726/2004. Doc. Ref: EMEA/27170/2006.

Abbreviations

AA arachidonic acid
AEDs antiepileptic drugs
ALLO allourinol
AMPAalfa-amino-3-hydroxy-5-methyl-4-isoxazole-propionic acid
BBB blood-brain barrier
CBF cerebral blood flow
CBFV cerebral blood flow velocity
CAT catalase
CSF cerebrospinal fluid
CNS central nervous system
CP cerebral palsy
DC dendritic cell
DEX dexamethasone
EAA excitatory amino acid
EPO erythropoietin
FDA Food and Drug Administration (USA)
FFA free fatty acids
FR free radical
GPX glutathione peroxidase
H-I hypoxic–ischemic
HIE Hypoxic ischemic encephalopathy
IL interleukins
I/R ischemia/reperfusion
IV intravenous
IVH intraventricular hemorrhage
LPS lipopolysaccharide
MBP myelin basic protein
NAC N- acetyl-l-cysteine
NICU Neonatal Intensive Care Unit
NMDA N methyl-D-aspartate
NO nitric oxide
NOS Nitric oxide synthase
NT's neurotransmitters
OL oligodendrocyte
PAF platelet activating factor
PB Phenobarbital
PLA2 phospholipase A2
Pre-OL premyelinating oligodendrocyte
PUFAs polyunsaturated fatty acids
PVL periventricular leukomalacia
PWMD periventricular white matter damage
PCW postconceptional weeks
RNS reactive nitrogen species
ROS reactive oxygen species
SCs Stem cell
SNN selective neuronal necrosis
SOD superoxide dismutase
SPZ subplate zone
SVZ subventricular zone
TLR toll-like receptor
TPM Topiramate
VLBW very low birth weight
WM white matter

Author details

Mª Carmen Carrascosa-Romero[1] and Carlos de Cabo-de la Vega[2]

*Address all correspondence to: carlosd@sescam.jccm.es

1 Department of Neuropediatrics, Albacete General Hospital, Albacete, Spain

2 Neuropsychopharmacology Unit, Albacete General Hospital, Albacete, Spain

References

[1] Abbracchio MP, Cattabeni F. Brain adenosine receptors as targets for therapeutic intervention in neurodegenerative diseases. *Ann N Y Acad Sci*. 1999; 890:79-92.

[2] Adibhatla RM, Hatcher JF. Citicoline mechanisms and clinical efficacy in cerebral ischemia. *J Neurosci Res* 2002; 15: 133-139.

[3] Adibhatla RM, Hatcher JF. Citicoline decreases phospholipase A2 stimulation and hydroxyl radical generation in transient cerebral ischemia. *J Neurosci Res* 2003; 73: 308-315.

[4] Ahola T, Lapatto R, Raivio KO, Selander B, Stigson L, Jonsson B, et al. N-acetylcysteine does not prevent bronchopulmonary dysplasia in immature infants: a randomized controlled trial. J. Pediatr. 2003; 143:713–719.

[5] Aicardi, J. Overview: neonatal seizures. In: Engel, JJ.; Pedley, TA., editors. Epilepsy: a comprehensive textbook. Philadelphia: Lippincott Williams & Wilkins; 2008. p. 2283-2285.

[6] Aly H, Abd-Rabboh L, El-Dib M, Nawwar F, Hassan H, Aaref M, et al. Ascorbic acid combined with ibuprofen in hypoxic ischemic encephalopathy: a randomized controlled trial. *J Perinatol* 2009; 29:438-43.

[7] Ambalavanan N, Carlo WA, Shankaran S, Bann CM, Emrich SL, Higgins RD, et al. Predicting outcomes of neonates diagnosed with hypoxemic-ischemic encephalopathy. *Pediatrics*. 2006; 118:2084–93.

[8] American College of Obstetricians and Gynecologists and the American Academy of Pediatrics. Criteria required to define an acute intrapartum hypoxic event as sufficient to cause cerebral palsy. Neonatal encephalopathy and cerebral palsy: defining the pathogenesis and pathophysiology. ACOG 2003:73–80.

[9] Andón FT, Pazinatto DB, Fadeel B. Perinatal nanomedicine: dendrimer-based therapy for neuroinflammatory disorders. *Nanomedicine* (Lond). 2012; 7(9):1294.

[10] Ara J, De Montpellier S. Hypoxic-preconditioning enhances the regenerative capacity of neural stem/progenitors in subventricular zone of newborn piglet brain. . *Stem Cell Res.* 2013; 11(2):669-86.

[11] Arvin KL, Han BH, Du Y, Lin S-Z, Paul SM, Holtzman DM. Minocycline markedly protects the neonatal brain against hypoxic-ischemic injury. *Ann Neurol.* 2002; 52:54–61.

[12] Azzopardi DV, Strohm B, Edwards AD, Dyet L, Halliday HL, Juszczak E, et al. TOBY Study Group. Moderate hypothermia to treat perinatal asphyxial encephalopathy. N Engl J Med. 2009; 361:1349–58.

[13] Ashkenazi A, Dixit VM. Death receptors: signaling and modulation. Science 1998; 281:1305 – 8.

[14] Badawi N, Felix JF, Kurubczuk JJ, Dixon G, Watson L, Keogh JM, et al. Cerebral palsy following term newborn encephalopathy: a population-based study. *Dev Med Child Neurol.* 2005; 47:293–98.

[15] Back SA, Gan X, Li Y, Rosenberg PR, Volpe JJ. Maturation-dependent vulnerability of oligodendrocytes to oxidative stress-induced death caused by glutathione depletion. *J Neurosci.* 1998; 18(16):6241–6253.

[16] Ballin A, Brown EJ, Koren G, Zipursky A. Vitamin C-induced erythrocyte damage in premature infants. *J Pediatr* 1988; 113:114–20.

[17] Barks JD.Current controversies in hypothermic neuroprotection. Seminars in Fetal and Neonatal Medicine. 2008; 13 (1): 30–34.

[18] Barks JD, Liu YQ, Shangguan Y, Silverstein FS.: Phenobarbital augments hypothermic neuroprotection. Pediatr Res. 2010; 67(5):532-7.

[19] Barks JD, Silverstein FS. Excitatory amino acids contribute to the pathogenesis of perinatal hypoxic–ischemic brain injury. *Brain Pathol* 1992; 2: 235–43.

[20] Barinaga, M. (1998) Stroke-damaged neurons may commit cellular suicide. *Science* 281, 1302–1303.

[21] Baud O. Antenatal corticosteroid therapy: benefits and risks. Acta Paediatr Suppl 2004; 93:6–10.

[22] Ben-Ari Y.: Basic developmental rules and their implications for epilepsy in the immature brain. *Epileptic Disord.* 2006; 8(2):91-102.

[23] Benders MJ, Bos AF, Rademaker CM, Rijken M, Torrance HL, Groenendaal F, et al. Early postnatal allopurinol does not improve short term outcome after severe birth asphyxia. Arch Dis Child Fetal Neonatal Ed 2006, 91:163-165.

[24] Bebarta VS, Kao L, Froberg B, Clark RF, Lavonas E, Qi M, et al: A multicenter comparison of the safety of oral versus intravenous acetylcysteine for treatment of acetaminophen overdose. *Clin Toxicol (Phila)*. 2010; 48(5):424-30.

[25] Bennet L, Tan S, Van den Heuij L, Derrick M, Groenendaal F, van Bel F, et al. Cell therapy for neonatal hypoxia-ischemia and cerebral palsy. *Ann Neurol*. 2012; 71(5): 589-600.

[26] Biban P, Filipovic-Grcic B, Biarent D, Manzoni P. International Liaison Committee on Resuscitation (ILCOR); European Resuscitation Council (ERC); American Heart Association (AHA); American Academy of Pediatrics (AAP). New cardiopulmonary resuscitation guidelines 2010: managing the newly born in delivery room. *Early Hum Dev*. 2011 Mar;87 Suppl 1:S9-11.

[27] Binder DK, Scharfman HE. Brain-derived neurotrophic factor. *Growth Factors*. 2004; 22: 123–131.

[28] Bittigau P, Sifringer M, Pohl D. Apoptotic neurodegeneration following trauma is markedly enhanced in the immature brain. *Ann Neurol* 1999; 45: 724–35.

[29] Bittigau P, Sifringer M, Ikonomidou C. Antiepileptic drugs and apoptosis in the developing brain. Ann NY Acad Sci 2003; 993:103–114.

[30] Blair E, Stanley FJ: Intrapartum asphyxia: A rare cause of cerebral palsy. *The Journal of Pediatrics*, Volume 112, Issue 4: 515-519.

[31] Blaschke AJ, Staley K, Chun J. Widespread programmed cell death in proliferative and postmitotic regions of the fetal cerebral cortex. Development 1996; 122:1165–74.

[32] Brener P, Ballardo M, Mariani G, Ceriani Cernadas JM. Medication error in an extremely low birth weight infant: paracetamol overdose. Arch Argent Pediatr. 2013; 111(1):53-5.

[33] Buettner GR, Jurkiewicz BA. Catalytic metals, ascorbate and free radicals: combinations to avoid. Radiat Res 1996; 145:532–41.

[34] Buller KM, Carty ML, Reinebrant HE, Wixey JA. Minocycline: a neuroprotective agent for hypoxic-ischemic brain injury in the neonate? *J Neurosci Res*. 2009; 87:599–608.

[35] Buonocore G, Perrone S, Turrisi G, Kramer BW, Balduini W. New pharmacological approaches in infants with hypoxic-ischemic encephalopathy. Curr Pharm Des. 2012; 18 (21):3086-100.

[36] Calabresi P, Cupini LM, Centonze D, Pisani F, Bernardi G. Antiepileptic drugs as a possible neuroprotective strategy in brain ischemia. *Ann Neurol* 2003; 53: 693–703.

[37] Carrascosa MC, Martínez Gutiérrez A, Onsurbe I, Vázquez MS, Catalan B, Tébar R.: "Neonatal convulsions in health care. Incidence, etiology and clinical aspects". *Rev Neurol* 1996: 24(134):1258-62.

[38] Carrascosa-Romero MC, Suela J, Alfaro-Ponce B, Cepillo-Boluda AJ. Ictiosis ligada al cromosoma X asociada a epilepsia, hiperactividad, autismo y retraso mental, por microdeleción Xp22.31.*Rev Neurol* 2012; 54: 241-8.

[39] Cayre M, Canoll P, Goldman JE. Cell migration in the normal and pathological postnatal mammalian brain. Prog Neurobiol. 2009; 88(1):41-63.

[40] Chakkarapani E, Thoresen M, Liu X, Walloe L, Dingley J. Xenon offers stable haemodynamics independent of induced hypothermia after hypoxia-ischaemia in newborn pigs. Intensive Care Med. 2012; 38(2):316-23.

[41] Chen J, Tu Y, Moon C, Matarazzo V, Palmer A, Ronnett G. The localization of neuronal nitric oxide synthase may influence its role in neuronal precursor proliferation and synaptic maintenance. Dev Biol 2004; 269:165-82.

[42] Cho H.J., Kim Y.J. Efficacy and safety of oral citicoline in acute ischemic stroke: drug surveillance study in 4,191 cases.Methods Find Exp Clin Pharmacol. 2009; 31: 171–176.

[43] Choi DW, Rothman SW. The role of glutamate neurotoxicity in hypoxic–ischemic neuronal death. Annu Rev Neurosci. 1990;13:171–82.

[44] Cilio M, Ferriero D. Synergistic neuroprotective therapies with hypothermia. Semin Fetal Neonatal Med 2010; 15(5):293-8.

[45] Clarke PGH, Puyal J, Vaslin A, Ginet V, Truttmann A: Multiple types of programmed cell death and their relevance to perinatal brain damage.In perinatal brain damage: from pathogenesis to neuroprotection, chp 3. Edited by LA Ramenghi, P Evrard, and E. Mercuri. Mariani Foundation Paediatric Neurology Series 19, John Libbey Eurotext, Montrouge, France, 2008, pp. 23–35.

[46] Clinical Practice Guidelines on Magnesium Sulphate Prior to Preterm Birth for Neuroprotection of the Fetus, Infant and Child. [Last cited on 15 Spe 2011]. Available from: http://www.nhmrc.gov.au/guidelines/publications/cp128.

[47] Committee opinion no. 573: magnesium sulfate use in obstetrics. Obstet Gynecol. 2013 Sep; 122(3):727-8.

[48] Costa C, Martella G, Picconi B, Prosperetti C, Pisani A, Di Filippo M, et al. Multiple mechanisms underlying the neuroprotective effects of antiepileptic drugs against in vitro ischemia. Stroke 2006; 37:1319–1326.

[49] Covey M, Murphy M, Hobbs C, Smith R, Oorschot D. Effect of the mitochondrial antioxidant, Mito Vitamin E, on hypoxic-ischemic striatal injury in neonatal rats: a dose-response and stereological study. Exp Neurol. 2006 Jun; 199(2):513-9.

[50] Crowther CA, Hiller JE, Doyle LW, Haslam RR. Effect of magnesium sulfate given for neuroprotection before preterm birth: a randomized controlled trial. JAMA. 2003; 290(20):2669-76.

[51] Crunkhorn S. Neurological disorders: Nanoparticle opens door to cerebral palsy treatment. Nat Rev Drug Discov. 2012; 11(6):440-1.

[52] Dammann O, Ferriero D, Gressens P: Neonatal Encephalopathy or Hypoxic-Ischemic Encephalopathy? Appropriate Terminology Matters. Pediatric Research. 2011; 70 (1): 1-2.

[53] Daneyemez M, Kurt E, Cosar A, Yuce E, Ide T. Methylprednisolone and vitamin E therapy in perinatal hypoxic-ischemic brain damage in rats. Neuroscience 1999; 92: 693-7.

[54] Dávalos A, Alvarez-Sabín J, Castillo J, Díez-Tejedor E, Ferro J, Martínez-Vila E, et al. International Citicoline Trial on acUte Stroke (ICTUS) trial investigators. Citicoline in the treatment of acute ischaemic stroke: an international, randomised, multicentre, placebo-controlled study (ICTUS trial). Lancet. 2012 Jul 28; 380(9839):349-57.

[55] Davalos A, Castillo J, Alvarez-Sabin J, et al. Oral citicoline in acute ischemic stroke: an individual patient data pooling analysis of clinical trials. *Stroke* 2002; 33: 2850–57.

[56] De Cabo-de la Vega C, Villanueva-Hernandez P, Prieto-Martin A. The neurochemistry of epilepsy, inhibitory neurotransmission and experimental models: new perspectives. *Rev Neurol*. 2006;42(3):159-68.

[57] Degos V, Loron G, Mantz J, Gressens P: Neuroprotective Strategies for the Neonatal Brain. Anesth Analg 2008; 106:1670 –80.

[58] Derks JB, Oudijk MA, Rosén KG, Thakor AS, Visser GHA, van Bel F,et al. Allopurinol maintains umbilical blood flow following repeated episodes of ischaemia-reperfusion in fetal sheep during late gestation [abstract]. J Soc Gynecol Invest 2006, 13:s35.

[59] De Vries LS, Jongmans MJ. Long-term outcome after neonatal hypoxic-ischemic encephalopathy. Arch Dis Child Fetal Neonatal Ed 2010; 95:F220–4.

[60] Diederich K, Frauenknecht K, Minnerup J, Schneider BK, Schmidt A, Altach E. Citicoline Enhances Neuroregenerative Processes After Experimental Stroke in Rats. *Stroke*. 2012; 43:1931-1940.

[61] Digicaylioglu M., Lipton S.A.Erythropoietin-mediated neuroprotection involves cross-talk between Jak2 and NF-kappaB signalling cascades. *Nature*.2001; 412: 641–647.

[62] Dingley J, Tooley J, Porter H, Thoresen M. Xenon provides short-term neuroprotection in neonatal rats when administered after hypoxia-ischemia. Stroke 2006; 37:501–6.

[63] Distefano G, Praticò AD. Actualities on molecular pathogenesis and repairing processes of cerebral damage in perinatal hypoxic-ischemic encephalopathy. Ital *J Pediatr*. 2010 Sep 16;36:63.

[64] Dommergues MA, Gallego J, Evrard P, Gressens P. Iron supplementation aggravates periventricular cystic white matter lesions in newborn mice. Eur J Paediatr Neurol 1998; 2:313–8.

[65] Dommergues MA, Plaisant F, Verney C, Gressens P. Early microglial activation following neonatal excitotoxic brain damage in mice: a potential target for neuroprotection. Neuroscience. 2003; 121:619–628.

[66] Donega V, van Velthoven CT, Nijboer CH, van Bel F, Kas MJ, Kavelaars A, et al. Intranasal mesenchymal stem cell treatment for neonatal brain damage: long-term cognitive and sensorimotor improvement. *PLoS One*. 2013; 8(1):e512-53.

[67] Douglas-Escobar M, Rossignol C, Steindler D, Zheng T, Weiss MD. Neurotrophin-induced migration and neuronal differentiation of multipotent astrocytic stem cells in vitro. PLoS One. 2012; 7(12):e51706.

[68] Doyle LW, Crowther CA, Middleton P, Marret S, Rouse D. Magnesium sulphate for women at risk of preterm birth for neuroprotection of the fetus. Cochrane Database Syst Rev. 2009 Jan 21; (1):CD004661.

[69] Dzhala VI, Brumback AC, Staley KJ. Bumetanide enhances phenobarbital efficacy in a neonatal seizure model. *Ann Neurol*. 2008; 63(2):222-35.),

[70] Edwards AD, Brocklehurst P, Gunn AJ, Halliday H, Juszczak E, Levene M, et al. Neurological outcomes at 18 months of age after moderate hypothermia for perinatal hypoxic ischaemic encephalopathy: synthesis and meta-analysis of trial data. BMJ 2010; 340:c363.

[71] Ehrenreich H, Weissenborn K, Prange H, Schneider D, Weimar C, Wartenberg K., et al. Recombinant human erythropoietin in the treatment of acute ischemic stroke.*Stroke*. 2009; 40: e647-e656.

[72] Elmahdy H, El-Mashad AR, El-Bahrawy H, El-Gohary T, El-Barbary A, Aly H. Human recombinant erythropoietin in asphyxia neonatorum: pilot trial. Pediatrics 2010; 125:e1135-42.

[73] Elterman RD, Glauser TA, Wyllie E, Reife R, Wu SC, Pledger G. A double-blind, randomized trial of topiramate as adjunctive therapy for partial-onset seizures in children. Topiramate YP Study Group. Neurology 1999; 52:1338–44.

[74] Esplugues JV. NO as a signalling molecule in the nervous system. Br J Pharmacol 2002; 135: 1079 – 95.

[75] Evans DJ, Levene MI, and Tsakmakis M. Anticonvulsants for preventing mortality and morbidity in full term newborns with perinatal asphyxia. *Cochrane Database of Systematic Reviews*, 2007; no. 3, Article ID CD001240,

[76] Fabian RH, Perez-Polo J, Kent TA. Perivascular nitric oxide and superoxide in neonatal cerebral hypoxia-ischemia. Am J Physiol Heart Circ Physiol 2010; 295: h1809-14.

[77] Fan LW, Lin S, Pang Y, Rhodes PG, Cai Z. Minocycline attenuates hypoxia-ischemia-induced neurological dysfunctionand brain injury in the juvenile rat. Eur J Neurosci 2006; 24:341–50.

[78] Fan LW, Pang Y, Lin SY, Tien LT, Ma TG, Rhodes PG, et al. Minocycline reduces lipopolysaccharide-induced neurological dysfunction and brain injury in the neonatal rat.*J Neurosci Res.* 2005; 82:71–82.

[79] Fatemi A, Wilson MA, Johnston MV. Hypoxic-Ischemic Encephalopathy in the Term Infant. Clin Perinatol. 2009; 36 835–858.

[80] Fauchere J, Dame C, Vonthein R, Koller B, Arri S, Wolf M, et al. An approach to using recombinant erythropoietin for neuroprotection in very preterm infants. Pediatrics 2008; 122:375-82.

[81] Fehlings, D; Hunt C. & Rosenbaum P. (2007). Cerebral Palsy. In: A Comprehensive Guide to Intellectual & Developmental Disabilities. Brown, I. & Percy, M.287- 295 ISBN-13: 978- 1-55766-700-7 Brookes Publishing Co. Baltimore.

[82] Ferriero DM, Holtzman DM, Black SM, Sheldon RA. Neonatal mice lacking neuronal nitric oxide synthase are less vulnerable to hypoxic-ischemic injury. Neurobiol Dis 1996; 3:64– 71. 60.

[83] Fiedorowicz M, Makarewicz D, Stańczak-Mrozek KI, Grieb P. CDP-choline (citicoline) attenuates brain damage in a rat model of birth asphyxia. Acta Neurobiol Exp 2008, 68: 389-397.

[84] Filippi L, Fiorini P, Daniotti M, Catarzi S, Savelli S, Fonda C, et al. Safety and efficacy of topiramate in neonates with hypoxic ischemic encephalopathy treated with hypothermia (NeoNATI). BMC Pediatr. 2012; 5 (12):144.

[85] Filippi L, Poggi C, la Marca G, Furlanetto S, Fiorini P, Cavallaro G, et al. Oral topiramate in neonates with hypoxic ischemic encephalopathy treated with hypothermia: a safety study. *J Pediatr.* 2010;157(3):361-6.

[86] FineSmith RB, Roche K, Yellin PB, Walsh KK, Shen C, Zeglis M, et al. Effect of magnesium sulfate on the development of cystic periventricular leukomalacia in preterm infants. Am *J Perinatol.* 1997; 14(5):303-7.

[87] Follett PL, Deng W, Dai W, Talos DM, Massillon LJ, Rosenberg PA, Volpe JJ, Jensen FE. Glutamate receptor-mediated oligodendrocyte toxicity in periventricular leukomalacia: a protective role for topiramate. J Neurosci 2004; 24: 4412–20.

[88] Follett PL, Rosenberg PA, Volpe JJ, et al. NBQX attenuates excitotoxic injury in developing white matter. J Neurosci 2000; 20:9235–41.

[89] Fulia F, Gitto E, Cuzzocrea S, Reiter R, Dugo L, Gitto P, et al. Increased levels of malondialdehyde and nitrite/nitrate in the blood of asphyxiated newborns: reduction by melatonin. J Pineal Res 2005; 31:343-9.

[90] Fujioka H, Shintaku H, Nakanishi H, Kim TJ, Kusuda S, Yamano T. Biopterin in the acute phase of hypoxia-ischemia in a neonatal pig model. *Brain Dev.* 2008; 30(1):1-6

[91] Glier C, Dzietko M, Bittigau P, Jarosz B, Korobowicz E, Ikonomidou C. Therapeutic doses of topiramate are not toxic to the developing rat brain. Exp Neurol. 2004; 187:403–9.

[92] Gitto E, Karbownik M, Reiter R, Tan D, Cuzzocrea S, Chiurazzi P, et al. Effects of melatonin treatment in septic newborns. Pediatr Res. 2001; 50: 756-60.

[93] Gitto E, Pellegrino S, Gitto P, Barberi I, Reiter RJ. Oxidative stress of the newborn in the pre- and postnatal period and the clinical utility of melatonin. J. Pineal Res. 2009; 46:128–139.).

[94] Gitto E, Reiter RJ, Cordaro SP, La Rosa M, Chiurazzi P, Trimarchi G, et al. Oxidative and inflammatory parameters in respiratory distress syndrome of preterm newborns: beneficial effects of melatonin. Am *J Perinatol* 2004; 21(4):209–16.

[95] Gitto E, Reiter RJ, Sabatino G, Buonocore G, Romeo C, Gitto P, et al. Correlation among cytokines, bronchopulmonary dysplasia and modality of ventilation in preterm newborns: improvement with melatonin treatment. J Pineal Res 2005; 39(3):287–93.].

[96] Gluckman PD, Wyatt J, Azzopardi DV, Ballard R, Edwards, Ferriero DM, et al. Selective head cooling with mild systemic hypothermia after neonatal encephalopathy: Multicenter Randomised Trial. Lancet.2005; 365:663–70.

[97] Gonzalez FF, Ferriero DM: Neuroprotection in the Newborn Infant. Clin Perinatol. 2009; 36 859–880.

[98] Gonzalez FF, Miller SP. Does perinatal asphyxia impair cognitive function without cerebral palsy? Arch Dis Child Fetal Neonatal Ed 2006; 91:454–9.

[99] Green JL, Heard KJ, Reynolds KM, Albert D. Oral and Intravenous Acetylcysteine for Treatment of Acetaminophen Toxicity: A Systematic Review and Meta-analysis. West J Emerg Med. 2013; 14(3):218-26.

[100] Group, EAIS, Effect of a novel free radical scavenger, edaravone (MCI-186), on acute brain infarction. Randomized, placebo-controlled, double-blind study at multicenters. Cerebrovasc Dis 2003; 15: 222–229.

[101] Grow J, Barks JD: Pathogenesis of hypoxic–ischemic cerebral injury in the term infant: current concepts. Clin. Perinatol. 2002; 29, 585–602.

[102] Gunes T, Ozturk MA, Koklu E, Kose K, Gunes I. Effect of allopurinol supplementation on nitric oxide levels in asphyxiated newborns. Pediatr Neurol 2007; 36:17-24.

[103] Gunn AJ, Bennet L. Fetal Hypoxia Insults and Patterns of Brain Injury: Insights from Animal Models. Clin Perinatol. 2009; 36 579–593.

[104] Gunn, A.J, Gunn TR, de Haan HH, Williams CE, Gluckman PD. Dramatic neuronal rescue with prolonged selective head cooling after ischemia in fetal lambs. J. Clin. Invest. 1997; 99: 248-256.

[105] Hall RT, Hall FK, and Daily DK. High dose phenobarbital therapy in term newborn with severe perinatal asphyxia: a randomized, prospective study with three-year follow-up J. Pediat.1998; 132 (2): 345–348.

[106] Hagberg H, Gilland E, Bona E, Hanson LA, Hahn-Zoric M, Blennow M, et al. Enhanced expression of interleukin (IL)-1 and IL-6 messenger RNA and bioactive protein after hypoxia-ischemia in neonatal rats. *Pediatr Res.* 1996; 40:603–609.

[107] Hammerman C, Kaplan M: Ischemia and reperfusion injury. The ultimate pathophysiologic paradox. Clin Perinatol 1998, 25:757-777.

[108] Hanrahan JD, Cox IJ, Azzopardi D, Cowan FM, Sargentoni J, Bell JD, et al. Relation between proton magnetic resonance spectroscopy within 18 hours of birth asphyxia and neurodevelopment at 1 year of age. *Dev Med Child Neurol* 1999; 41:76 – 82.

[109] Haynes RL, Baud O, Li J Kinney HC, Volpe JJ, Folkerth DR. Oxidative and nitrative injury in periventricular leukomalacia: a review. *Brain Pathol* 2005; 15:225–33.

[110] Haynes RL, Folkerth RD, Keefe RJ, Sung I, Swzeda LI, Rosenberg PA, et al. Nitrosative and oxidative injury to premyelinating oligodendrocytes in periventricular leukomalacia. J. Neuropathol. Exp. Neurol. 2003; 62:441–450.

[111] Hazinski MF, Nolan JP, Billi JE, Böttiger BW, Bossaert L, de Caen AR, et al. Part 1: Executive summary: 2010. International Consensus on Cardiopulmonary Resuscitation and Emergency Cardiovascular Care Science With Treatment Recommendations. Circulation. 2010; 122(16 Suppl 2):S250-75.

[112] Holliday M. A. Body composition and energy needs during growth, in *Human Growth: A Comprehensive Treatise*, (1986). Vol. 2, 2nd Edn. eds Falkner F., Tanner J. M., editors. (New York, NY: Plenum); 101–117.

[113] Hossain MA. Molecular mediators of hypoxic-ischemic injury and implications for epilepsy in the developing brain. Epilepsy Behav. 2005 Sep; 7(2):204-13.

[114] Hossain MA, Fielding KE, Trescher WH, Ho T, Wilson MA, Laterra J. Human FGF-1 gene delivery protects against quinolinate- induced striatal and hippocampal injury in neonatal rats. Eur J Neurosci 1998; 10:2490–9.

[115] Hu BR, Liu CL, Ouyang Y, Blomgren K, Siesjo BK. Involvement of caspase-3 in cell death after hypoxia–ischemia declines during brain maturation. J Cereb Blood Flow Metab 2000; 20: 1294–300.

[116] Huang J, Agus DB, Winfree CJ, Kiss S, Mack WJ, McTaggart RA, Choudhri TF, Kim LJ, Mocco J, Pinsky DJ, Fox WD, Israel RJ, Boyd TA, Golde DW, Connolly ES Jr.: Dehydroascorbic acid, a blood–brain barrier transportable form of vitamin C, mediates

potent cerebroprotection in experimental stroke. *Proc Natl Acad Sci U S A*. 2001; 98(20):11720-4.

[117] Hudome S, Palmer C, Roberts RL, Mauger D, Housman C, Towfighi J. The role of neutrophils in the production of hypoxic-ischemic brain injury in the neonatal rat.*Pediatr Res.* 1997; 41: 607–616.

[118] Hurtado O, Moro MA, Cardenas A, Sanchez V, Fernandez-Tome P, Leza JC, et al. Neuroprotection afforded by prior citicoline administration in experimental brain ischemia: effects on glutamate transport. Neurobiol Dis 2005; 18: 336-345.

[119] Ikonomidou C, Turski L. Why did NMDA receptor antagonists fail clinical trials for stroke and traumatic brain injury? Lancet Neurol 2002; 1:383–6.

[120] Inder TE, Volpe JJ: Mechanisms of perinatal brain injury. Semin Neonatol.2000, 5:3-16.

[121] Itoh T, Satou T, Nishida S, Tsubaki M, Hashimoto S, Ito H. The novel free radical scavenger, edaravone, increases neural stem cell number around the area of damage following rat traumatic brain injury. Neurotox Res. 2009; 16(4):378-89.

[122] Jackson T, Xu A, Vita J, Keaney JJ. Ascorbate prevents the interaction of superoxide and nitric oxide only at very high physiological concentrations. Circ Res 1998; 83:916-22.

[123] Jacobs SE, Berg M, Hunt R, Tarnow-Mordi WO, Inder TE, Davis PG. Cooling for newborns with hypoxic ischaemic encephalopathy. Cochrane Database Syst Rev. 2013 Jan 31; 1:CD003311

[124] Jacobs SE, Morley CJ, Inder TE, et al. Whole-body hypothermia for term and near-term newborns with hypoxic-ischemic encephalopathy: a randomized controlled trial. Arch Pediatr Adolesc Med 2011; 165:692–700.

[125] Jaffer H, Morris VB, Stewart D, and Labhasetwar V: Advances in Stroke Therapy. Drug Deliv Transl Res. 2011; 1(6): 409–419.

[126] Jan J, O'Donnell M. Use of melatonin in the treatment of paediatric sleep disorders. J Pineal Res 1996; 21: 193-9.

[127] Jatana M, Singh I, Singh A, Jenkins D. Combination of systemic hypothermia and N-acetylcysteine attenuates hypoxic-ischemic brain injury in neonatal rats. Pediatr Res 2006; 59:684-9.

[128] Ji H, Tan S, Igarashi J, Li H, Derrick M, Martasek P, et al. Selective neuronal nitric oxide synthase inhibitors and the prevention of cerebral palsy. *Ann Neurol* 2009; 65:209-17.

[129] Johnson MT, McCammon CA, Mullins ME, Halcomb SE. Evaluation of a simplified N-acetylcysteine dosing regimen for the treatment of acetaminophen toxicity. Ann Pharmacother. 2011; 45(6):713-20.

[130] Johnston MV. Excitotoxicity in neonatal hypoxia. Ment Retard Dev Disabil Res Rev 2001; 7:229–34.

[131] Johnston MV, Trescher WH, Ishida A, Nakajima W. Neurobiology of hypoxic–ischemic injury in the developing brain. Pediatr Res 2001; 49:735–41.

[132] Judaš M, Sedmak G, Pletikos M, Jovanov-Milošević N. Populations of subplate and interstitial neurons in fetal and adult human telencephalon. J Anat. 2010; 217(4): 381-99.

[133] Juul S. Erythropoietin in the central nervous system, and its use to prevent hypoxic-ischemic brain damage. *Acta Paediatr Suppl.* 2002; 91: 36–42.

[134] Juul S. Neuroprotective role of erythropoietin in neonates. J Matern Fetal Neonatal Med. 2012 Oct;25 Suppl 4:105-7.

[135] Juul SE, McPherson, RJ, MR, Bauer LA, Ledbetter KJ, Gleason CA, Mayock DE. A phase I/II trial of high-dose erythropoietin in extremely low birth weight infants: pharmacokinetics and safety. Pediatrics 2008; 122:383-91.

[136] Kaandorp JJ, BendersMJ, Rademaker CM, Torrance HL, Martijn A Oudijk MA et al: Antenatal allopurinol for reduction of birth asphyxia induced brain damage (ALLO-Trial);a randomized double blind placebo controlled multicenter study. BMC Pregnancy and Childbirth 2010, 10:8. http://www.biomedcentral.com/1471-2393/10/8.

[137] Kahle KT, Barnett SM, Sassower KC, Staley KJ. Decreased seizure activity in a human neonate treated with bumetanide, an inhibitor of the NaC-KC- 2Cl- cotransporter NKCC1. J. Child Neurol.. 2009; 24:572–576.

[138] Kang SK, Lee DH, Bae YC, Kim HK, Baik SY, Jung JS. Improvement of neurological deficits by intracerebral transplantation of human adipose tissue-derived stromal cells after cerebral ischemia in rats. Exp Neurol 2003, 183: 355-366.

[139] Kannan S, Dai H, Navath RS, Balakrishnan B, Jyoti A, Janisse J, et al. Dendrimer-based postnatal therapy for neuroinflammation and cerebral palsy in a rabbit model. Sci Transl Med. 2012; 4(130):130-46.

[140] Kattwinkel J, Perlman JM, Aziz K, et al. Neonatal resuscitation: 2010 American Heart Association Guidelines for cardiopulmonary resuscitation and emergency cardiovascular care. Pediatrics 2010; 126(5):e1400–13.

[141] Keda T, Xia YX, Kaneko M, Sameshima H, Ikenoue T. Effect of the free radical scavenger, 3-methyl-1-phenyl-2-pyrazolin-5-one (MCI-186), on hypoxia-ischemia-induced brain injury in neonatal rats. *Neuroscience Letters.* 2002; 329(1):33–36.].

[142] Kelen D, Robertson NJ. Experimental treatments for hypoxic ischaemic encephalopathy. Early Hum Dev 2010; 86:369-77.

[143] Khanna A, Walcott BP, Kahle KT.: Limitations of Current GABA Agonists in Neonatal Seizures: Toward GABA Modulation Via the Targeting of Neuronal Cl(-) Transport. *Front Neurol.* 2013; 4:78.

[144] Kidd GA, Hong H, Majid A, Kaufman DI, Chen AF. Inhibition of brain GTP cyclohydrolase I and tetrahydrobiopterin attenuates cerebral infarction via reducing inducible NO synthase and peroxynitrite in ischemic stroke. Stroke 2005; 36:2705–11.

[145] Koike M, Shibata M, Tadakoshi M, Gotoh K, Komatsu M, Waguri S, et al. Inhibition of autophagy prevents hippocampal pyramidal neuron death after hypoxic-ischemic injury. Am J Pathol 2008, 172:454–469.

[146] Kostovic I, Judas M. Prolonged coexistence of transient and permanent circuitry elements in the developing cerebral cortex of fetuses and preterm infants. *Dev Med Child Neurol.* 2006 May; 48(5):388-93.

[147] Kroncke KD. Nitrosative stress and transcription. Biol Chem 2003; 384:1365–77.

[148] Knudsen TT, Thorsen S, Jensen SA, Dalhoff K, Schmidt LE, Becker U, et al. Effect of intravenous N-acetylcysteine infusion on haemostatic parameters in healthy subjects. Gut. 2005; 54(4):515-21.

[149] Kumral A, Tüzün F, Oner MG, Genç S, Duman N, Ozkan H. Erythropoietin in neonatal brain protection: the past, the present and the future. Brain Dev. 2011; 33(8): 632-43.

[150] Kurinczuk JJ, White-Koning M, Badawi N. Epidemiology of neonatal encephalopathy and hypoxic-ischaemic encephalopathy. Early Hum Dev. 2010; 86:329–38.

[151] Labiche LA & Grotta JC: Clinical Trials for Cytoprotection In Stroke.*NeuroRx; 2004; 1 (1): 46-70.*

[152] Lakic N, Surlan K, Jerin A, Meglic B, Curk N, Bunc M. Importance of erythropoietin in brain protection after cardiac surgery: a pilot study. Heart Surg Forum 2010; 13:e185-9

[153] Lechpammer M, Manning SM, Samonte F, Nelligan J, Sabo E, Talos DM, et al. Minocycline treatment following hypoxic-ischemic injury attenuates white matter injury in a rodent model of periventricular leukomalacia. *Neuropathol Appl Neurobiol.* 2008; 34: 379–393.

[154] Lee TF, Tymafichuk CN, Bigam DL, Cheung PY. Effects of postresuscitation N-acetylcysteine on cerebral free radical production and perfusion during reoxygenation of hypoxic newborn piglets. Pediatr Res. 2008; 64(3):256-61.

[155] Leonardo CC and Pennypacker KR: Neuroinflammation and MMPs: potential therapeutic targets in neonatal hypoxic-ischemic injury. Journal of Neuroinflammation 2009, 6-13.

[156] Levene MI. Cool treatment for birth asphyxia, but what's next? Archives of Disease in Childhood: Fetal and Neonatal Edition, 2010; 95 (3): F154–F157.

[157] Levene MI, Gibson NA, Fenton AC, Wyatt JS, Penrice J, Edwards AD, et al. The use of a calcium-channel blocker, nicardipine, for severely asphyxiated newborn infants. *Dev Med Child Neurol* 1990; 32:567– 74.

[158] Leviton A, Allred EN, Kuban KC, Hecht JL. Onderdonk AB, O'shea TM, Paneth N. Microbiologic and histologic characteristics of the extremely preterm infant's placenta predict white matter damage and later cerebral palsy. the ELGAN study. Pediatr Res. 2010; 67:95–101.

[159] Li J, Ramenaden ER, Peng J, Koito H, Volpe JJ, Rosenberg PA. Tumor necrosis factor alpha mediates lipopolysaccharide-induced microglial toxicity to developing oligodendrocytes when astrocytes are present. J. Neurosci. 2008; 28:5321–5330.

[160] Lippman-Bell JJ, Rakhade SN, Klein PM, Obeid M, Jackson MC, Joseph A, Jensen FE. AMPA Receptor antagonist NBQX attenuates later-life epileptic seizures and autistic-like social deficits following neonatal seizures. Epilepsia 2013. Article first published online: 1 OCT 2013. DOI: 10.1111/epi.12378.

[161] Lipton ST. Neuronal protection and destruction by NO. Cell Death Differ 1995;6:943–51.

[162] Liu J, Lee T, Chen C, Bagim D, Cheung P. N-acetylcysteine improves hemodynamics and reduces oxidative stress in the brains of newborn piglets with hypoxia-reoxygenation injury. J Neurotrauma 2010; 27: 1865-73.

[163] Liu W, Khatibi N, Sridharan A, Zhang JH. Application of medical gases in the field of neurobiology. Med Gas Res. 2011; 1(1):13.

[164] Liu X-H, Eun B-L, Silverstein S, Barks JD. The platelet-activating factor antagonist BN 52021 attenuates hypoxic-ischemic brain injury in the immature rat. *Pediatr Res.* 1996; 6: 797–803.

[165] Lo EH. A new penumbra: transitioning from injury into repair after stroke. Nat Med. 2008; 14: 497-500.

[166] Lorek A, Takei Y, Cady EB, Wyatt JS, Penrice J, Edwards AD, et al. Delayed ("secondary") cerebral energy failure after acute hypoxia-ischemia in the newborn piglet: Continuous 48-hour studies by phosphorus magnetic resonance spectroscopy. *Pediatr Res.* 1994; 36: 699-706.

[167] Lujan R, Shigemoto R, Lopez-Bendito G. Glutamate and GABA receptor signalling in the developing brain. Neuroscience 2005; 130:567–80.

[168] Llado J, Haenggeli C, Maragakis NJ, Snyder EY, Rothstein JD. Neural stem cells protect against glutamate-induced excitotoxicity and promote survival of injured motor

neurons through the secretion of neurotrophic factors. *Mol Cell Neurosci.* 2004; 27: 322–331].

[169] Madsen JT, Jansen P, Hesslinger C, Meyer M, Zimmer J, Gramsbergen JB. Tetrahydrobiopterin precursor sepiapterin provides protection against neurotoxicity of 1-methyl-4-phenylpyridinium in nigral slice cultures. J Neurochem 2003; 85:214-23.

[170] Magee L, Sawchuck D, Synnes A, von Dadelszen P. SOGC Clinical Practice Guideline. Magnesium sulphate for fetal neuroprotection. J Obstet Gynaecol Can. 2011; 33:516–29.

[171] Maiese K, Chong ZZ, Li F, Y. C. Shang YC. Erythropoietin: elucidating new cellular targets that broaden therapeutic strategies. *Progress in Neurobiology*. 2008; 85 (2): 194–213.

[172] Manning SM, Boll G, Fitzgerald E, Selip D, Volpe JJ, Jensen F. The clinically available NMDA receptor antagonist, memantine, exhibits relative safety in the developing rat brain. *J Neurosci.* 2010;

[173] Maragakis NJ, Rothstein JD. Mechanisms of Disease: astrocytes in neurodegenerative disease. Nat. Clin. Pract. Neurol. 2006; 2:679–689

[174] Marks KA, Mallard CE, Roberts I, et al. Nitric oxide synthase inhibition attenuates delayed vasodilation and increases injury after cerebral ischemia in fetal sheep. Pediatr Res 1996; 40(2):185–91.

[175] Marlow N, Rose AS, Rands CE, et al. Neuropsychological and educational problems at school age associated with neonatal encephalopathy. Arch Dis Child Fetal Neonatal Ed 2005; 90: F380–7.

[176] Martin LJ, Al-Abdulla NA, Brambrink AM, Kirsch JR, Sieber FE, Portera-Cailliau C. Neurodegeneration in excitotoxicity, global cerebral ischemia, and target deprivation: a perspective on the contributions of apoptosis and necrosis. Brain Res Bull. 1998; 46:281–309.

[177] Martinez, J.; Moreno, J. J. Role of Ca^2-independent phospholipase A2 on arachidonic acid release induced by reactive oxygen species. *Arch. Biochem. Biophys.* 392:257–262; 2001.

[178] Massey LK, Liebman M, Kynast-Gales SA. Ascorbate increases human oxaluria and kidney stone risk. J Nutr 2005; 135:1673–7

[179] Matés JM, Pérez-Gómez C, Núñez de Castro I.: Antioxidant enzymes and human diseases. Clin Biochem. 1999; 32 (8):595-603.

[180] Matyja E, Taraszewska A, Nagańska E, Grieb P, Rafałowska J: CDP-choline protects motor neurons against apoptotic changes in a model of chronic glutamate excitotoxicity *in vitro. Folia Neuropathol 2008; 46 (2): 139-148.*

[181] Medja F, Lelievre V, Fontaine RH, Lebas F, Leroux P, Ouimet T, et al. Thiorphan, a neutral endopeptidase inhibitor used for diarrhoea, is neuroprotective in newborn mice. Brain 2006; 129:3209–23.

[182] Mehmet H, Yue X, Squier MV, Lorek A, Cady E, Penrice J, et al. Increased apoptosis in the cingulate sulcus of newborn piglets following transient hypoxia-ischaemia is related to the degree of high energy phosphate depletion during the insult. Neurosci Lett 1994; 181:121–5.

[183] Ment LR, Vohr BR, Makuch RW, Westerveld M, Katz KH, Schneider KC, et al. Prevention of intraventricular hemorrhage by indomethacin in male preterm infants. *J Pediatr*; 2004 145(6):832–834.

[184] McDonald JW, Silverstein FS, Johnston MV. Neurotoxicity of N-methyl-D-aspartate is markedly enhanced in developing rat central nervous system. Brain Res 1988; 459:200–3.

[185] McDonald JW, Behrens MI, Chung C, Bhattacharyya T, Choi DW. Susceptibility to apoptosis is enhanced in immature cortical neurons. Brain Res 1997;759:228–32

[186] McQuillen PS, Ferriero DM. Selective vulnerability in the developing central nervous system. Pediatr Neurol 2004; 30:227–35.

[187] Miao H, Jiang L, Huang L. Effects of simvastatin on the expression of intercellular adhesion molecule-1 mRNA in neonatal brain withhypoxic-ischemic damage. J Nanosci Nanotechnol.2005; 5(8):1261-5.

[188] Miller LJ, Marx J: Apoptosis-Special Section, Introduction. Science 1998; 281:1301.

[189] Mir C, Clotet J, Aledo R, Durany N, Argemi J, Lozano R, et al. CDP-choline prevents glutamate-mediated cell death in cerebellar granule neurons. J Mol Neurosci 2003; 20: 53-60.

[190] Mishra OP, Delivoria-Papadopoulos M. Cellular mechanisms of hypoxic injury in the developing brain. Brain Res Bull 1999; 48:233– 8.

[191] Mokra D, Drgova A, Kopincova J, Pullmann R, Calkovska A: Anti-inflammatory treatment in dysfunction of pulmonary surfactant in meconium-induced acute lung injury. Adv Exp Med Biol. 2013; 756:189-96.

[192] Moncada S, Palmer RM, Higgs EA. Nitric oxide: physiology, pathophysiology, and pharmacology. Pharmacol Rev 1991; 43:109–42.

[193] Muir KW, Lees KR, Ford I, Davis S; Intravenous Magnesium Efficacy in Stroke (IMAGES) Study Investigators. Magnesium for acute stroke (Intravenous Magnesium Efficacy in Stroke trial): randomised controlled trial. Lancet. 2004; 363(9407):439-45.

[194] Muramatsu K, Sheldon RA, Black SM, Tauber M, Ferriero DM. Nitric oxide synthase activity and inhibition after neonatal hypoxia ischemia in the mouse brain. Brain Res Dev Brain Res.2000; 123(2):119–12.

[195] Nakai A, Shibazaki Y, Taniuchi Y, Oya A, Asakura H, Koshino T, et al Vitamins ameliorate secondary mitochondrial failure in neonatal rat brain. Pediatr Neurol 2002; 27:30–5.

[196] Nakajima K, Kohsaka S: Microglia: Neuroprotective and neurotrophic cells in the central nervous system. Curr Drug Targets Cardiovasc Haematol Disord 2004; 4: 65–84.

[197] Nijboer CH, Groenendaal F, Kavelaars A, Hagberg HH, van Bel F, Heijnen CJ. Gender-specific neuroprotection by 2-iminobiotin after hypoxia-ischemia in the neonatal rat via a nitric oxide independent pathway. J Cereb Blood Flow Metab 2007; 27:282-92.

[198] Noh MR, Kim SK, Sun W, Park SK, Choi HC, Lim JH, et al. Neuroprotective effect of topiramate on hypoxic ischemic brain injury in neonatal rats. Exp Neurol. 2006;201:470–478

[199] Nolana JP, Soarb J, Zidemanc DA, Biarentd D,Bossaerte L L, Deakinf C: European Resuscitation Council Guidelines for Resuscitation 2010. Section 1. Executive summary. Resuscitation. 2010; 81:1219–1276

[200] Northington FJ, Ferriero DM, Graham EM, Traystman RJ, Martin LJ. Early neurodegeneration after hypoxia-ischemia in neonatal rat Is necrosis while delayed neuronal death Is apoptosis. Neurobiol Dis 2001; 8: 207–19.

[201] Nozaki K, Finklestein SP, Beal MF. Basic fibroblast growth factor protects against hypoxia–ischemia and NMDA neurotoxicity in neonatal rats. J Cereb Blood Flow Metab. 1993; 13:221–8.

[202] Olney JW, Wozniak DF, Jevtovic-Todorovic V, Farber NB, Bittigau P, Ikonomidou C. Drug-induced apoptotic neurodegeneration in the developing brain. *Brain Pathol* 2002; 12:488–498.

[203] Olsson B, Johansson M, Gabrielsson J, Bolme P. Pharmacokinetics and bioavailability of reduced and oxidized N-acetylcysteine. Eur. J. Clin. Pharmacol. 1988; 34:77–82.

[204] O'Shea TM, Doyle LW. Perinatal glucocorticoid therapy and neurodevelopmental outcome: an epidemiologic perspective. Semin Neonatol 2001;6:293–307

[205] Otten U, Marz P, Heese K, Hock C, Kunz D, Rose-John S. Cytokines and neurotrophins interact in normal and diseased states. *Ann N Y Acad Sci* 2000; 917:322–30.

[206] Pabon MM, Borlongan CV Advances in the cell-based treatment of neonatal hypoxic-ischemic brain injury: *Future Neurol*. 2013; 8(2):193-203.

[207] Painter MJ: Animal Models of Perinatal Asphyxia: Contributions, Contradictions, Clinical Relevance. Seminars in Pediatric Neurology, Vol 2 (1), 1995: pp 37-56.

[208] Painter MJ, Scher MS, Stein AD, Armatti S, Wang Z, Gardiner JC, et al. Phenobarbital compared with phenytoin for the treatment of eonatal seizures. N. Engl. J. Med. 1999; 341, 485–489.

[209] Paintlia MK, Paintlia AS, Barbosa E, Singh I, Singh AK. N-acetylcysteine prevents endotoxin-induced degeneration of oligodendrocyte progenitors and hypomyelination in developing rat brain. *J Neurosci Res* 2004; 78: 347-61.

[210] Palmer C, Towfighi J, Roberts RL, et al. Allopurinol administered after inducing hypoxiaischemia reduces brain injury in 7-day-old rats. Pediatr Res 1993; 33:405–11.

[211] Palmer C. Hypoxic-ischemic encephalopathy. Therapeutic approaches against microvascular injury, and role of neutrophils, PAF, and free radicals. *Clin Perinatol.* 1995; 22: 481-517.

[212] Palmer C, Roberts RL, Bero C. Deferoxamine posttreatment reduces ischemic brain injury in neonatal rats. Stroke 1994; 25: 1039–45.

[213] Palsdottir K, Dagbjartsson A, Thorkelsson T, Hardardottir H: Birth asphyxia and hypoxic ischemic encephalopathy, incidence and obstetric risk factors. Laeknabladid. 2007; 93(9):595-601.

[214] Peeters-Scholte C, Braun K, Koster J, Kops N, Blomgren K, Buonocore G, et al. F: Effects of allopurinol and deferoxamine on reperfusion injury of the brain in newborn piglets after neonatal hypoxia-ischemia. Pediatr Res 2003; 54:516-522.

[215] Perlman JM, Risser R. Relationship of uric acid concentrations and severe intraventricular hemorrhage/leukomalacia in the premature infant. *J Pediatr* 1998; 132:436– 9.

[216] Perlman JM, Wyllie J, Kattwinkel J, Atkins DL, Chameides L, Goldsmith JP, et al. Neonatal resuscitation: 2010 International Consensus on cardiopulmonary resuscitation and emergency cardiovascular care science with treatment recommendations. Pediatrics 2010; 126(5):e1319–44.

[217] Perlman M, and Shah PS: Hypoxic-Ischemic Encephalopathy: Challenges in Outcome and Prediction. *J Pediatr* 2011; 158:e51-4.

[218] Pimentel-Coelho PM, Mendez-Otero R. Cell therapy for neonatal hypoxic-ischemic encephalopathy. Stem Cells Dev. 2010; 19:299-310.

[219] Pimentel-Coelho PM, Rosado-de-Castro PH, da Fonseca LM, Mendez-Otero R. Umbilical cord blood mononuclear cell transplantation for neonatal hypoxic-ischemic encephalopathy. Pediatr Res 2012; 71(4 Pt 2):464–73.

[220] Portera-Cailliau C, Price DL, Martin LJ. Excitotoxic neuronal death in the immature brain is an apoptosis–necrosis morphological continuum. J Comp Neurol 1997; 378:70–87.

[221] Puyal J, Vaslin A, Mottier V, Clarke PGH. Postischemic Treatment of Neonatal Cerebral Ischemia Should Target Autophagy. *Ann Neurol* 2009; 66: 378–389.

[222] Rees S, Richard Harding R, WalkerD. The biological basis of injury and neuroprotection in the fetal and neonatal brain. Int J Dev Neurosci. 2011; 29(6): 551–563.

[223] Robertson CM. Long term follow-up of term infants with perinatal asphyxia. In: Stevenson DK, Benitz WE, Sunshine P, editors. Fetal and neonatal brain injury. 3rd edition. New York, NY: Cambridge University; 2003. p. 829–58.

[224] Robertson CM, Finer NN. Long-term follow-up of term neonates with perinatal asphyxia. Clin Perinatol 1993; 20: 483 – 500.

[225] Robertson NJ, Faulkner S, Fleiss B, Bainbridge A, Andorka C, Price D,et al: Melatonin augments hypothermic neuroprotection in a perinatal asphyxia model. Brain. 2013; 136(Pt 1):90-105.

[226] Robertson NJ, Tan S, Groenendaal F, van Bel F, Juul SE, Bennet L, et al. Which neuroprotective agents are ready for bench to bedside translation in the newborn infant? .J Peds. 2012; Vol. 160 (4): 544-552.e4.

[227] Russell GA, Cooke RW. Randomised controlled trial of allopurinol prophylaxis in very preterm infants. Arch Dis Child Fetal Neonatal Ed. 1995 Jul; 73(1):F27-31.

[228] Rutherford M, Ramenghi LA, Edwards AD, Brocklehurst P, Halliday H, Levene M, et al. Assessment of brain tissue injury after moderate hypothermia in neonates with hypoxic-ischaemic encephalopathy: a nested substudy of a randomised controlled trial. Lancet Neurol 2010; 9: 39–45.

[229] Rybakowski C, Mohar B, Wohlers S, Leichtweiss H, Schreoder H. The transport of vitamin C in the isolated human near-term placenta..Eur J Obstet Gynecol Reprod Biol 1995; 62:107-14.

[230] Sámano C, Nasrabady SE, Nistri A. A study of the potential neuroprotective effect of riluzole on locomotor networks of the neonatal rat spinal cord in vitro damaged by excitotoxicity. Neuroscience. 2012; 11(222):356-65.

[231] Samanta S, Perkinton MS. Morgan M. Williams RJ. Hydrogen peroxide enhances signal-responsive arachidonic acid release from neurons: role of mitogen-activated protein kinase. J. Neurochem. 70:2082–2090; 1998.

[232] Sameshima H, Ota A, Ikenoue T. Pretreatment with magnesium sulfate protects against hypoxic-ischemic brain injury but postasphyxial treatment worsens brain damage in seven-day- old rats. *American Journal of Obstetrics and Gynecology*.1999; 180(3):725–730.

[233] Sandilands EA, Bateman DN. Adverse reactions associated with acetylcysteine. Clin Toxicol (Phila). 2009; 47(2):81-8.

[234] Sarco DP, Becker J, Palmer C, Sheldon RA, Ferriero DM. The neuroprotective effect of deferoxamine in the hypoxic-ischemic immature mouse brain. Neurosci Lett 2000; 282(1–2): 113–6.

[235] Sarkar S, Barks JD, Bapuraj JR, Bhagat I, Dechert RE, Schumacher RE, et al. Does phenobarbital improve the effectiveness of therapeutic hypothermia in infants with hypoxic-ischemic encephalopathy?.*J Perinatol*. 2012; 32(1):15-20.

[236] Sarnat HB, Sarnat MS: Neonatal encephalopathy following fetal distress. A clinical and electroencephalographic study. Arch Neurol 1976; 33: 696-705.

[237] Saver JL. Citicoline: update on a promising and widely available agent for neuroprotection and neurorepair. *Rev Neurol* Dis 2008; 5: 167–77.

[238] Sävman K, Blennow M, Gustafson K, Tarkowski E, Hagberg H.Cytokine response in cerebrospinal fluid after birth asphyxia. Pediatr Res 1998; 43:746– 51.

[239] Sato Y, Nakanishi K, Hayakawa M, Kakizawa H, Saito A, Kuroda Y et al. Reduction of brain injury in neonatal hypoxic-ischemic rats by intracerebroventricular injection of neural stem/progenitor cells together with chondroitinase ABC. *Reprod Sci.* 2008; 15: 613–620.

[240] Scafidi J, Gallo V: New concepts in perinatal hypoxia ischemia encephalopathy. Curr Neurol Neurosci Rep 2008, 8:130-138.

[241] Schaper M, Gergely S, Lykkesfeldt J, Zb€aren J, Leib S, T€auber M, et al. Cerebral vasculature is the major target of oxidative protein alterations in bacterial meningitis. J Neuropathol Exp Neurol 2002; 61:605-13.

[242] Schwartz JC. Racecadotril: a new approach to the treatment of diarrhoea. Int J Antimicrob Agents 2000; 14: 75–9.

[243] Secades JJ, Lorenzo JL. Citicoline: pharmacological and clinical review, 2006 update. Methods Find Exp Clin Pharmacol 2006; 28 Suppl B: 1-56.

[244] Shalak L, Perlman JM. Hypoxic-ischemic brain injury in the term infant-current concepts. Early Hum Dev. 2004; 80:125–141.

[245] Shankaran S: Hypoxic-ischemic. Encephalopathy and Novel Strategies for Neuroprotection. Clin Perinatol. 2012, 39: 919–929.

[246] Shankaran S, Barnes PD, Hintz SR, Laptook AR, Zaterka-Baxter KM, McDonald SA, et al. Brain injury following trial of hypothermia for neonatal hypoxic-ischaemic encephalopathy. Arch Dis Child Fetal Neonatal Ed. 2012; 97(6):F398-404

[247] Shankaran S, Laptook AR, Ehrenkranz RA, Tyson JE, McDonald SA, Donovan EF, et al. Whole-body hypothermia for neonates with hypoxic-ischemic encephalopathy. N Engl J Med. 2005; 353:1574–84.

[248] Shankaran S, Pappas A, Laptook AR, McDonald SA, Ehrenkranz RA, Tyson JE et al. Outcomes of safety and effectiveness in a multicenter randomized, controlled trial of whole-body hypothermia for neonatal hypoxic-ischemic encephalopathy. Pediatrics 2008;122:e791–

[249] Shankaran S, Woldt E, Koepke T, Bedard MP, Nandyal R. Acute neonatal morbidity and long-term central nervous system sequelae of perinatal asphyxia in term infants. Early Hum Dev 1991; 25:135–48.

[250] Shelov, SP. American Academy of Pediatrics, American Heart Association - *Neonatal Resuscitation.* Editor American Academy of Pediatrics, 2011, sixth edition. ISBN: 1581105002, 9781581105001.

[251] Shen M, Wu RX, Zhao L, Li J, Guo HT, Fan R, et al. Resveratrol attenuates ischemia/reperfusion injury in neonatal cardiomyocytes and its underlying mechanism. PLoS One. 2012; 7(12):e51223.

[252] Shimizu K, Rajapakse N, Horiguchi T, Payne RM, Busija DW. Neuroprotection against hypoxia-ischemia in neonatal rat brain by novel superoxide dismutase mimetics. Negrosci Lett. 2003; 346: 41-44.

[253] Shingo T, Sorokan ST, Shimazaki T, Weiss S. Erythropoietin regulates the in vitro and in vivo production of neuronal progenitors by mammalian forebrain neural stem cells. *J Neurosci.* 2001; 21: 9733–9743

[254] Shim SY, Kim HS.: Oxidative stress and the antioxidant enzyme system in the developing brain. Korean *J Pediatr.* 2013; 56(3):107-11.

[255] Simbruner G, Mittal RA, Rohlmann F, Muche R. neo. nEURO.network Trial Participants. Systemic hypothermia after neonatal encephalopathy: outcomes of neo.nEURO. network RCT. *Pediatrics.*2010;126:e771–8.

[256] Sinha S, Davies J, Toner N, Bogle S, Chiswick M. Vitamin E supplementation reduces frequency of periventricular haemorrhage in very preterm babies. Lancet 1987; i: 466-71.

[257] Sinor A.D., Greenberg D.A. Erythropoietin protects cultured cortical neurons, but not astroglia, from hypoxia and AMPA toxicity. *Neurosci Lett.* 2000; 90: 213–215.

[258] Sizonenko SV, Bednarek N, Gressens P. Growth factors and plasticity. *Semin Fetal Neonatal Med.* 2007; 12: 241–249.

[259] Sfaello I, Baud O, Arzimanoglou A, Gressens P. Topiramate prevents excitotoxic damage in the newborn rodent brain. Neurobiol Dis 2005; 20:837–48.

[260] Slaughter LA, Anup D. Patel AD, and Slaughter JL, Pharmacological Treatment of Neonatal Seizures: A Systematic Review. Journal of Child Neurology 2013; 28(3) 351-364.

[261] Subirós N, Del Barco DG, Coro-Antich RM Erythropoietin: still on the neuroprotection road. Ther Adv Neurol Disord. 2012; 5:161-73.

[262] Tagin MA, Woolcott CG, Vincer MJ, Whyte RK, Stinson DA. Hypothermia for neonatal hypoxic ischemic encephalopathy: an updated systematic review and meta-analysis. Arch Pediatr Adolesc Med 2012; 166:558–66.

[263] Torrance HL, Benders MJ, Derks JB, Rademaker CM, Bos AF, Berg Van Den P, et al. Maternal allopurinol during fetal hypoxia lowers cord blood levels of the brain injury marker S-100B. Pediatrics 2009, 124:350-357.

[264] Trovarelli G, de Medio GE, Dorman RV, Piccinin GL, Horrocks LA, Porcellati G. Effect of cytidine diphosphate choline (CDP-choline) on ischemia-induced alterations of brain lipid in the gerbil. Neurochem Res 1981; 6: 821-833.

[265] Tsuji M, Wilson MA, Lange MS, Johnston MV. Minocycline worsens hypoxic-ischemic brain injury in a neonatal mouse model. Exp Neurol 2004; 189:58–65.

[266] Turrens JF, Alexandre A, Lehninger AL. Ubisemiquinone is the electron donor for superoxide formation by complex III of heart mitochondria. Arch Biochem Biophys 1985; 237: 408 – 14.

[267] Valls I Soler A, Sanjurjo P, Vázquez Cordero C. Controlled study of the administration of CDP-choline to preterm newborn infants with respiratory distress syndrome].*An Esp Pediatr*. 1988 Jun;28(6):493-6.

[268] Vallat-Decouvelaere AV, Chretien F, Gras G, Le Pavec G, Dormont D. Expression of excitatory amino acid transporter-1 in brain macrophages and microglia of HIVinfected patients. A neuroprotective role for activated microglia? J Neuropathol Exp Neurol 2003; 62:475–85.

[269] Van Bel F, Shadid M, Moison RM, Dorrepaal CA, Fontijn J, Monteiro L, et al. Effect of allopurinol on postasphyxial free radical formation, cerebral hemodynamics, and electrical brain activity. *Pediatrics*.1998, 101:185-193.

[270] van den Tweel ER, van Bel F, Kavelaars A, Peeters-Scholte CM, Haumann J, Nijboer CH, et al. Long-term neuroprotection with 2-iminobiotin, an inhibitor of neuronal and inducible nitric oxide synthase, after cerebral hypoxia-ischemia in neonatal rats. J Cereb Blood Flow Metab. 2005; 25(1):67-74.

[271] van Velthoven CT, Kavelaars A, Heijnen CJ. Mesenchymal stem cells as a treatment for neonatal ischemic brain damage. Pediatr Res. 2012;71(4 Pt 2):474-81

[272] Vasquez-Vivar J, Whitsett J, Derrick M, Ji X, Yu L, Tan S. Tetrahydrobiopterin in the prevention of hypertonia in hypoxic fetal brain. *Ann Neurol* 2009;66:323-31.

[273] Volpe JJ, Neurology of the Newborn, 5th edition, 2008.

[274] Volpe JJ. Cerebral white matter injury of the premature infantmore common than you think. Pediatrics 2003; 112:176–80.

[275] Volpe JJ, Kinney HC, Jensen FE, Rosenberg PA.The developing oligodendrocyte: key cellular target in brain injury in the premature infant. *Int J Dev Neurosci*. 2011 Jun; 29(4):423-40.

[276] Wang B, Navath RS, Romero R, Kannan S, Kannan R. Anti-inflammatory and anti-oxidant activity of anionic dendrimer-N-acetyl cysteine conjugates in activated microglial cells. Int J Pharm. 2009 Jul 30;377(1-2):159-68.

[277] Wang X, Feuerstein GZ. Induced expression of adhesion molecules following focal brain ischemia. J Neurotrauma. 1995; 12:825-32.

[278] Wang X, Svedin P, Nie C, Lapatto R, Zhu C, Gustavsson M, et al. N-acetylcysteine reduces lipopolysaccharide-sensitized hypoxicischemic brain injury. *Ann Neurol* Mar; 2007 61(3):263–271.

[279] Wang XL, Yu SL, Yu T, Li JH, Guo Ar, Liang HT. Treatment of neonatal hypoxicischaemic encephalopathy (HIE) with compound salvia miltiorrhizae and citicoline: a comparative study in China. Singapore Paediatr J 1997; 39: 120–123.

[280] Wang Y, Cao M, Liu A, Di W, Zhao F, et al.Changes of inflammatory cytokines and neurotrophins emphasized their roles in hypoxic-ischemic brain damage. Int J Neurosci. 2013; 123(3):191-5].

[281] Welin AK, Svedin P, Lapatto R, Sultan B, Hagberg H, Gressens P, et al. Melatonin reduces inflammation and cell death in white matter in the mid-gestation fetal sheep following umbilical cord occlusion. *Pediatr Res.* 2007; 61:153–158.

[282] Whitelaw A, Thoresen M. Antenatal steroids and the developing brain Arch Dis Child Fetal Neonatal Ed 2000; 83:F154–7.

[283] Widness J, Schmidt R, Sawyer S. Erythropoietin transplacental passagereview of animal studies. J Perinat Med 1995; 23:61-70.

[284] Willmot M, Gibson C, Gray L, Murphy S, Bath P. Nitric oxide synthase inhibitors in experimental ischemic stroke and their effects on infarct size and cerebral blood flow: a systematic review. Free Radic BiolMed 2005; 39:412-25.

[285] Xu J, Yu S, Sun AY, Sun GY. Oxidant-mediated AA release from astrocytes involves cPLA2 AND iPLA2. Free Radical Biology & Medicine. 2003; 34 (12): 1531–1543.

[286] Yrjanheikki J, Tikka T, Keinanen R, Goldsteins G, Chan PH, Koistinaho J. A tetracycline derivative, minocycline, reduces inflammation and protects against focal cerebral ischemia with a wide therapeutic window. Proc Natl Acad Sci USA. 1999; 96:13496–500.

[287] Yu X., Shacka JJ, Eells JB, Suarez-Quian C, Przygodzki RM, Beleslin-Cokic B, et al. Erythropoietin receptor signalling is required for normal brain development.*Development*. 2002; 129: 505–516

[288] Zhang RI, Chopp M, Li Y, Zalonga C, Jiang M, Jones M, et al. Anti-ICAM-1 antibody reduces ischemic cell damage after transient middle cerebral artery occlusion in the rat. *Neurology*. 1994; 44:1747–1751.

[289] Zhou WH, Cheng GQ, Shao XM, Liu XZ, Shan RB, Zhuang DY, et al. Selective head cooling with mild systemic hypothermia after neonatal hypoxic-ischemic encephal-

opathy: a multicenter randomized controlled trial in China. *J Pediatr*. 2010; 157:367–72.

[290] Zhu C, Kang W, Xu F, Cheng X, Zhang Z, Jia L, et al. Erythropoietin improved neurologic outcomes in newborns with hypoxic-ischemic encephalopathy. Pediatrics 2009; 124:e218-26.

[291] Zhu C, Wang X, Huang Z, Qiu L, Xu F, Vahsen N, et al. Apoptosis-inducing factor is a major contributor to neuronal loss induced by neonatal cerebral hypoxia-ischemia. Cell Death Differ. 2007; 14(4):775-84.

Chapter 6

Neuromusculoskeletal Rehabilitation of Cerebral Palsy Using SEMLARASS

Deepak Sharan

Additional information is available at the end of the chapter

http://dx.doi.org/10.5772/57957

1. Introduction

The primary aim of motor treatment of Cerebral Palsy (CP) is to control and correct spasticity, abnormal movement patterns and lever arm dysfunction, since these problems directly affect the activities of daily living, ability to walk and developmental milestones.

1.1. Disabilities in CP

1.1.1. Primary disabilities

Primary disabilities are due to the direct effect from the brain lesion, and includes spasticity, delayed milestones, sensory problems, cognitive impairment, imbalance, lack of muscle strength, abnormal movements, etc. The injury to the Central Nervous System affects critical inputs to the reticulospinal and corticospinal tract, which in turn affects the motor units and results in abnormal control and weakness. Loss of descending inhibitory input through the reticulospinal tract and other system increases the excitability of gama and alpha neurons, producing spasticity.

1.1.2. Secondary disabilities

Secondary disabilities are due to the after effects of primary disabilities. These include contractures, lever arm dysfunctions and scoliosis. The contractures are the after effects of spasticity. Common contractures in CP are hip flexion (psoas), knee flexion (hamstring) and equinus ankle (gastrocnemius) in the lower extremities, and elbow flexion (biceps and pronator) and wrist flexion (forearm flexors) contractures in the upper extremities. The concept of lever arm dysfunction was proposed by Dr. James Gage. Lever is a rigid structure that transmits and modifies force or motion when forces are applied at one point and is able to rotate about another. In the human body, joints function as levers to transmit forces. Lever arm (also known as a moment arm) is the perpendicular distance from the line of application of a force to the axis of joint rotation. Lever arm dysfunction is the disruption in the moment generation of a muscle joint complex because of an ineffective lever or moment arm despite normal muscle force. The types of lever arm dysfunction are: Short lever-arm (coxa valga), Flexible lever-arm (pes valgus), Malrotated lever-arm (external tibial torsion), An abnormal pivot or action point (hip subluxation or dislocation), and/or Positional lever–arm dysfunction (crouch gait).

Positional lever–arm dysfunction (crouch gait). The result of lever arm dysfunction is functional weakness and decreased power generation, due to abnormal direction of pull of muscles (Gage JR, Novacheck TF, 2001; Novacheck TF, Gage JR, 2007)

Children with CP are unable to perform sufficient movement to adequately stretch their muscles, particularly when those muscles are spastic. The stretch is the stimulus that facilitates muscular growth. The growing bones often become plastically deformed or twisted because of the uneven/abnormal muscle forces present during growth in a child with CP and the spastic muscles cause joint displacement by pulling on the bones. This leads to lever arm dysfunction in the femur as increased femoral anteversion, in the tibia as external tibial torsion and in the heel as valgus deformity. Another important lever arm dysfunction is hip joint dislocation or subluxation.

1.1.3. Tertiary disabilities

Tertiary disabilities are coping responses to primary or secondary disabilities and must be carefully identified and left alone. An example is circumduction of hip in co-spasticity of rectus femoris and hamstring.

1.2. Progression of gait abnormalities in CP

The gait in an ambulant child with CP is characterised by crouched position (flexion of the hip and knees), intoeing (internal rotation of femur), and walking on the toes (equinus foot placement). This gait is extremely inefficient and result in increased energy expenditure due to the presence of simultaneous contraction of agonist and antagonist muscles. If the child continues walking in the presence of contractures and lever arm dysfunction the inevitable result is over lengthening of tendons (e.g., calcaneus deformity due to over lengthened tendoachilles) and joint decompensation, after which further ambulation becomes increasingly difficult.

1.3. Overview of current treatment options for cerebral palsy and their limitations

a. **Oral Muscle Relaxants** [Gracies JM, et. al, 1997; Hattab JR, 1980; Ryan DM, Blumenthal FS, 1993] Baclofen, Diazepam, Dantrolene, Tizanidine, Clonidine, Vigabatrin, Idroclimide etc. are available for the management of spasticity. The benefits of these drugs are severely limited by neurological effects such as drowsiness, daytime sedation, lassitude, exhaustion, light headedness, ataxia, confusion, dizziness, headache, insomnia, myalgia, muscle weakness and euphoria, hallucinations, nightmares, depression and dyskinesia. Other reported side effects include withdrawal symptoms, gastrointestinal disturbances such as a dry mouth, nausea, vomiting, constipation or diarrhoea.

b. **Phenol & Alcohol Injection** [Kolaski K, et. al., 2008] Phenol or carbolic acid (3–7%) and Alcohol (45–100%) is injected directly to the spastic muscles or to the corresponding motor nerve. Phenol denatures protein and causes non-selective tissue destruction in the injected area (including coagulation of nerve and muscle) followed by Wallerian degeneration of neurons, that can occur for several weeks. Because diffusion of both is limited, the area of effective denervation extends just a few millimeters from the injection site. The duration of denervation is 3-6 months for alcohol and 4-8 months for phenol. It should be reserved only for motor nerves such as the obturator nerve and musculocutaneous nerve. Disadvantages are pain on injection requiring a general anaesthetic or heavy sedation, non-selective protein denaturation and possible permanent muscle fibrosis, dysaesthesias and motor weakness, e.g., foot drop.

c. **Botulinum Toxin Type A** [Jianjun L, et.al, 2013] The limitations include high expense, transient effect lasting a few weeks or months, lack of efficacy in the presence of generalised spasticity, muscle contractures or lever arm disease, and life threatening complications (permanent paralysis, respiratory paralysis and death in hundreds of children) have been reported leading to FDA issuing a warning in January 2008 against its use in CP.

d. **Neurodevelopmental Therapy** The policy statement from the American Academy of Cerebral Palsy & Developmental Medicine concluded that the long-term benefits of Neurodevelopmental Therapy were marginal and/or not measurable. Except for an immediate improvement in the range of motion of joints, no other functional parameters showed any long-term improvement in a review of the available literature [Butler C, Darrah J, 2001]. A child with mild CP shows improvement with therapy, whereas the more severe cases progressively develop contractures and lever arm dysfunctions around the age of 4-7 years, after which no further improvement occurs with continuing therapy.

e. **Intrathecal baclofen (ITB) therapy.** A few studies have reported positive and statistically significant results for lower-extremity muscle tone. From ITB the quality of life improvement estimates were 0.27 for a bed-bound patient not in pain, to 0.5 for a bed-bound patient experiencing severe spasm-related pain. Over five years, the total discounted cost was £15,400. The cost per quality adjusted life year ranged from £6,900 to £12,790 [Sampson FC, et. al., 2002]. Several life threatening complications have been reported with ITB therapy such as complications related to pump replacement and implantation (19%), infection (10%), CSF leakage (17%), catheter drawing out, disconnection or breakage (10.5%). Drowsiness, nausea, headache, muscle weakness, light-headedness and return of pretreatment spasticity can be caused by intrathecal pump delivering an incorrect dose of baclofen [Motta F, et.al., 2007]. Intrathecal baclofen withdrawal syndrome is a very rare, potentially life-threatening complication of baclofen pump caused by an abrupt cessation of ITB [Ross JC, et.al., 2011].

f. **Selective Dorsal Rhizotomy** [Lundkvist A, Hagglund G, 2006; Crawford K, et. al., 1996; Mooney JF 3rd, Millis MB, 1999; Turi M, Kalen V, 2000] This is associated with adverse permanent effects (sensory disturbance, bladder dysfunction, scoliosis, lordosis, hip dislocations and foot deformities), loss of antigravity stability and worsening of motor function, and has no efficacy in patients with contractures, lever arm disease or upper limb involvement.

g. **Conventional Orthopaedic Surgery.** Orthopaedic surgery has been used to restore normal alignment, correcting contractures and deformities, and achieving stability by arthrodesis. Earlier it was believed that adductor spasticity was the predominant risk factor for hip dislocation, which frequently required surgical intervention [Spiegel DA, et. al., 2004]. It is now evident, after the advent of computerised gait analysis that what was called "scissoring" gait in the past is actually due to spasticity of medial hamstrings and increased femoral anteversion, in most instances [Scrutton D, et. al., 2001]. Adductor tenotomy and obturator neurectomy (a commonly performed operation) in this situation will convert an assisted ambulator into a non-ambulator, because of denervation of adductor brevis (an important hip flexor and antigravity muscle), besides producing an abduction deformity of hips. Matsuo T, et. al. [1986] compared a group of nineteen patients who had concomitant adductor tenotomy and anterior branch obturator neurectomy with a group of twenty-three patients who had only myotomy of the adductor longus and gracilis muscles. The nineteen patients in the neurectomy group had an unacceptable broad-based gait with hyperabduction of the hips. Iliopsoas lengthening or tenotomy at its insertion at lesser trochanter of femur is frequently done for a fixed flexion deformity of the hip. However, it is now known that this causes loss of power in hip flexion, and hence selective release of the psoas muscle at the pelvic brim has been recommended. [Matsuo T, et. al., 1987] The gluteus medius is a major internal rotator of the hip, and when it is spastic it can cause the objectionable in-toed gait in spastic diplegia. Steel's procedure of transfer of the insertion of the gluteus medius and minimus to the anterolateral aspect of the proximal part of the femur was effective in a few patients. But the patients in whom the procedure failed had a severe Trendelenburg limp postoperatively [Steel HH, 1980]. The correction of contractures of one joint without concomitant correction of spasticity of another joint may result in irreparable over lengthening of tendons. The commonest clinical scenario is that of a child with spastic diplegia who walks on the toes. The immediate response of many surgeons is to perform Z lengthening of the tendoachilles, a simple procedure that takes only a few minutes to perform, but the functional effects to the child may be devastating and permanent. Most children with spastic diplegia have hip and knee flexion contractures, and tiptoe in order to shift the center of gravity close to the body. It is extremely rare to encounter a true tendoachilles contracture in diplegics, and inappropriate lengthening of tendoachilles in the presence of hip and knee contractures inevitably leads to an unstable calcaneus deformity, which cannot be braced nor salvaged surgically. Besides, the crouch at the knee persists. The problems with Conventional Orthopaedic Surgery are: 1) Lengthening of monoarticular muscles or tendons (adductor brevis, iliopsoas, tendoachilles) leading to loss of antigravity action and severe weakness, 2) Over lengthening of tendons is common, 3) Muscle transfers (e.g., Eggers) often lead to reverse deformity, e.g., genu recurvatum, 4) It does not help the severely

http://dx.doi.org/10.5772/57957

involved: quadriplegics, athetoid, dystonia, 5) Joint fusion (Grice fusion) leads to degeneration of surrounding joints, 6) Lever arm dysfunction is rarely corrected simultaneously or early enough leading to recurrence of contractures and 7) It cannot control spasticity, produce reciprocal movements to facilitate antigravity muscles, and improve functional skills and voluntary movement of the hand.

Studies have shown that the GMFCS remains unchanged or "stable" in the vast majority of children with currently available treatment modalities, including single-event multilevel surgery in bilateral spastic CP [Rutz E, et. al., 2012]. Conventional treatment for CP fails to simultaneously and effectively address spasticity, abnormal movement patterns and lever arm dysfunction, leading to a need for better rehabilitation strategies to address the functional problems in a person with CP.

2. SEMLARASS — A new management approach for CP

With the help of experience of treating CP for a decade and extensive research, Dr. Deepak Sharan conceptualised Single Event Multilevel Lever Arm Restoration and Anti Spasticity Surgery (SEMLARASS) at RECOUP Neuromusculoskeletal Rehabilitation Centre, Bangalore, India in 2001.

SEMLARASS has the following components:

1. **Single Event:** All surgeries are completed under a single anaesthetic, requiring only one hospital admission and one period of rehabilitation,
2. **Multilevel:** All the affected regions and all orthopaedic deformities (soft tissue and bony) are corrected simultaneously in view of interdependence of joints,
3. **Lever Arm Restoration:** Simultaneous correction of lever arm dysfunctions to improve the direction of pull of muscles and to facilitate muscle strengthening postoperatively,
4. **Anti Spasticity:** using Orthopaedic Selective Spasticity Control Surgery (OSSCS), developed by Dr. Takashi Matsuo (Tokyo, Japan) based on the concept that multiarticular muscles, which have less antigravity activity, are hyperactive in CP. Therefore, spasticity and athetotic movements can be controlled by releasing them selectively. The monoarticular muscles, which have antigravity activity and are responsible for maintaining an upright posture, are carefully preserved [Matsuo T, 2002].

The unique features of SEMLARASS [Sharan D, 2005] are:

a. Operating between the ages of 4 years to 6 years (preferably), to avoid joint decompensation and over lengthening of tendons that happen due to continued usage of deformed joints,

b. Minimally invasive procedures using image intensification that do not require large skin incisions and consequent risk of blood loss and infection,

c. Use of only External Fixators that do not require a second operation for removal,

d. All bony operations done to restore deformed lever arms are extra-articular to allow for the maximum growth potential of children's bones,

e. Simultaneous lever arm restoration is essential for spasticity and contracture correction as well as to reduce chances of recurrence of deformities and repeat surgery at a later stage, and to improve the direction of pull of muscles and facilitating strengthening,

f. Tendon lengthening or transfers are avoided to reduce weakness or overcorrection, and

g. The surgery is followed by a time bound, structured and intensive physician directed rehabilitation protocol developed by Dr. Sharan.

SEMLARASS has been successfully carried out in over 1000 children from across the world. Several children have completed a follow up of 12 years and no child has had to reoperated due to recurrence of contracture or deformity.

2.1. Recommended age for SEMLARASS

The best functional outcomes are achieved between the age group of 4 to 6 years. The child develops a mature gait pattern by the age of 4 years and is better able to cooperate with an intensive post-operative physiotherapy programme. Once this window of opportunity is lost (usually due to reluctance of physiotherapists or physicians to let go or the insistence of the family in exploring non-operative options at any cost) and complex decompensated joint pathology has developed, the results of operation are less gratifying, though functional improvements still occur in older children and adults. Unstable lever arm disease must be operated irrespective of age if there is to be any hope of preserving ambulation.

2.2. Advantages of SEMLARASS

i. There is no loss of antigravity activity and weakness of the muscles because mono-articular muscles are preserved,

ii. There is no loss of sensation or sense of stereognosis,

iii. There is no increase in the occurrence of dislocations,

iv. There is little risk of recurrence of contractures and deformities with continued growth,

v. Only one hospital admission and post operative rehabilitation period,

vi. Even non ambulatory cases improve functionally and achieve a better quality of life, and

vii. More cost effective compared to other treatment options (Sharan D, 2005).

2.3. Functional results of SEMLARASS

The study included 314 children with CP with mean age 9.7±4.8 years. The types of CP were spastic diplegia (58%), spastic quadriplegia (35%) and spastic/athetoid/dystonic quadriplegia

http://dx.doi.org/10.5772/57957

(7%). The results showed a significant improvement after a 1 year post-surgical rehabilitation. Correlation studies showed median value of Functional Mobility Scale (FMS) of 3 before surgery and 5 after surgery. Before surgery the median value of Gross Motor Functional Classification System (GMFCS) was level 4 and after surgery it was level 2. The GMFCS improved 2 levels on average. Before surgery the mean value of Pediatric QOL (PQOL) was 39.64±17.49; after surgery the mean value was 23.11 ±14.02. Before surgery median value of Manual Ability Classification System (MACS) was 3 and after surgery it was 1. Children with severe CP (GMFCS IV and V) showed more positive correlation than mild to moderate cases. No child was wheel chair bound at the end of the rehabilitation and all the children were able to walk at least with help of a walking aid. A significant improvement was noted in their participation levels, motivation and a significant improvement in the over all quality of life. Over 50 patients have been followed up beyond 10 years and there have been no recurrences (Sharan D, 2012).

2.4. Outcome of SEMLARASS in severe cerebral palsy (GMFCS IV AND V)

Gross Motor Function Classification System (GMFCS) is a 5 level classification system with clinically meaningful distinctions in motor function between levels and its emphasis on self-initiated movement with particular emphasis on sitting (trunk control) and walking. Persons with CP at GMFCS levels IV and V are non-ambulatory and at a greater risk of complications like hip subluxation/dislocation, Osteopenia/Osteoporosis and Low Energy Fractures. Prevention of these complications requires that these persons are made ambulant with or without support. However, the recommended rehabilitation strategy at present for these groups is wheel chair aided mobility leading to a "Catch 22" situation. The purpose of the study was to find out the functional outcome of SEMLARASS and rehabilitation in persons with cerebral palsy (GMFCS levels IV and V). In this study 170 children with GMFCS V&IV were participated. Mean age of the participants was 9.68±4.77. The follow up ranged from 1 year to 3 years (mean= 1 year). The outcome measures such as component of Gross Motor Function Measure (GMFM-88), Functional Mobility Scale (FMS), Physicians Rating Scale (PRS), Manual Ability Classification System (MACS) were used to compare the functional status of the child before and after SEMLARASS. The results showed a significant improvement in all GMFM-88 components and the values were Lying and Rolling (A); GMFM V: t-9.77 (P<0.001), GMFM IV t-8.56 (P<0.001), Sitting (B); GMFM V: t-20.01 (P<0.001), GMFM IV: t-12.61 (P<0.001), Crawling and Kneeling (C); GMFM V: t-22.26 (P<0.001), GMFM IV: t-21.01(P<0.001); Standing (D); GMFM V: t-20.01 (P<0.001), GMFM IV: t-22.64 (P<0.001),Walking, Running and Jumping (E); GMFM V: t-12.71 (P<0.001), GMFM IV t-15.65 (P<0.001), and total GMFM-88; GMFM V t-31.55 (P<0.001), GMFM IV: t-32.86 (P<0.001), respectively. The result of Pre-Post PRS evaluation showed a significant improvement for both sides (Right: t-8.60, (P<0.001); Left: t-9.21, (P<0.001). The improvement in the MACS (Right: t-4.05 (P<0.001); Left: t-5.74 (P<0.001) and FMS (t-5.46 (P<0.001) were also significant among both GMFCS V and IV (Sharan, 2012).

2.5. Complications of SEMLARASS

A study was done to quantify the complications encountered during rehabilitation following SEMLARASS in 463 consecutive patients. The following complications were reported: Myofascial Pain Syndrome (149, 32.60%), Prolonged Articular Stiffness beyond 4 weeks (111, 24.23%), Patellofemoral Pain Syndrome (38, 8.13%), Osteopenia (36, 7.88%), Meralgia Paresthetica (26, 5.69%), Pressure Ulcers (19, 4.10%), Hypertrophic Scar (18, 3.94%), Low Energy Fractures (19, 4.06%), Superficial Pin Tract Infection (12,2.56%), Wound Dehiscence (9,1.92%), Patellar Tendinitis (8,1.71%), Myositis Ossificans (7,1.51%), Complex Regional Pain Syndrome (5, 1.07%), Rickets (3, 0.6%), Osteomyelitis (2, 0.43%), Transient Common Peroneal Nerve Palsy (2, 0.43%), Transient Axillary Nerve Palsy (2, 0.43%), Skin Hypersensitivity (1, 0.21%), and IT Band Friction Syndrome (1, 0.21%). There was a significant association between the anatomical distribution of abnormality and osteopenia ($\chi 2 - 8.01$, $p<0.05$). A preoperative GMFCS level IV and V was associated with a higher prevalence of complications like Osteopenia, Low Energy Fractures and Myositis Ossificans. However, none of the complications were life threatening, permanent or affecting the long term outcome of surgery. To minimise the rate of complications we recommend a structured rehabilitation protocol carried out by an experienced multidisciplinary medical team. Before the surgery, the patients, parents and care givers should be counselled regarding the prevalence of these complications, along with the available prevention and treatment options (Sharan D, 2013).

2.6. Surgical Techniques used in SEMLARASS

SEMLARASS differs from conventional orthopaedic surgery in the following ways:

a. **Technique of surgery:** conventional orthopaedic surgery relies on tendon lengthening, aponeurotic releases and tendon transfers, while SEMLARASS employs the techniques of selective spasticity control surgery, e.g., intramuscular myofascial lengthening and controlled sliding tendon lengthening. Tendon transfers are avoided in SEMLARASS.

b. **Selective release of multiarticular muscles:** In conventional orthopaedic surgery, a 5 year old child with spastic diplegia who walks on the toes, intoed, crouched at hips and knees, with equinovalgus would be treated by adductor tenotomy (and possibly obturator neurectomy), iliopsoas lengthening, distal hamstring lengthening or advancement (Eggers procedure), tendoachilles lengthening and Grice fusion of subtalar joint. In SEMLARASS, the procedures would be OSSCS of psoas, gracilis, medial hamstrings, gastrocnemius, femoral rotational osteotomy and medial displacement sliding calcaneal osteotomy. Hence, in conventional orthopaedic surgery several monoarticular muscles are released while in SEMLARASS only multiarticular muscles are selectively released.

c. **Simultaneous release of flexor and extensor muscle groups in each joint:** in SEMLARASS (except at wrists, hands and feet), while in conventional orthopaedic surgery only the flexor or the extensor muscle group is usually released.

d. **Timing of surgery:** conventional orthopaedic surgery is recommended in adolescence to avoid the risk of postoperative recurrence of contractures and deformities. Since SEM-

LARASS corrects lever arm dysfunction at a much younger age the risk of recurrence is minimal.

e. **Simultaneous lever arm restoration:** in conventional orthopaedic surgery, lever arm restoration is left till adolescence and is often not combined with soft tissue surgery. It is frequently not addressed at all. Unlike SEMLARASS, joint fusions are often used and internal fixation (plates and screws) is used to stabilise osteotomies that requires a second operation to remove the implants.

2.7. Assessment in SEMLARASS

In SEMLARASS the assessment is a "bridge" between the patient, surgery, rehabilitation and prognosis. This bridge carries the patient from one phase of the treatment to the next phase. Assessments in SEMLARASS are mainly divided into: 1) Pre-operative assessment, 2) Post-operative assessment during rehabilitation and 3) Functional outcome measurement.

2.7.1. Pre-operative assessment

Pre operative assessment starts from first consultation with the Paediatric Orthopaedic Surgeon followed by other assessments by paediatric neurologist/developmental paediatrician, physical and occupational therapists, child psychologist, speech therapist and medical social worker.

The main objectives of the pre operative assessment are:

1. To make an accurate diagnosis, whether CP or other progressive neuromuscular disorder
2. To classify the type of CP and level of impairment
3. Screening any associated risk factors in the patient
4. To determine whether SEMLARASS is indicated
5. If SEMLARASS indicated, then to decide the nature of surgical procedures
6. For predicting the expected functional outcome

The following methods are used at RECOUP for pre-operative assessment:

1. Detailed history,
2. General assessment,
3. Assessment of sensory deficit and function,
4. Neurological assessment,
5. Musculoskeletal assessment, and
6. Instrumented gait analysis

2.7.2. Musculoskeletal assessment and surgical prescription

The primary objective of the musculoskeletal assessment is to assess the musculoskeletal problems in the extremities and prescribe the appropriate surgical procedures. The assessment includes measurement of spasticity along with associated fixed deformities and lever arm dysfunctions.

Assessment	Surgical Prescription
Staheli test (hip flexion deformity)	OSSCS psoas
Duncan Ely's test (rectus femoris contracture)	OSSCS rectus femoris
Adductor test (hip adductor contracture)	OSSCS gracilis, distal adductor magnus
Popliteal angle (hamstring contracture)	OSSCS distal and/or proximal hamstring
Silfverskiold test (gastrocnemius contracture)	OSSCS gastrocnemius
Craig test (increased femoral anteversion)	Distal femoral derotation osteotomy
Hip subluxation or dislocation	Proximal femoral varus derotation osteotomy
Hip dislocation with dysplastic (shallow) acetabulum	Proximal femoral varus derotation osteotomy and tectoplasty
Tibial torsion	Proximal tibial derotation osteotomy
Hind food valgus	Medial displacement calcaneal osteotomy
Plano-valgus foot with midfoot break	Calcaneal lengthening osteotomy

Table 1. Surgical prescription based on pre-operative assessment

2.7.3. Assessment during postoperative rehabilitation

Purpose of the assessment:

1. To analyse the progress from one phase to another
2. To assess effectiveness of the treatment interventions
3. To determine the need to modify the interventions and
4. To judge the prognosis during the rehabilitation.

The main assessment methods used are:

1. Goniometry for hip, knee and ankle range of motion
2. Muscle power, spasticity and/or deformity
3. Soft tissue integrity
4. Balance and antigravity control
5. Mobility status and

6. Over all functional status.

This periodic assessment is usually performed by the Paediatric Orthopaedics Surgeon, who makes immediate changes in the treatment protocol in discussion with the rehabilitation team during the rehabilitation to ensure the optimal functional outcome.

2.7.4. Functional outcome measurement

The functional outcome measurement are performed one day before the surgery, 3rd postoperative month, 6th month, 1 year and after 2 years. The objectives of this assessment is: 1) to find out the overall outcome of SEMLARASS, 2) to compare the effect of SEMLARASS in different types of CP (e.g., based on GMFCS and MACS), 3) to document the results and publish research studies related to SEMLARASS.

At RECOUP, the following outcome measures are commonly used:

1. Gross Motor Functional Measure (GMFM),
2. Modified Ashworth Scale,
3. The Amsterdam Gait Classification,
4. Melbourne Assessment of Unilateral Upper Limb Function and House Classification for upper extremity use,
5. Functional Mobility Scale (FMS),
6. Pediatric Outcomes Data Collection Instrument (PODCI),
7. The Fahn-Marsden dystonia scales, and
8. Pediatric Quality of Life Inventory (PEDS QL)

2.8. Post-SEMLARASS rehabilitation

A successful functional outcome after SEMLARASS requires intensive rehabilitation for atleast 2 years after the operation. Children and their parents need to be well prepared before the operation, as well as strongly motivated after the operation in order to endure the prolonged rehabilitation process. A Paediatric Physiotherapist is designated as the programme coordinator.

The entire Post-SEMLARASS Rehabilitation is divided into four phases:

a. Non-ambulatory phase: 0-2 weeks
b. Weight bearing phase: 2-4 weeks
c. Ambulatory phase: 1-6 months
d. Maintenance phase: 6-24 months

Stages	Time Period	Treatment Choice	Intervention	Outcome Measure
Phase 1 Non Ambulatory Phase	0-2 weeks	Soft Tissue Techniques	Oedema Control, Scar Tissue Mobilisation, MFR, Stretching	MMT, ROM
		Pain Management Techniques	Wax Bath, IFT,TENS, UST, Desensitisation, PRT, MTrP Therapy (Cross Friction Massage, Ischaemic Compression), Relaxation Exercises, CBT, Imagery, Breathing Exercises	
		Joint Mobilisation Techniques	Mulligan MWM: Hip, Knee, Ankle	
		Basic Strengthening Exercises	FES, Isometric Exercises, PRE with Theraband, MET, PNF, Aquatic Therapy	
		Balance Training	Ball Exercises	
Phase II Weight Bearing Phase	2 – 4 weeks	Advanced Strengthening Exercises	FES, PRE with Weight Cuffs, Core Stabilisation, Hippotherapy, Aquatic Therapy, PNF, Ball Exercises, EMG Biofeedback	MMT, ROM
		Advanced Balance Training	Ball Exercises, Exercises on Balance Board/ Wobble Board/Trampoline, Anti Gravity Muscle Strengthening, Virtual Reality Based Therapy	
		Mobility Exercises	Mat Activities, Static Cycling	
		Gait Training	Body Weight Supported Treadmill Training, Parallel Bar Walking (with Assistive Devices)	
Phase III Ambulatory Phase	1 – 6 Months	Advanced Balance Training	Ball Exercise, Balance Board Training with/without Assistive Devices, Dynamic Mobility Exercises, Virtual Reality Based Therapy	MMT GMFCS GMFM FMS PedsQL (Child & Parent Report), MACS The Amsterdam Gait Classification
		Advanced Strengthening Exercises	Core Stabilisation, PNF, Mat Activities, EMG Biofeedback, Hippotherapy, Yoga, Tai Chi	
		Gait Training		
		Sensory Re-education	Sensory Integration	
		Upper Limb Rehabilitation	MFR, Stretches, CIMT, Fine Motor Activities, Virtual Reality Based Therapy	
Phase IV Maintenance Phase	6 - 24 Months	Gross Motor Functions and Fine Motor Activities	Cycling, Running on Treadmill, Uphill Running, Threading a Needle, Painting, Drawing, Shoelace tying, Writing	GMFCS GMFM FMS

http://dx.doi.org/10.5772/57957

Stages	Time Period	Treatment Choice	Intervention	Outcome Measure
		Sporting Activities	Badminton, Basketball and Football	MMT PedsQL (Child & Parent Report) PODCI The Amsterdam Gait Classification
		Community Based Rehabilitation	Community Walking	

Table 2. Post SEMLARASS Rehabilitation Protocol. SEMLARASS Post-Surgical Rehabilitation Intervention Pathway

Abbreviations:

PRT – Positional Release Techniques

TrP – Trigger Point Therapy

MFR – Myofascial Release

MMT – Manual Muscle Testing

ROM – Range of Motion, measured by a goniometer

MWM – Mobilisation with Movement

CBT – Cognitive Behavioural Therapy

IFT – Interferential Therapy

TENS – Transcutaneous Electrical Nerve Stimulation

UST – Ultrasonic Therapy

FES – Functional Electrical Stimulation

PRE – Progressive Resistive Exercises

MET – Muscle Energy Techniques

PNF – Proprioceptive Neuromuscular Facilitation

EMG – Electromyograph

MACS - Manual Ability Classification System

GMFCS - Gross Motor Function Classification System

GMFM - Gross Motor Function Measure

PedsQL – Paediatric Quality of Life Inventory

PODCI - Paediatric Outcome Disability Classification Index

PRS - Physician Rating Scale

FMS - Functional Mobility Scale

CHQ - Child Health Questionnaire

CIMT – Constraint Induced Movement Therapy

2.9. Prognosis following SEMLARASS

The DEEPAK SHARAN'S Prognostic Score for CP is currently undergoing validation studies. The following factors have been found to determine the final functional results following SEMLARASS:

- Dysfunction of lever arm in the lower extremities
- Epileptic episodes in last two years
- Excessively lengthened tendons (e.g., due to previous Botulinum Toxin injections or Tendon lengthening surgery)
- Post operative complications in relation with severity and its duration
- Abnormalities of bone mineralisation
- Known cognitive deficits
- Sensory problems
- Hand function
- Age at the time of surgery
- Rehabilitation in terms of intensity and quality
- Achieved level of gross motor function
- Neurosurgery in the past
- Socioeconomic background

3. Representative case studies

3.1. Case study 1

Prop Status

H, a 7 year old boy with spastic diplegia could barely take few steps when held by an adult with severe crouching at knee and toe walking. The GMFCS level was 4.

Treatment

SEMLARASS, with OSSCS Psoas, proximal and distal Semimembranosus, Semitendinosus, Gastrocnemius, along with bilateral femoral external rotation osteotomies followed by 6 months of post operative rehabilitation.

http://dx.doi.org/10.5772/57957

Current functional status

At a follow up of 6 years, he was walking and performing all the activities of daily living independently. Now he carries his bag to school by himself, climbs stairs and also participates in running, football and cycling. The current GMFCS level was 1.

"Our son, born with Spastic Diplegia, till the age of five could only walk on toes with constant support and now after the surgery and aggressive rehabilitation at RECOUP started walking independently with lot of confidence and enthusiasm."
Parents of H

Figure 1. Before and after SEMLARASS

3.2. Case study 2

Preop status

S, a 10 year boy, was a known case of spastic diplegia. He was an independent ambulator with severe crouching at knee, intoeing and toe walking, with frequent falls. The GMFCS level was 2.

Treatment

SEMLARASS, with OSSCS Psoas, distal Semimembranosus, Semitendinosus, Gastrocnemius, along with bilateral femoral external rotation osteotomies followed by 3 months of post operative rehabilitation.

Current functional status

At a follow up of 5 years, he was walking independently with a near normal gait and doing all the activities of daily living himself. Now he can sprint and play sports at a competitive level. The current GMFCS level was 1.

"My child made quick improvements following SEMLARASS. Now, he walks and runs independently. When he first attained physical stability after the surgery he would just run around nonstop to express his joy. He now calls himself Mohan Lal (after the famous Malayalam Cine Star) since he believes he is now a different person altogether. We are thankful to the RECOUP team for giving our child a life which we once thought impossible." **Parents of S**

Figure 2. Before and after SEMLARASS

3.3. Case study 3

Preop status

A, was a 12 year old girl and could barely take a few steps with extremely crouched knee with the support of an adult (GMFCS level 4). She was studying in a normal school but had to be carried by her parents to and back from the school, and could not participate in any outdoor activities.

Treatment

SEMLARASS, with OSSCS Psoas, proximal and distal Semimembranosus, Semitendinosus, Biceps Femoris, Gastrocnemius, along with bilateral femoral external rotation osteotomies, tibial internal rotation osteotomies and medial displacement calcaneal osteotomies, followed by 8 months of post operative rehabilitation.

Current functional status

At 5 year follow up, she was walking independently, doing all the activities of daily living herself, travelling to school by school bus without any assistance. The current GMFCS level was 2.

http://dx.doi.org/10.5772/57957

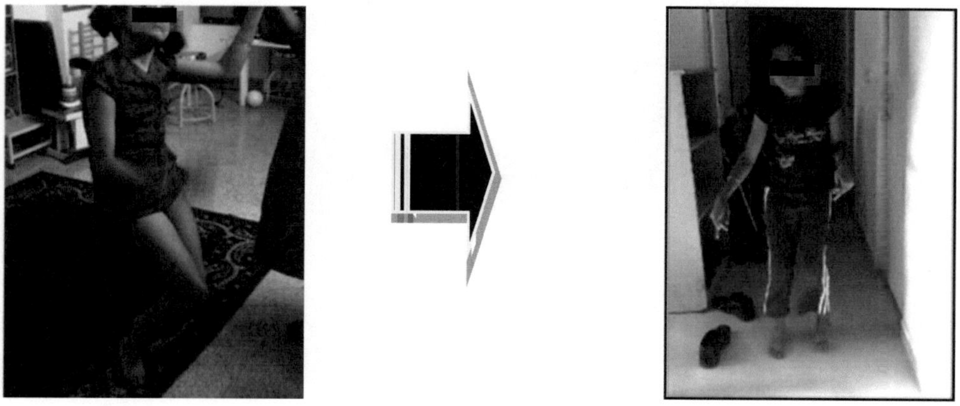

"I was 100 percent confident that, I would be able to walk independently. Finally I reached that goal. This is a rare and unexpected experience in life. Now I am enjoying my life with my friends and no barrier in front me." **A**

Figure 3. Before and after SEMLARASS

3.4. Case study 4

Preop status

A, was a 19 year old girl with spastic athetoid hemiplegia (GMFCS level 2) with severe and disfiguring upper limb deformities, including swan neck deformities of all fingers. She had undergone failed tendon transfers in the forearm and wrist earlier.

Treatment

OSSCS right forearm flexors and pronators and hand intrinsics, distal semimembranosus, gastrocnemius, tibialis posterior and external rotation tibial osteotomy, followed by 3 months of post operative rehabilitation.

Current functional status

At a follow up of 8 years she had a cosmetically acceptable and functional upper extremity and was an independent community ambulator (GMFCS level 1).

3.5. Case study 5

Preop status

A, an 11 year old boy was diagnosed to have cerebral palsy (spastic quadriplegia, GMFCS level 5), was always carried by his father. He had not attained sitting balance and had severe spasticity and contractures in all 4 extremities.

Treatment

SEMLARASS, with OSSCS Psoas, Gracilis, proximal and distal Semimembranosus, Semitendinosus, Biceps Femoris, Gastrocnemius, along with bilateral femoral external rotation

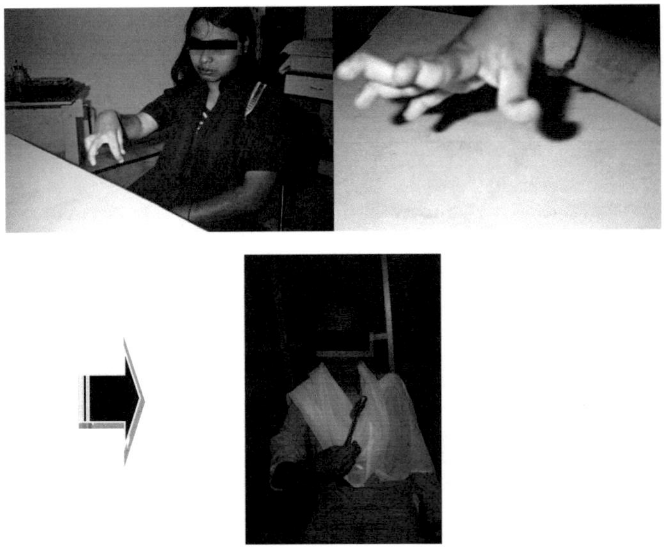

Figure 4. Before and after SEMLARASS

osteotomies, tibial internal rotation osteotomies and medial displacement calcaneal osteotomies, followed by 4 months of post operative rehabilitation.

Current functional status

At a follow up of 6 years, he was walking in the community with help of elbow crutches and a few steps independently and doing all the activities of daily living herself. Now he was studying in normal school and walked to school himself. The current GMFCS level was 2.

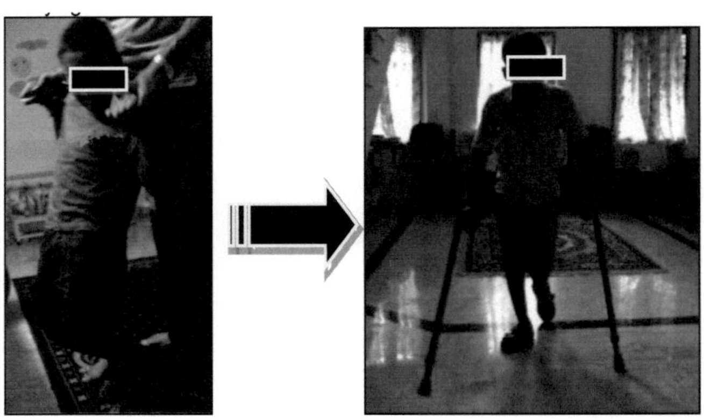

"While critically analysing the progress accomplished by my son through SEMLARASS I would like to mention that the improvement shown by him was dramatic and beyond my expectations. Before surgery I had to carry him to the school due to marked crouching and tip toeing. Now these problems are corrected to a great extent and he is walking with just one hand support and short distances independently." **Parents of A.**

Figure 5. Before and after SEMLARASS

http://dx.doi.org/10.5772/57957

3.6. Case study 6

Preop status

B, an 18 year old male and a known case of spastic quadriplegia was a non ambulator, an assisted sitter with severe spasticity in all 4 extremities (GMFCS level 5). He needed two caregivers to help him transfer.

Treatment

SEMLARASS, with OSSCS Psoas, Gracilis, proximal and distal Semimembranosus, Semitendinosus, Biceps Femoris, Gastrocnemius, along with bilateral femoral external rotation osteotomies, tibial internal rotation osteotomies and medial displacement calcaneal osteotomies, followed by 8 months of post operative rehabilitation.

Current functional status

At a follow up of 5 years, he was walking with help of single stick or one hand support and doing all the activities of daily living himself. He could stand independently for short durations and was now studying in a special school. The current GMFCS level was 3.

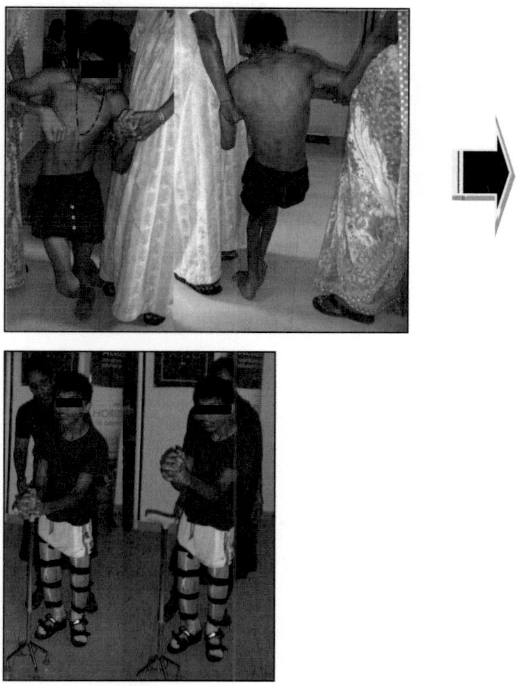

*"I am extremely thankful to the staff of RECOUP for their immense efforts to help me achieve a good quality life. Now I have the ability to operate a computer and I am studying computers in my school. My future plan is to advertise the experiences and miracles in my life for helping other children with disabilities"***B.**

Figure 6. Before and after SEMLARASS

3.7. Case study 7

Preop status

O, a 10 year boy had spastic athetoid quadriplegia with generalised dystonia. He was a non ambulator and an assisted sitter with non-functional upper limbs (GMFCS level 5).

Treatment

He underwent SEMLARASS, with OSSCS Psoas, distal Semimembranosus, Semitendinosus, Gastrocnemius, along with bilateral femoral external rotation osteotomies. He also underwent OSSCS of proximal biceps femoris, triceps, distal teres major, latissimus dorsi, forearm flexors, pronators, hand intrinsics and adductor pollicis, followed by 9 months of rehabilitation.

Current functional status

At a follow up of 5 years, he could walk inside the house independently and used a single stick for walking in the community. He had good hand function on both sides and was independently going to school. The current GMFCS level was 2.

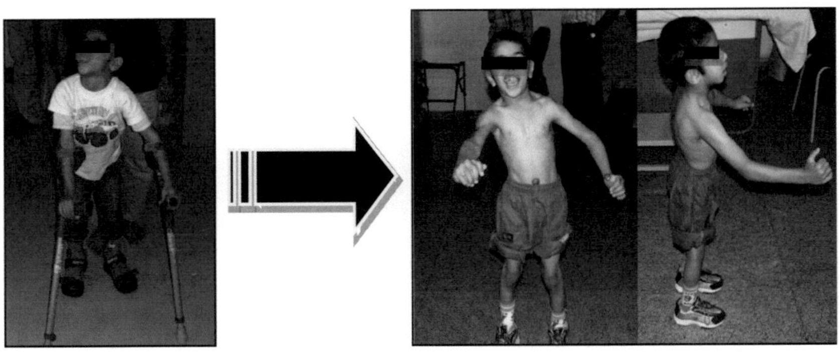

Figure 7. Before and after SEMLARASS

3.8. Case study 8

Preop status

S, a 10 year old girl, and a known case of spastic athetoid quadriplegia, did not have head control and was unable to sit or lie down in bed due to the severe spasticity and abnormal movements. The GMFCS level was 5.

Treatment

She underwent SEMLARASS, with OSSCS Psoas, distal Semimembranosus, Semitendinosus, Gastrocnemius, along with bilateral femoral external rotation osteotomies. She also underwent OSSCS of proximal biceps femoris, triceps, distal teres major, latissimus dorsi, forearm flexors, pronators, hand intrinsics and adductor pollicis in the second stage, followed by 9 months of rehabilitation.

http://dx.doi.org/10.5772/57957

Current functional status

At a follow up of 5 years, she was walking with rollator and sitting without support. She had achieved complete head control and good hand function. Also the abnormal movements were greatly reduced. The current GMFCS level was 3.

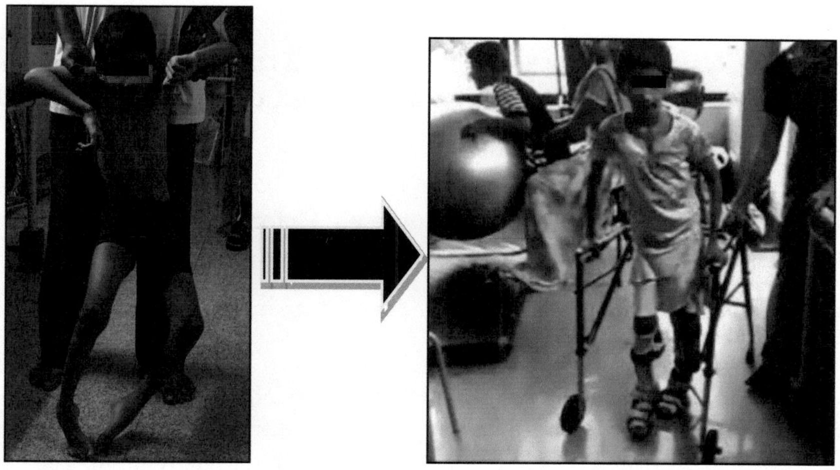

Figure 8. Before and after SEMLARASS

4. Conclusion

SEMLARASS is the first and only treatment modality for CP to systematically, simultaneously and effectively address all the major problems restricting the child's functional activities, i.e., spasticity or movement dysfunction and lever arm dysfunction. Currently, a well-planned and executed SEMLARASS, in the context of a multi-disciplinary team, provides the child with CP with the best hope for a dramatic, predictable and lasting functional improvement (e.g., improved GMFCS level). This is especially true of spastic quadriplegia, dystonia and athetosis that account for over two-thirds of cases of CP, where available treatment options are extremely limited.

Author details

Deepak Sharan*

Address all correspondence to: deepak.sharan@recoup.in

Dept. of Paediatric Orthopaedics and Rehabilitation, RECOUP Neuromusculoskeletal Rehabilitation Centre, Bangalore, Karnataka, India

References

[1] Gage JR, Novacheck TF. An update on the treatment of gait problems in cerebral palsy. J Pediatr Orthop B. 2001 Oct;10(4):265-74.

[2] Novacheck TF, Gage JR. Orthopedic management of spasticity in cerebral palsy. Childs Nerv Syst. 2007 Sep; 23(9):1015-31.

[3] Gracies JM, Nance P, Elovic E, McGuire J, Simpson DM. Traditional pharmacological treatments for spasticity. Part II: General and regional treatments. Muscle Nerve Suppl. 1997; 6: S92-120.

[4] Hattab JR. Review of European clinical trials with baclofen. In: Feldman RG, Young RR, Koella WP (eds.), Spasticity: Disordered Motor Control. Chicago: Year Book Medical Publishers 1980, pp. 71–86.

[5] Ryan DM, Blumenthal FS. Baclofen-induced dyskinesia. Arch Phys Med Rehabil. 1993 Jul; 74(7): 766-7.

[6] Kolaski K, Ajizian SJ, Passmore L, Pasutharnchat N, Koman LA, Smith BP. Safety profile of multilevel chemical denervation procedures using phenol or botulinum toxin or both in a pediatric population. Am J Phys Med Rehabil. 2008 Jul; 87(7): 556-66.

[7] Jianjun L, Shurong J, Weihong W, Yan Z, Fanyong Z, Nanling L. Botulinum toxin-A with and without rehabilitation for the treatment of spastic cerebral palsy. J Int Med Res. 2013 Jun; 41(3): 636-41.

[8] Butler C, Darrah J. Effects of neurodevelopmental treatment (NDT) for cerebral palsy: an AACPDM evidence report. Dev Med Child Neurol. 2001 Nov; 43(11): 778-90.

[9] Sampson FC, Hayward A, Evans G, Morton R, Collett B. Functional benefits and cost/benefit analysis of continuous intrathecal baclofen infusion for the management of severe spasticity. J Neurosurg. 2002 Jun; 96(6): 1052-7.

[10] Motta F, Buonaguro V, Stignani C. The use of intrathecal baclofen pump implants in children and adolescents: safety and complications in 200 consecutive cases. J Neurosurg. 2007 Jul; 107(1 Suppl): 32-5.

[11] Ross JC, Cook AM, Stewart GL, Fahy BG. Acute intrathecal baclofen withdrawal: a brief review of treatment options. Neurocrit Care. 2011 Feb; 14(1): 103-8

[12] Lundkvist A, Hagglund G. Orthopaedic surgery after selective dorsal rhizotomy. J Pediatr Orthop B 2006, 15: 244–246.

[13] Crawford K, Karol LA, Herring JA. Severe lordosis after dorsal rhizotomy. J Pediatr Orthop. 1996 May-Jun; 16(3): 336-9..

http://dx.doi.org/10.5772/57957

[14] Mooney JF 3rd, Millis MB. Spinal deformity after selective dorsal rhizotomy in patients with cerebral palsy. Clin Orthop 1999; 364: 48-52.

[15] Turi M, Kalen V. The risk of spinal deformities after selective dorsal rhizotomy. J Pediatr Orthop 2000; 20(1):104-7.

[16] Spiegel DA, Loder RT, Alley KA, Rowley S, Gutknecht S, Smith-Wright DL, Dunn ME. Spinal Deformity Following Selective Dorsal Rhizotomy; J Pediatr Orthop 2004; 24: 30-36.

[17] Scrutton D, Baird G, Smeeton N. Hip dysplasia in bilateral cerebral palsy: incidence and natural history in children aged 18 months to 5 years. Dev Med Child Neurol 2001; 43: 586–600.

[18] Matsuo T, Tada S, Hajime T. Insufficiency of the hip adductor after anterior obturator neurectomy in 42 children with cerebral palsy. J Pediatr Orthop. 1986 Nov-Dec; 6(6): 686-92.

[19] Matsuo T, Hara H, Tada S. Selective lengthening of the psoas and rectus femoris and preservation of the iliacus for flexion deformity of the hip in cerebral palsy patients. J Pediatr Orthop. 1987 Nov-Dec; 7(6): 690-8.

[20] Steel HH. Gluteus medius and minimus insertion advancement for correction of internal rotation gait in spastic cerebral palsy. J Bone Joint Surg Am. 1980 Sep; 62(6): 919-27.

[21] Rutz E, Tirosh O, Thomason P, Barg A, Graham HK. Stability of the Gross Motor Function Classification System after single-event multilevel surgery in children with cerebral palsy. Dev Med Child Neurol. 2012 Dec; 54(12): 1109-13.

[22] Matsuo T. Cerebral Palsy: Spasticity control and Orthopaedics: An Introduction to Orthopaedic Selective Spasticity Control Surgery. Soufusha, Tokyo, 2002.

[23] Sharan D. Recent advances in the management of Cerebral Palsy. Indian Journal of Pediatrics 2005; 72(11): 969-973.

[24] Sharan D. Functional Outcome of a new Single Event Multilevel Lever Arm Restoration and Anti Spasticity Surgery for Cerebral Palsy. 33rd Orthopaedic World Conference, Dubai, United Arab Emirates (November 28-30, 2012).

[25] Sharan D. Functional Outcome of a New Surgical Approach in Severe Cerebral Palsy (GMFCS IV and V), 4th International Cerebral Palsy Conference, Pisa, Italy (October 10-13, 2012).

[26] Sharan D. Complications during Post-Surgical Rehabilitation Following Single Event Multilevel Surgery in Cerebral Palsy, 14th International Society for Prosthetics and Orthotics 2013 World Congress, Hyderabad, India (February 4-7, 2013).

Chapter 7

Stem Cell Therapy for Cerebral Palsy – A Novel Option

Alok Sharma, Hemangi Sane,
Nandini Gokulchandran, Prerna Badhe,
Pooja Kulkarni and Amruta Paranjape

Additional information is available at the end of the chapter

http://dx.doi.org/10.5772/57152

1. Introduction

Discovery of stem cells by James Till and Ernest McCulloch in 1961, stands as one of the most remarkable medical-research achievements of the 20th century. This discovery provided a foundation for further breakthroughs in the field of stem cells. Sir Martin J. Evans along with Mario R. Capecchi, and Oliver Smithies were jointly awarded a Nobel Prize in 2007 for their contribution in introducing specific gene modifications in mice by the use of embryonic stem cells. Later in 2012, John B. Gurdon and Shinya Yamanaka were also jointly awarded a Nobel Prize for discovering that mature cells can be reprogrammed to become pluripotent cells. [1]

Ramon y Cajal in 1926 stated "Once the development was ended, the founts of growth and regeneration of the axons and dendrites dried up irrevocably. In the adult centers the nerve paths are something fixed, ended, and immutable. Everything may die, nothing may be regenerated. It is for the science of the future to change, if possible, this harsh decree." [2]. It was a long-standing belief that cells of the central nervous system once damaged cannot be regenerated. The medical science of stem cells has finally made restoration of CNS possible which has changed the old concept of medicine. Not too long ago, this therapy was hamstrung by various controversies, ethical and moral issues. But, tremendous progress of research in this field has finally led to its translation from laboratory to innovative cellular therapies.

A variety of cells including embryonic stem cells, adult stem cells, umbilical cord blood cells and induced pluripotent stem cells have been explored as a therapeutic alternative for treating a broad spectrum of neurologic disorders including stroke, Alzheimer's, Parkinson's, spinal cord injury, cerebral palsy etc. amongst others. It is essential to select suitable cells depending on the nature and status of neurological dysfunctions to achieve optimal therapeutic efficacy. Along with the selection of cells, the route of administration also plays an important role to maximize the clinical therapeutic effect of the cell therapy. Numerous preclinical studies have been carried out to study the safety of intrathecal, intravenous and direct cerebral implantation. A plethora of published literature is also available to provide evidence of stem cells initiating functional restoration of CNS. The postulated mechanisms of action involved are neuromodulation, neuroprotection, axon sprouting, neural circuit reconstruction, neurogenesis, neuroregeneration, neurorepair, and neuroreplacement.

In view of the fact that stem cell therapy has a promising therapeutic potential in the treatment of neurological disorders, it is important for all the professionals in the medical field to understand the concepts of this upcoming therapeutic strategy.

In this chapter, we have focused on stem cell therapy for Cerebral Palsy (CP) which is a heterogeneous group of neurological disorders mainly observed in infants. It results due to a static brain lesion at the time of pregnancy or early life. The survival of CP children has increased due to advanced modern medicine which has led to their growing population. CP involves impairment of movement, muscle function, and cognitive functioning and the effects range from mild to severe. [3] Chronic motor disability along with intellectual disability, epilepsy, behavioral disorders, and sensory and perceptual disturbances are few of the complications seen in these patients. No biological intervention has been effective for CP and the standard approach is limited to supportive management strategies which do not address the core issue of neural tissue damage. Currently, stem cell based strategies have garnered attention due to their ability of neuroregeneration and neuroprotection in CP.

We have discussed the clinical aspects of stem cell therapy in cerebral palsy supported by various human case studies and clinical trials. We have also enumerated our experience and results wherein our subjects were administered autologous bone marrow mononuclear cells. In our study it was found to be safe, feasible and efficacious and may be used as a combinatorial strategy with the currently available standard treatments.

Here, we try to summarize the current vast knowledge available for stem cell therapy in cerebral palsy.

2. Stem cells

Stem cells are defined as "cells that have the ability to renew them continuously and possess pluripotent or multipotent ability to differentiate into many cell types." [4]These cells exhibit a unique property of "plasticity" where in cells isolated from one tissue convert to cells of different tissues by crossing lineage barriers and adopting the expression profile and phenotype of cells that are unique to other tissues. [5]

Stem cells are categorized based on their potential to differentiate into other types of cells.

1. Totipotent cells: These cells have the ability to differentiate into all possible cell types of the human body including extra embryonic and placental cells.

http://dx.doi.org/10.5772/57152

2. Pluripotent cells: These cells have ability to differentiate into any of the three germ layers viz. endoderm, mesoderm and ectoderm.

3. Multipotent cells: These cells have the ability to differentiate into specialized cells.

4. Oligopotent cells: These cells have the ability to differentiate into a few cell types.

5. Unipotent cells: These cells have the ability to produce cells only of their own type, but are capable of self-renewal to be classified as a stem cell.

2.1. Types of stem cells

Stem cells are broadly classified based on their source, as follows:

i. **Embryonic stem cells (ESCs):** These cells are pluripotent cells derived from a 4-7 day old blastocyst stage embryo. The cells are harvested from the inner cell mass (ICM) of the blastocyst. Invitro, they can be indefinitely maintained and expanded as pure populations of undifferentiated cells. [6] Inspite of their great potency in tissue repair, these cells have triggered various ethical and moral issues due to the destruction of human embryos involved. [7] They also have tumorigenic side effects as ESCs and tumor cells share cellular and molecular phenotypes such as rapid proliferation rate, lack of contact inhibition, a susceptibility to genomic instability, high activity of telomerase, high expression of oncogenes and epigenetic status amongst others. They form teratomas which have the potential to degenerate into malignant teratocarcinomas [8] The likelihood of development of tumors in children cannot be overlooked as they have many years of life ahead of them for the tumor formation to occur.

ii. **Fetal Stem Cells:** These cells are isolated either from the aborted fetus or from the extra embryonic structures of the fetal origin such as the amniotic fluid and placenta. Fetal blood is a rich source of haemopoietic stem cells (HSC). Non-haemopoietic mesenchymal stem cells (MSC) are also found in the first trimester fetal blood. [9] These cells have better homing capacity, greater multipotentiality and differentiation potential and lower immunogenicity as compared to the allogeneic adult stem cells. [10] Although these cells have a greater therapeutic potential they may also be associated with infections. These cells are prone to KS-associated herpes virus (KSHV) infections. [11] As the safety is not yet substantiated, fetal cells are not often used for transplantation. Use of fetal cells is ethically controversial as it is associated with abortion. Research on fetal cells is permitted only after the decision of abortion is made.

iii. **Umbilical cord:** Umbilical cord contains a heterogeneous mixture of stem/progenitor cells at different lineage commitment stages. Cells are isolated either from the cord blood or the Wharton jelly. They consist of embryonic stem cell-like and other pluripotential stem cells, which can give rise to hematopoietic, epithelial, endothelial, and neural tissues. [12] Various banks have evolved to collect and preserve the umbilical cord blood. But the utility of these centers is still questionable. The protocols and guidelines for collection and retrieval of cells are still being standardized. Other disadvantages of use of UCBCs are limited by the fact that minimum necessary dosage of cells for cell engraftment is 1×10^7 cells per kilogram which includes the total nucleated cell fraction along with stem cells. Thus, the available dose of autologous cells obtained at birth may be insufficient for transplantation at an older age of the child [13].

Amount of stem cells found in the cord blood is 10% less than that obtained from the bone marrow. [14] There are some reports of associated Herpes virus and JC virus infection by allogeneic UCBCs transplantation [15,16]

iv. **Induced pluripotent stem cells (iPSC):** To circumvent the ethical issues involved in the use of embryonic stem cells, pluripotent cells were generated directly from the patients' own cells. Induced pluripotent stem cells are non-pluripotent adult cells (somatic cells) which have been genetically reprogrammed to form pluripotent cells. The iPSC technology develops patient-specific cell therapy protocols. The availability of iPSCs is particularly advantageous for research involving neurological diseases, since it is difficult to obtain diseased tissue sample for study from living patients. [17]

v. **Adult stem cells:** These cells are multipotent stem cells, isolated from adult tissues (i.e. after birth). They include hematopoietic stem cells, bone marrow derived stem cells, adipose tissue-derived stem cells, neural stem cells amongst others [18] Adult stem cells are found in almost all the tissues of the body and help to maintain and repair organs and tissues throughout a person's life. These cells are majorly found in the bone marrow, brain, skeletal muscle, liver, pancreas, fat, skin and skeletal muscle. The different types of adult stem cells include multipotent adult progenitor cells, oligodendrocyte progenitor cells neural stem cells, glial progenitor.

2.2. Major sources of adult stem cells

i. **Bone marrow:** Anterior or posterior superior iliac crest is the preferred site for the bone marrow aspiration. If bone marrow cannot be obtained from the iliac crest due to positioning difficulties or obesity, sternum may be used in adults. However, aspiration from sternum poses a great risk of complication. [19]

Bone marrow is a proficient source of autologous cells with distinct regenerative properties, which can be quickly harvested and are thus applicable for both chronic and acute diseases. Cells isolated from the bone marrow not only differentiate into blood cells but also into neural tissues. [20] The mononuclear cell fraction derived from the bone marrow is a heterogeneous population containing differentially matured B-cells, T-cells and monocytes, as well as rare progenitor cells such as hematopoietic stem cells (HSC), mesenchymal stromal cells (MSC), endothelial progenitor cells (EPC) and very small embryonic-like cells (VSEL). The hematopoietic stem cells (HSCs) are the blood cells which give rise to the myeloid and lymphoid lineages. HSCs also have a potential to transdifferentiate into various nonhematopoietic cell lineages especially neural lineage. [21] [Figure1] Bone marrow mesenchymal stem cells (BMMSCs) give rise to mesodermal lineage cells such as osteoblasts, chondrocytes, adipocytes, and muscle cells along with neuroectodermal cells. [22] BMMSCs express a unique surface

http://dx.doi.org/10.5772/57152

molecule profile, including expression of CD13, CD29, CD44, CD49e, CD73, CD90, CD105, CD146, CD166, CD271, STRO-1, Octamer-4 (Oct4), and stage-specific embryonic antigen-4 (SSEA4). It is generally believed that BMMSCs are negative for hematopoietic cell markers such as CD14, CD34, c-kit, SCA1. [23]

It has been observed that use of cell mixture is more efficacious than individual sub fractionated cells of the bone marrow. They promote angiogenesis, mediate vascular repair, and express several cytoprotective growth factors and cytokines. These cells are also safe and due to its easy availability they are most preferred for cellular therapy. These cells are used for the treatment of various neurological disorders such as cerebral palsy, stroke, Parkinson's, Spinal cord injury, etc. along with diabetes, orthopedic conditions, cancers and wound healing. [24,25]

In brain injury, stem cells induce neuroprotection and neural repair by inflammatory suppression, causing tissue reconstruction of completely or partially damaged cells and prevent cell death. [26] On administration, these cells migrate to the injured tissue and initiate host repair and recovery through direct and indirect cell-cell signaling. [27]The safety of use of bone marrow derived cells has been well established as they have an autologous origin and do not result in tumor formation.

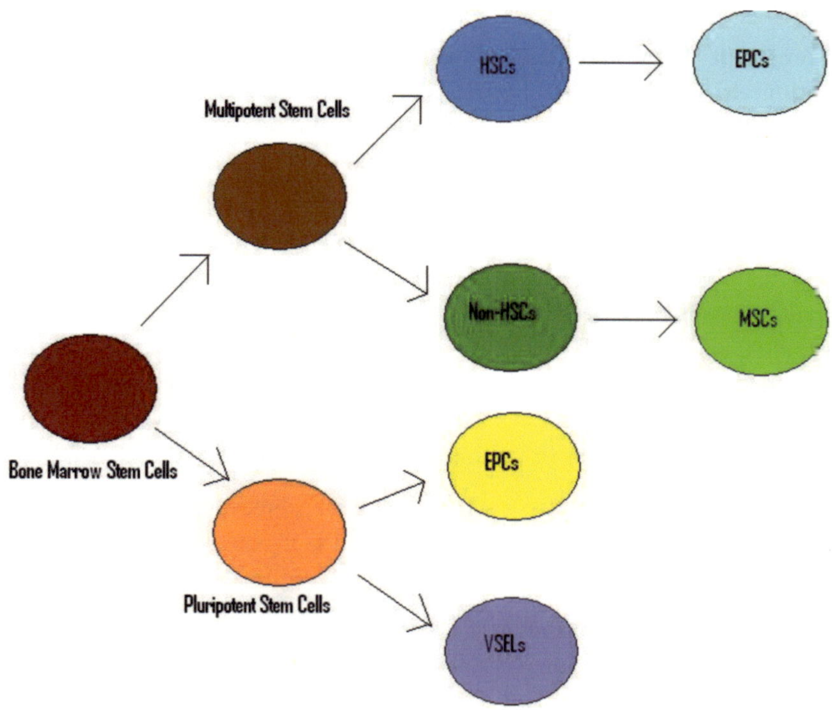

Figure 1. Bone marrow stem cells

ii. **Adipose tissue:** Adipose tissue derived stem cells (ASCs) are multipotent cells, found abundantly in fat tissue. They can differentiate into several lineages, including adipose cells, chondrocytes, osteoblasts, neuronal cells, endothelial cells, and cardiomyocytes. These cells are obtained either through liposuction or lipectomy. Mesenchymal stem cells make up the majority of the adipose derived stem cells. [28] Due to their plasticity, they are a preferred alternative to the BMSCs. [29] One of the major disadvantages of adipose derived stem cell is the isolating procedure. Therefore, a professional technician is required for cell isolation. In experimental cerebral palsy models, infusion of adipose derived stem cells has shown to improve physical activities and cognitive deficits. They have the ability to replace damaged oligodendrocytes and neurons without forming glial scars.[30]

iii. **Dental pulp:** A population of stem cells has been isolated from the human dental pulp known as dental pulp stem cells (DPSCs). They have an ability to regenerate a dentin-pulp-like tissue. [31] DPSCs are a heterogeneous population of cells as they are composed of both mesenchymal and ectodermic cells. These cells are readily obtained from routine dental procedures such as removal of impacted third molars, deciduous teeth and have been shown to possess properties similar to neural stem cells and mesenchymal stem cells. [32] Under appropriate conditions, these cells also undergo neuronal differentiation. [33] One of the disadvantages of DPSCs is that it takes longer to culture mesenchymal stem cells from teeth active tissue. Also, it is difficult to harvest a large quantity of stem cells from teeth.

iv. **Menstrual Blood stem cells:** Recently, stem cells have been identified from the endometrial tissue. The endometrial lining of the uterus has tremendous capacity of regeneration. The menstrual blood consists of a heterogeneous population of cells. However, isolation of these cells is a very invasive procedure. [34] Their potential in CP is not studied.

v. **Peripheral blood stem cells:** Peripheral blood consists of circulating stem cells capable of restoring hematopoiesis. Hematopoietic stem cells obtained from PB by leukapheresis have been used for transplantation as an alternative to bone marrow-derived stem cells (BM-stem cells). The haemopoietic "potential" of PBSCs is equivalent to that of BM- stem cells [35] Their use in CP is not well established.

2.3. Various routes of administration of stem cells for cerebral palsy

The appropriate route of cell administration is essential prerequisite for the success of cellular therapy. For the treatment of cerebral palsy, cells are injected via various routes such as intrathecal, intravenous and intracerebral.

i. **Intrathecal administration:** Intrathecal administration of cells involves delivery of cells via lumbar puncture. It is a minimally invasive procedure as compared to other routes of administration. This mode of injection allows efficient delivery of cells and the possibility of migration of cells to the tissues other than the damaged ones is avoided. [36]

This procedure needs to be done under aseptic condition to avoid any sort of infections. In case of cerebral palsy, it is considered to be the safe, feasible and efficacious route of administration. In cerebral palsy, studies have shown that this route of administration results in positive functional outcomes. [37]

ii. **Intravenous administration:** Intravenous system of delivery of cells has been widely used in cellular therapy due to its broad bio distribution and easy access. It is one of the most minimally invasive and safe modes of administration. Studies have shown that on administering cells intravenously, few cells reach the damaged site while a majority of cells get trapped in the lungs. Pulmonary passage could be one of the major hindrances for intravenous administration of stem cells. Hence, for effective result of intravenous stem cell transplantation, it is necessary to increase the number of cells injected. [38]

iii. **Intracerebral administration:** In cerebral palsy, intracerebral transplantation of stem cells is carried out by injecting cells in the subventricular zone. This leads to migration of cells to the areas of ischemia but the outcome is not as remarkable as compared to the other minimally invasive procedures. Additionally, in CP the injury of the brain is diffused and a local injection could be focused on a particular area which might not be as effective. Intracerebral injection also increases the risk of bleeding and neural tissue injury. [39] Since, it is yet to establish cellular therapy as a cure, an invasive transplantation is not recommended due to the risks involved. Hence, a safer route of administration should be used.

2.4. Mechanism of action of stem cells in cerebral palsy

To understand the mechanism of action of stem cells in the treatment of cerebral palsy, it is important to understand the empirical neuropathophysiology. In spite of the vast and varied etiology; underlying cellular mechanisms, that cause the morbidity or mortality associated with cerebral palsy, are tissue damage caused by hypoxia and ischemia. The clinical manifestations of this cellular damage, depends on a range of factors including the time of insult, the severity of insult and cause of the insult. Brain tissue is heterogeneous and responds differently to hypoxia and ischemia. Therefore, a certain type of brain tissue is implicated to cause cerebral palsy. Recent preclinical, immunohistochemical and imaging evidence suggests periventricular white matter injury (PWMI), particularly damage to oligodendrocytes (OLs) as a primary cause of cerebral palsy [40,41,42]. PWMI is a spectrum ranging from cystic focal necrotic lesions, periventricular leuckomalacia (PVL) to specific cortical scarring in the deep regions of sulci, Ulegyria to diffuse myelination disturbances. Oligodendrocyte progenitors are abundantly present in the subventricular and periventricualr zones, therefore damage to these cells is seen as PVL in neuroimaging investigations. The extent of the damage to the white matter and its consequences are dependent on the developmental stage at which the damage occurred, brain vascularization and the type of tissue[43].

Vascularization of the brain begins as early as 28th day of gestation with the formation of carotid arteries, followed by the large arteries, their branches, communicating arteries, long penetrating arteries and ends with the formation of short penetrating arteries in the post term period. Damage at pre term leads to focal cystic necrosis in the vascular end zones of the long

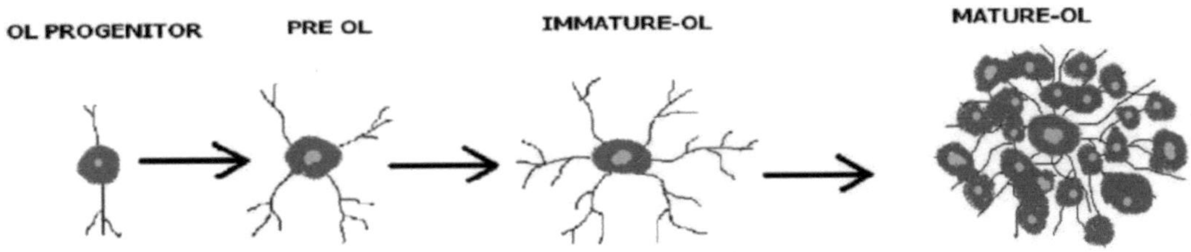

Figure 2. Phases of Oligodendrocyte development

penetrating arteries causing PVL, damage at term leads to tissue injury at the border zones of the long and short penetrating arteries giving rise to Ulegyria and damage at post term leads to diffuse myelination disturbances caused at the vascular end zones of short penetrating arteries [44]. Subsequently most vulnerable cells, precursors of Oligodendrocytes (OLs), undergo necrosis through apoptosis. This leads to myelination disturbances. Oligodendrocytes evolve through an established lineage of OL progenitors to pre OLs to immature OLs to mature OLs. [Figure 2] Hypoxic ischemia as observed in cerebral palsy leads to death of pre OLs and subsequent deficiency of mature OLs and myelination. Other cell types and mechanisms that contribute to pathophysiology of CP are axonal damage and microglial activation [45]. Following this primary insult to the nervous tissue, activation of glial cells leads to secretion of various chemical mediators of tissue necrosis in the neural microenvironment, leading to secondary white matter injury. These mediators are Reactive oxygen and nitrogen species, glutamates, adenosine and inflammatory cytokines like Tumor necrosis factor alpha (TNF-α), interferon gamma (INF-γ), Interleukin -1 beta (IL-1β) and superoxide radicals [46].

Cellular therapy regulates all of the above cellular mechanisms. Neuroplasticity of the brain is maximal during childhood. Hence, stem cell intervention is more successful in these children as the integration of new cells in the brain to carry out the repair process is more effective [47]. Stem cell possess the capacity to home onto the injured sites of brain, as guided by chemo attractant pathway [48]. The effects of cellular therapy are twofold, enhancing the brain tissue repair caused by various paracrine mechanisms and regeneration of neural tissue. Stem cells help in modifying the microglial response by exhibiting immunomodulatory, neurotrophic properties and enhance axonal sprouting. Various neurotrophic factors secreted by the stem cells are connective tissue growth factor, fibroblast growth factor 2 and 7 that are responsible for cell proliferation, interleukins responsible for cytoprotection [49,50,51]. Stem cell therapy restores lost myelin by replacing dead cells with new oligodendrocytes and their progenitors. Indirectly, it may also support their survival by introducing other cell types able to restore missing enzymes to an otherwise deficient environment [47]. Stem cell therapy also has an anti-inflammatory effect on the neural microenvironment as they reduce the levels of TNF-α,IL-1β, IL-1α, IL-6 and increased levels of IL-10 [52]; therefore, enhancing the endogenous brain repair. Stem cells also secret various growth factors like vascular endothelial growth factor (VEGF), fibroblast growth factor (FGF), brain fibroblast growth factor (bFGF). These growth factors initiate neoangiogenesis and induce secretion of hormones like erythropoietin. The cascade events triggered due to these lead to formation of new vessels as well increased bold flow. Improved blood circulation of the brain thus helps retrieving the lost tissue functions [50].

3. Published literature

Several preclinical experiments on animal models of cerebral palsy have been carried out to demonstrate the potential of cell transplantation to minimize damage and promote recovery. However, limited clinical trials have been initiated to study the effect cell therapy in humans.

Human umbilical cord blood cells (hUCBCs) have been explored to a great extent in cerebral palsy. hUCBCs have been administered in rat models of neonatal hypoxia/ ischemia. They protect the mature neurons in the neocortex from injury, bring about near-normalization of brain damage in the subventricular zone (SVZ) leading to significant improvement in behavioral functions. The long lasting effect of these cells is due to the paracrine effects of hUCBCs which stimulate recovery in the injured brain and protect against further brain damage. [53] On transplantation, hUCBCs have shown to ameliorate neurological and motor deficits in CP model by reducing the levels of pro-inflammatory cytokines (Interleukin-1α (IL-1α), Interleukin-1β (IL-1β), and Tumor necrosis factor α (TNFα)) [54,55]

They also alleviate spastic paresis in neonatal rat models resulting in normal walking [56] Studies have also shown that these cells promote neural stem cell proliferation via Sonic hedgehog (Shh) signaling pathway improving the brain damage. [57] Human umbilical cord blood (hUCB) cells have also shown to reduce sensorimotor deficits after hypoxic ischemic brain injury in neonatal rats. The dimensions of cortical maps and receptive fields, which are significantly altered after injury, are largely restored. Additionally, the lesion induced hyperexcitability is no longer observed in treated animals compared to control animals. The results demonstrate that hUCB cells reinstall the way central neurons process information by normalizing inhibitory and excitatory processes. [58] it is also observed that these cells exhibit a neuroprotective effect in the striatum, and decrease the number of activated microglial cells in the cerebral cortex of treated animals, further resulting in better functional recovery. [59] Tanaka et al, observed that CD133+ cells derived from hUCBCs reduce the cortical damage and also promote axonal growth in neonatal rat organ co-cultures exposed to hypoxia. [60]

Various preclinical studies have shown that transplantation of stem cells in the CP models lead to survival, homing and differentiation of these cells into neurons, oligodendrocytes, astrocytes etc.

Park et al reported clonal neural stem cells (NSCs) when transplanted into brains of postnatal hypoxic-ischemic (HI) injury mice, home preferentially to and integrate extensively within the ischemic areas. They differentiate into neurons and oligodendrocytes, the cell types damaged due to HI. [61] A chinese study, wherein neural stem cells derived from human fetal brain (hNSCs) were transplanted into cerebral ventricle of HI injury neonatal rat, too demonstrated the survival, migration and differentiation capacity of these cells in rat brain. [62] Similarly, Zheng et al showed that Multipotent astrocytic stem cells (MASCs) from mice transplanted into a rat model of hypoxia-ischemia (HI) survive, migrate and differentiate into neurons and astrocytes. [63] In their study, Titomanlio et al implanted neurosphere-derived precursors in neonatal mouse model of cerebral palsy induced by excitotoxicity. They observed that cells migrated to the lesion site, remained undifferentiated at day 10, and differentiated into oligodendrocyte and neurons at day 42. Although grafted cells finally die there few weeks later, this procedure triggered a reduction in lesion size and an improvement in memory performance compared with untreated animals. [64] Chen et al, transplanted magnetically labeled mesenchymal stem cells in a model of perinatal brain injury. They found that these cells migrate to lesion sites and proliferate. They are neuroprotective and indirectly contribute to brain repair. [65]

Yasuhara et al, investigated the efficacy of intrahippocampal transplantation of bone marrow derived multipotent progenitor cells (MPCs) in HI injury. They found that transplanted MPCs ameliorated motor deficits associated with HI injury. [66] Webber et al in their study highlighted the protective effects of oligodendrocyte precursor cell transplantation in neonatal inflammation-induced rat model of periventricular leukomalacia. [67]

All the above preclinical studies have been carried out in animal models of acute hypoxic injury, hence showing significant results. But, similar results are difficult to replicate in human cases since the intervention always, cannot be carried out immediately post injury. Thus, more studies should be carried out in chronic injury models. Based on this observation, it can also be concluded that earlier the intervention, better is the outcome.

Below are few of the published studies carried out in human cases of cerebral palsy.

Luan et al carried out a study on 45 patients diagnosed with severe CP. They underwent transplantation of neural progenitor cell (NPC) derived from aborted fetal tissue. After 1 year, the developmental level in gross motor, fine motor, and cognition of the treatment group was significantly higher compared to the control group. These results suggested that NPC transplantation is a safe and effective therapeutic method for treating children with severe CP. [68]

Chen et al, injected neural stem cell-like (NSC-like) cells derived from autologous marrow mesenchymal stem cells in 30 cases of cerebral palsy. On follow up, they observed an increase in the GMFM scores and language quotients compared to the control group. No adverse events were recorded indicating that NSC-like cells are safe and effective for the treatment of motor deficits related to cerebral palsy. [69] Mink et al carried out a double blind, randomized, controlled trial in which they administered allogeneic umbilical cord blood cells potentiated with recombinant human erythropoietin (rhEPO) in CP patients. They observed improvement in motor and cognitive dysfunction in children with CP, accompanied by structural and metabolic changes in the brain. [70]

Papadopoulos et al administered autologous umbilical cord blood cells in 2 children diagnosed with spastic diplegic CP. They found that this therapy was safe, feasible and led to functional improvements in children which was seen by the change in GMFCS. [71]

Li et al, transplanted autologous bone marrow mesenchymal cells in an 11 year old CP case with visual impairment. On six month follow up, he could walk better and his vision had improved significantly. These findings were supported by the electrophysiological examinations. [72]

Jensen et al recently published a study wherein a 2 ½ year old boy received autologous umbilical cord blood mononuclear cells intravenously. At 2-months follow-up the boy's motor control improved, spastic paresis was largely reduced, and eyesight was recovered, as did the EEG. He smiled when played with, was able to sit and to speak simple words. At 40 months, independent eating, walking in gait trainer, crawling, and moving from prone position to free sitting were possible, and there was significantly improved receptive and expressive speech competence (four-word sentences, 200 words). This suggested that autologous cord blood transplantation could be a treatment alternative for cerebral palsy. [73]

Wang et al reported a case of a 5-year old girl with CP who underwent umbilical cord mesenchymal stem cells (MSCs) transplantation. She was treated with multiple times of intravenous and intrathecal administration of MSCs derived from her young sister and was followed up for 28 months. The gross motor dysfunction was improved. Immunity was enhanced, physical strength improved along with speech and comprehension. [74]

Purandare et al, reported a case of cerebral palsy who was administered with autologous bone marrow mononuclear cells. On follow up, they recorded a significant improvement in motor, sensory, cognitive, and speech. Bowel and bladder control was also achieved. On the GMFCS-E&R level, the patient was promoted from grade III to I. Hence, concluding that intrathecal infusion of autologous BMMNCs is feasible, effective, and safe in CP patients. [75]

In our previously published cases of cerebral palsy, one was with comorbid intellectual disability [76] while the other one was without any comorbidity. [77] These cases were administered with autologous bone marrow mononuclear cells intrathecally. Six months after the treatment, both cases showed significant functional outcomes which was supported by improvement in PET CT scan.

3.1. Ongoing trials

Currently, there are five clinical trials on stem cell therapy for cerebral palsy registered in clinicaltrials.gov. [http://clinicaltrials.gov/]

2 studies are from India studying the safety and efficacy of bone marrow MNCs in cerebral palsy in children below 15 years. One of the studies is a combination of phase 1 and phase 2 while the other is a combination of phase 2 and phase 3.

2 studies are from Iran, one evaluating the side effects of bone marrow derived CD133 cells transplantation in cerebral palsy patients and the other studying the safety of multiple intrathecal injections of bone marrow derived CD133 cells.

A study from USA is based on evaluating the safety and effectiveness of a single, autologous, cord blood stem cell infusion for the treatment of cerebral palsy in children. It is a randomized, controlled, blind, crossover study

4. Administration of autologous bone marrow derived mononuclear cells in children with cerebral palsy.

Sharma et al, carried out a study on 71 children, wherein they administered 20 cases of cerebral palsy with autologous bone marrow mononuclear cells, intrathecally. [78] These cases included dystonic and spastic CP. Symptoms commonly observed in them were delayed milestones, spasticity, motor impairment, ambulation deficits, cognitive impairment, swallowing and speech problems etc.

Autologous bone marrow MNCs were selected as they are easily obtainable, safe and do not involve any ethical issues. As discussed earlier, intrathecal route of administration is a minimally invasive, safe and an effective procedure as compared to other routes. Studies have also proved that a mixture of cells exhibits more benefits as compared to a single sub fraction of cells. [79] Hence, we chose to carry out intrathecal autologous BMMNC transplantation.

The patients were administered Granulocyte Colony Stimulating Factor (GCSF), 48 hours and 24 hours before the harvest and transplantation of BMMNC. On the day of the transplantation, bone marrow was aspirated under general anesthesia in the operation theatre with aseptic precautions. Approximately, 100 ml of bone marrow (varying between 80 ml and 100 ml, based on the age and body weight) was aspirated from the region of anterior superior iliac spine using the bone marrow aspiration needle and collected in the heparinized tubes.

The aspirate was then transferred to the laboratory where the mononuclear cells were separated by the density gradient method. CD34+ counting was done by Fluorescence activated cell sorting (FACS). The MNCs were checked for viability (Average viability count was found to be 97%).

The separated autologous BMMNCs were immediately injected on the same day, intrathecally using an 18G Touhy needle and epidural catheter at the level between fourth and fifth lumbar vertebrae. The average numbers of cells injected were 8.19×10^7. Simultaneously 20mg/kg body weight methyl prednisolone in 500 ml Ringer Lactate was given intravenously to enhance survival of the injected cells. Patient was monitored for any adverse events.

On mean follow up of 15 months ± 1 month post stem cell administration, improvement was observed in 85% cases. [Figure 3] Significant improvement was observed in spasticity, neck holding, drooling of saliva, muscle strength in upper and lower limbs, sitting and standing balance, gross and fine motor activities, speech, swallowing, ambulation, and cognition. There was also a reduction in dystonic movements. No major adverse events were recorded. Some minor side effects such as headache, nausea and vomiting were experienced by few children who were self-limiting (resolved within a week) and treated with medications. The improvements in these patients sustained even after the follow up period of the study. None of them showed any deterioration on the GMFCS [78]

We are also currently conducting a clinical study to assess the efficacy of autologous BMMNC in 64 patients with CP. These patients are being treated with a combination of cell therapy and rehabilitation. The unpublished data analysis have shown preliminary results as improvement

http://dx.doi.org/10.5772/57152

in oromotor skills, speech, neck holding, sitting, standing and walking balance and significant reduction in muscular tone and dystonic movements. These changes were observed in all types of cerebral palsy over 6 months with varied follow up periods. [Figure 4,5,6]

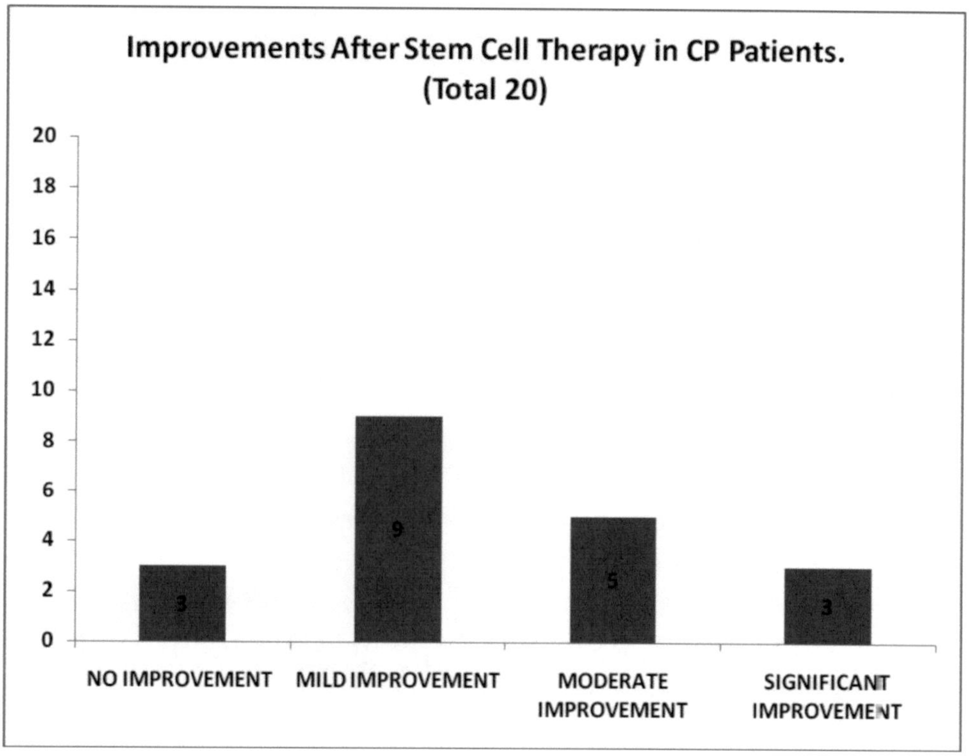

Figure 3. Graph demonstrating overall improvements in CP patients after stem cell therapy.

Percentage improvement noted in patients of diplegic cerebral palsy was as follows. Oromotor skills (75%), speech (64%), neck holding (100%),Sitting balance (67%), standing balance(67%), walking balance (67%), ambulation (30%), leg movements (54%), overhead movements (38%), distal hand movements (69%), upper limb spasticity (38%), Lower limb spasticity (38%), trunk muscle tone (36%), trunk dissociation (30%)

Percentage improvement noted in patients of quadriplegic cerebral palsy was as follows. Oromotor skills (58%), speech (40%), neck holding (94%),Sitting balance (48%), standing balance(27%), walking balance (21%), ambulation (13%), involuntary movements (25%), upper limb spasticity (51%), Lower limb spasticity (50%), trunk muscle tone (36%)

Percentage improvement noted in patients with other types of cerebral palsy was as follows. Oromotor skills (55%), speech (55%), neck holding (40%),Sitting balance (45%), standing balance(50%), walking balance (27%), involuntary movements (9%), upper limb spasticity (22%), lower limb spasticity (14%), dystonia of upper limbs (50%), dystonia of lower limbs (100%), dystonia of the trunk (50%)

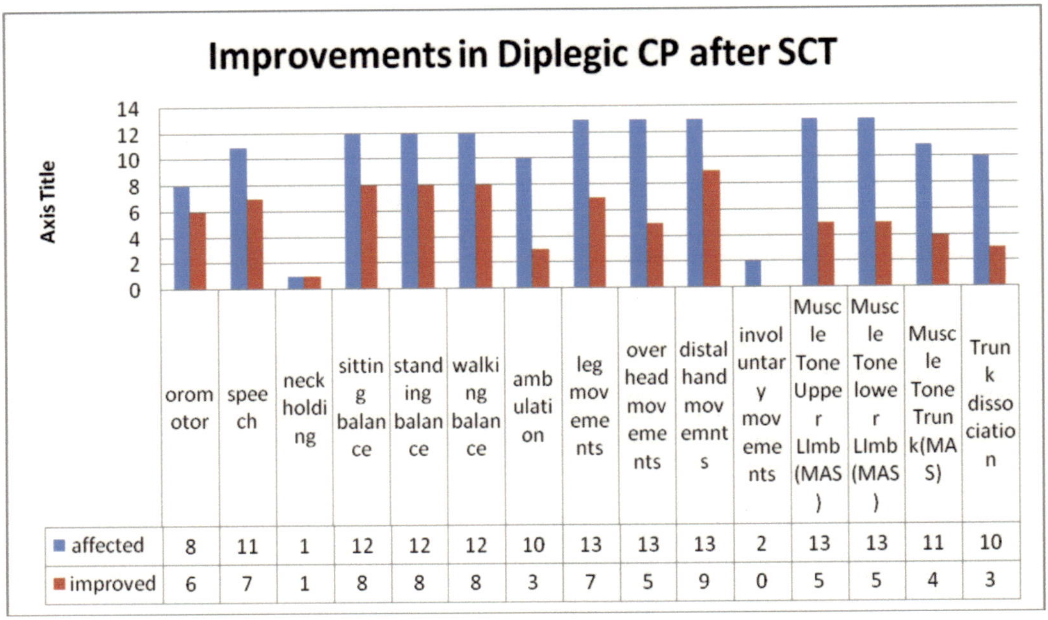

Figure 4. Graph demonstrating improvements in Diplegic CP patients after stem cell therapy

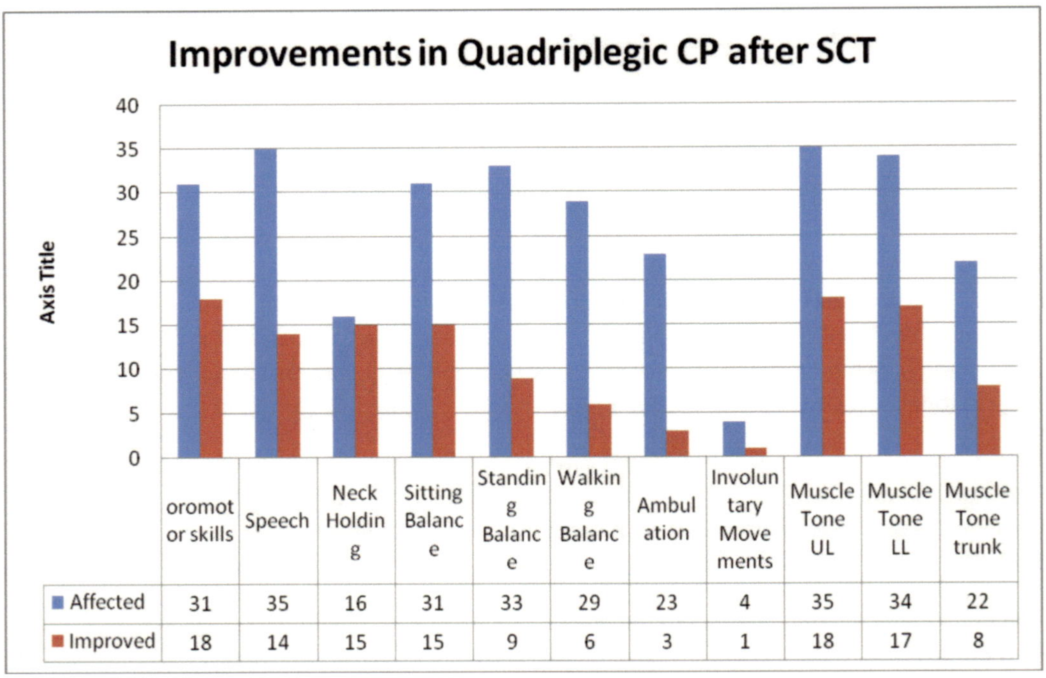

Figure 5. Graph demonstrating improvements in Quadriplegic CP patients after stem cell therapy

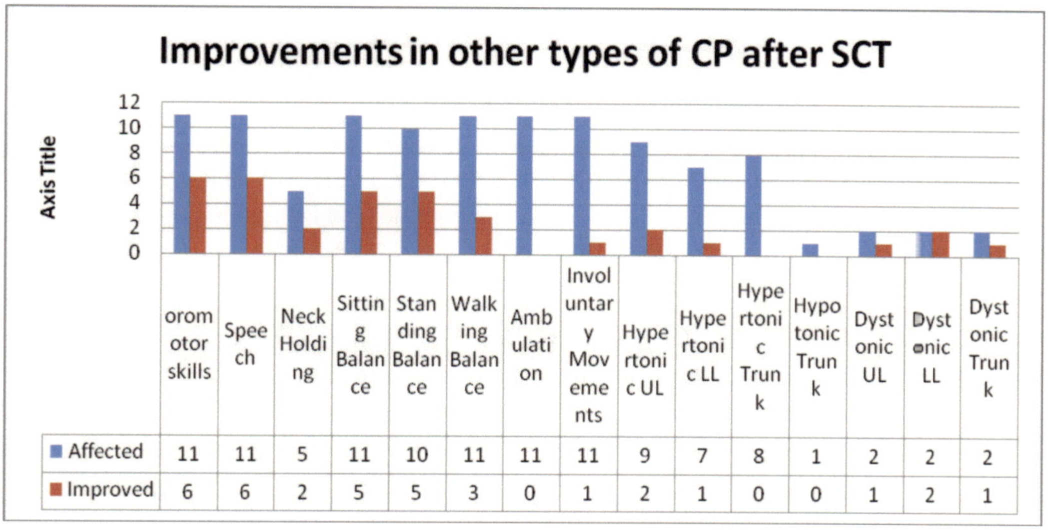

Figure 6. Graph demonstrating improvements in other types of CP patients after stem cell therapy

5. Objective imaging evidence

Various clinical outcome measures have been devised to measure changes in sensory, motor, cognitive, perceptual and Behaviour functions in CP. [80] It is however important to understand the underlying mechanisms behind these changes.

MRI scans not only help reveal the underlying pathology of CP, but it also correlates with the clinical findings. [81] It has been observed that, clinical trials using MRI as a primary outcome measure failed to fully identify the effects of the therapy on clinical measures. [82] MRI shows the structural malformations and grey and white matter lesions, these are only suggestive of underlying tissue mechanisms; but MRI is not sensitive to measure the changes at cellular level. Principle mechanisms underlying the benefits of cellular therapy are the changes brought about in the microenvironment of cells reducing cell necrosis, ischemia and hypoxia. These changes therefore cannot be measured on a plane MRI and hence it is not a sensitive tool to monitor the effects of stem cell therapy. Functional neuroimaging on the other hand may be an appropriate option to monitor the finer changes at cellular level.

The basic principle underlying the functional neuroimaging of the brain is that the cerebral blood flow and metabolism is associated with neuronal activity. [83] Stem cell therapy aims to modulate neuronal tissue function in the patients of cerebral palsy, through various paracrine mechanisms. Measurement of the tissue function is therefore a preferred outcome measure to monitor the effects of cellular therapy.

Positron Emission Tomography – Computed Tomography (PET – CT) is one of the techniques of functional neuroimaging that measures the metabolism of the nervous tissue in terms of

Fleuro-deoxy glucose (FDG) uptake. FDG is a radioactive glucose analogue that undergoes glycolysis in the same manner as that of glucose. Once it has been metabolized to FDG – 6 – Phosphate it cannot be further metabolized and is trapped inside the cell due to the impermeability of the cell membrane for this molecule. With increased glycolysis higher concentration of FDG-6-phosphate is observed. Photons emitted by this radioactive isotope are then measured to identify concentration of FDG in the nervous tissue [84]. This is expressed as a ratio of the actual uptake and the calculated presumed uptake of FDG, standard uptake value [85]. Because of its ability to measure the finer changes in tissue metabolism, FDG PET-CT holds a great potential as a monitoring tool. [86,87]

PET-CT scan is performed following a standard protocol. Various guidelines are available for appropriate standardization, image acquisition and interpretation during PET-CT scanning. Dosage of the radioactive dye is calculated based on the age and weight of the patient. Calculated units are then administered systemically 30 minutes before scanning. [88].

FDG, chemically expressed as 18F-FDG is radionuclide and therefore special safety concerns with its use need to be address. The primary safety concern is exposure to radiation. Incidence of developing any side effect is negligible with the use of 18F-FDG PET-CT. The half-life of 18F-FDG is 109.8 minutes and is excreted via urine. The tissue metabolism of 18F-FDG is the same in adult and children and the dose administered in children is "as low as reasonably achievable". Various adjustments with regards to scanning technology and measurement period are made to enhance the quality of the image with the administered dose [89]. PET-CT is sensitive to measure the cellular changes and it is a standardized imaging modality which makes it a good monitoring tool to assess the effects of cellular therapy.

In our previous case studies involving cerebral palsy patients treated with autologous bone marrow derived mononuclear cells (BMMNCs), the clinical outcome was correlated with changes in the PET scan. In one case of a 20 year old CP patient with co morbid intellectual disability, a repeat PET-CT scan showed significant increase in the FDG uptake in various affected areas of the brain, which correlated with the clinical improvement in social behavior, balance and motor control. However the MRI remained unchanged (76).

In another case of a 2 year old child with cerebral palsy, we observed similar correlation of clinical improvement with the PET-CT changes in metabolism. (77) Six months following cellular therapy she developed good neck control along with improvement in balance and speech. These clinical changes were synonymous with the increased FDG uptake in the bilateral mesial temporal structures, right basal ganglia, frontal, parietal, temporal and occipital lobes.

Functional MRI is also one of the emerging techniques to study the functional outcome of the intervention. The technique of fMRI is based on Blood oxygenation level dependent (BOLD) contrast between rest and activated states of human brain. (90) Activation of neuronal tissue leads to increased metabolism and increased oxygen demands that have a twofold effect of greater oxygen extraction and increased cerebral blood flow; both of which result in higher BOLD signals than that of resting tissue. In ischemic tissues the blood flow–metabolism couple is impaired. Stem cells enhance angiogenesis, increasing the cerebral blood flow. Hence, fMRI may be effectively used to monitor the therapeutic outcome of stem cell therapy and should be studied further. (91)

6. Role of rehabilitation in combination with stem cell therapy

For a long time, rehabilitation has been the standard approach for cerebral palsy. The goal of rehabilitation in cerebral palsy is to develop coordination, build strength, improve balance, maintain flexibility, optimize physical functions, manage spasticity and maximize independence. Rehabilitation is multidisciplinary. Various therapeutic regimens aim to enhance particular clinical, functional and psychosocial consequences of CP. Physiotherapy makes use physical modalities to muscle spasticity, increase flexibility, balance and co-ordination, build strength and enhance function. Physiotherapists also prescribe different assistive devices to gain higher functionality. Multiple medical and surgical regimens are also instituted to deal with these physical impairments. Botox injections are most commonly used to reduce spasticity of the muscles, but the effects are short lived. A variety of surgical techniques are utilized to correct deformities.

Occupational therapy is focused at therapeutic regimens to improve cognitive abilities of the child and increase participation in activities of daily and social living. Children with CP most often present with poor oromotor control and speech disorders, speech therapy aims at correcting these impairments. Psychiatric and psychological intervention is aimed at patient and caregiver wellbeing. It helps to improve the quality of life by addressing co-morbid psychological disturbances and cognitive impairments.

All of the rehabilitative modalities face the fundamental limitation of inability to repair the damage to the nervous tissue. Some studies have defined minor improvements in motor and social skills. Wright and Nicholson and Sommerfeld et al have demonstrated that physical therapy alone does not show a consistent benefit in cerebral palsy. [92]

However, rehabilitation in combination with stem cell therapy may augment its benefits. Exercise helps in enhancing the cell proliferation and neurogenesis. [93] Increased mobilization of hematopoietic stem cells and erythropoieticprogrnitor cells (EPCs) to peripheral blood is observed post exercise. It also increases angiogenesis and oxygen supply to the brain thereby improving the cognitive function. [94,95,96]. Regular exercise induces suppression of pro-inflammatory cytokines and up regulation of anti-inflammatory cytokines in various tissues of the body including brain. [97,98,99] One of the key mechanisms for homing of bone marrow mononuclear stem cells is the chemokine stromal derived factor -1 / CXCR4+ receptor pathway, exercise has also been found to up regulate expression of CXCR4+ receptors is ischemic tissue ensuring enhanced homing of stem cells. In addition, mobilization of stem cells enhanced homing, improved angiogenesis exercise also exerts immunomodulatory effects. [97] Benefits of regular exercise resonate with the cellular mechanisms of stem cell therapy and therefore it augments the therapeutic potential of stem cell therapy. Exercise and rehabilitation has a synergistic effect for the benefits of cell transplantation. [100]

7. Future direction

Stem cell therapy for cerebral palsy still remains in its infant stage. Although cellular therapy for cerebral palsy has moved from the preclinical studies to bedside therapy; evidence remains inconclusive regarding multitude of variables. These variables are pertaining to cerebral palsy and cellular therapy.

Cerebral palsy is a heterogeneous group of disorders. This inherently reduces the generalizability of the findings. Pre-clinical models of effects of cellular therapy in cerebral palsy are far from the ideal state and show benefits only in acute injury. Majority of the human application of stem cells in cerebral palsy is for individuals who already have established pathology, hence at a chronic stage. Animal models of chronic injury are therefore required to study the efficacy and mechanism of action of stem cells. The individuals suffering from cerebral palsy are from various age groups, and present with varied kinds and severities of clinical manifestations; there is only a limited evidence about which of these groups will benefit the most from cellular therapy.

Only preliminary evidence using basic research methodologies is available for the effects of cellular therapy in humans. Various factors limit the methodological robustness of the current trials. There are limited controlled trials in humans with cerebral palsy. We require more double blind, randomized, multicenter controlled clinical trials to prove the safety and feasibility of stem cells. The evidence available is heterogeneous in methodology, patient population, outcome measures and cellular therapy provided. Stem cells provide their beneficial effects through numerous mechanisms; it is difficult to underpin the exact mechanism of action of stem cells. Types, sources and numbers of cells administered, frequency of transplantation, time of transplantation are concerns which require attention imperatively. It is important to not only conduct more trials but also to standardize research protocols to allow comparison. Comparative studies will help in establishing the most effective cell based therapy for cerebral palsy.

Apart from these issues, development and validation of outcome measures to obtain evidence of the efficacy of intervention is very important. Different scales, monitoring tools need to be standardized. Modalities should be developed to study the effect of cell transplantation at a cellular level. Outcomes that can successfully assess these cellular changes are measuring the serum, plasma and cerebrospinal fluid biomarkers, which are invasive. Less invasive modalities would be functional imaging techniques. PET-CT scan has been used as an outcome to assess the effects of cellular therapy however it is required to further explore its various components in depth. fMRI also provides insights regarding the changes at cellular level however there is no evidence of its use in monitoring the changes post cellular therapy. It is therefore necessary to explore how functional imaging can provide us a better understanding of the cellular mechanisms.

8. Conclusion

Stem cell therapy has been extensively studied but still needs to be standardized before it becomes a definitive treatment modality. Autologous BMMNCs are safe and feasible option

http://dx.doi.org/10.5772/57152

but their effectiveness needs more clinical trials. Other types of stem cells need to establish safety and efficacy. Though not a cure, stem cell therapy has emerged as a novel therapeutic option to improve the quality of life.

Author details

Alok Sharma[1*], Hemangi Sane[2], Nandini Gokulchandran[1], Prerna Badhe[1], Pooja Kulkarni[2] and Amruta Paranjape[3]

*Address all correspondence to: alok276@gmail.com

1 Department of Medical Service and Clinical Research, Neurogen Brain and Spine Institute, Mumbai, India

2 Department of Research and Development, Neurogen Brain and Spine Institute, Mumbai, India

3 Department of Neurorehabilitation, Neurogen Brain and Spine Institute, Mumbai, India

References

[1] http://www.nobelprize.org/nobel_prizes/medicine/laureates/2012/press.html

[2] Cajal SRY. Degeneration and regeneration of the nervous system. In: May RM, editor. Trans. London: Oxford University Press; 1928

[3] Panteliadis CP., Strassburg HM. Cerebral Palsy: Principles and Management. Stuttgart, NY: Georg Thieme; 2004

[4] Potten CS, Loeffler M. Stem cells: attributes, cycles, spirals, pitfalls and uncertainties. Lessons for and from the crypt. Development. 1990;110(4):1001-20

[5] Lakshmipathy U, Verfaillie C. Stem cell plasticity. Blood Rev. 2005;19(1):29-38.

[6] Evans M, Kaufman M. Establishment in culture of pluripotent cells from mouse embryos. Nature 1981; 292: 154–156.

[7] De Wert G, Christine M. Human embryonic stem cells: research, ethics and policy. Human reproduction 2003; 18(4): 672-682.

[8] Ben-David U, Benvenisty N. The tumorigenicity of human embryonic and induced pluripotent stem cells. Nat Rev Cancer. 2011;11:268-77

[9] Pappa KI, Anagnou NP. Novel sources of fetal stem cells: where do they fit on the developmental continuum? Regen Med. 2009;4(3):423-33

[10] Ilancheran S, Michalska A, Peh G, Wallace EM, Pera M, Manuelpillai U. Stem cells derived from human fetal membranes display multilineage differentiation potential. Biol Reprod. 2007;77(3):577-88.

[11] Parsons C, Barbara S, Dean K. Susceptibility of human fetal mesencyhmal stem cells to Kaposi sarcoma-associated herpesvirus. Blood 2004; 104(9): 2736-2738

[12] Newcomb JD, Willing AE, Sanberg PR. Umbilical cord blood cells. Methods Mol Biol. 2009;549:119-36.

[13] Stanevsky A, Goldstein G, Nagler A. Umbilical cord blood transplantation: pros, cons and beyond. Blood Rev. 2009;23(5):199-204.

[14] Moise K.J. Umbilical cord stem cells. Obstetrics and Gynecology. 2005; 106(6), 1393–1407

[15] Chik KW, Chan PK, Li CK, et al. Human herpesvirus-6 encephalitis after unrelated umbilical cord blood transplant in children. Bone Marrow Transplantation. 2002;99:991–994

[16] El-Cheikh J, Fürst S, Casalonga F, Crocchiolo R, Castagna L, et al. JC Virus Leuko-Encephalopathy in Reduced Intensity Conditioning Cord Blood Transplant Recipient with a Review of the Literature. Mediterr J Hematol Infect Dis. 2012;4(1):e2012043

[17] Robbins RD, Prasain N, Maier BF, et al. Inducible pluropotent stem cells: not quite ready for prime time? Current Opinion in Organ Transplantation. 2010;15:61–67.

[18] Kim SU, de Vellis J. Stem cell-based cell therapy in neurological diseases: a review. Journal of Neuroscience Research, 2009; 87(10): 2183–2200

[19] Trewhitt, Kathryn G. Bone marrow aspiration and biopsy: collection and interpretation. Oncology nursing forum. Vol. 28. No. 9. Oncology nursing society, 2001.

[20] Bianco P, Riminucci M, Gronthos S, Robey PG, Bone marrow stromal stem cells: nature, biology, and potential applications. Stem Cells 2001; 19(3):180–192

[21] Bonilla S, Alarcón P, Villaverde R, Aparicio P, Silva A, Martínez S. Haematopoietic progenitor cells from adult bone marrow differentiate into cells that express oligodendroglial antigens in the neonatal mouse brain. Eur J Neurosci. 2002;15(3):575-82.

[22] Zhao LR, Duan WM, Reyes M, Keene CD, Verfaillie CM, Low WC. Human bone marrow stem cells exhibit neural phenotypes and ameliorate neurological deficits after grafting into the ischemic brain of rats. Exp Neurol. 2002;174(1):11 20.

[23] Boxall S, Jones E. Markers for Characterization of Bone Marrow Multipotential Stromal Cells. Stem Cells International, vol. 2012, Article ID 975871, 12 pages, 2012.

[24] Yoo J, Kim HS, Hwang DY. Stem cells as promising therapeutic options for neurological disorders. J Cell Biochem. 2013;114(4):743-53.

[25] Orozco L, Munar A, Soler R, Alberca M, Soler F, Huguet M, Sentís J, Sánchez A, García-Sancho J. Treatment of Knee Osteoarthritis With Autologous Mesenchymal Stem Cells: A Pilot Study. Transplantation. 2013 ;95(12):1535-1541.

[26] Dobrowolski S, Lepski G. Stem cells in traumatic brain injury. Am. J. Neurosci 2013; 4: 13-24.

[27] Ciara T, Case C. Transplanted Mesenchymal Stem Cells Aid the Injured Brain Through Trophic Support Mechanisms. Stem Cells and Cancer Stem Cells, Volume 4. Springer Netherlands, 2012. 297-303.

[28] Zuk PA, Zhu M, Mizuno H, Huang J, Futrell JW, Katz AJ, Benhaim P, Lorenz HP, Hedrick MH. Multilineage cells from human adipose tissue: implications for cell-based therapies. Tissue Eng. 2001;7(2):211-28.

[29] Kim M, Kim I, Lee SK, Bang SI, Lim SY. Clinical trial of autologous differentiated adipocytes from stem cells derived from human adipose tissue. Dermatol Surg. 2011;37(6):750-9.

[30] Nakama K, Choi SW, Yang PS, Song KC, Ko MS, Jo JY, Ra JC. Therapy of autologous human adipose tissue-derived mesenchymal stem cells for the cerebral palsy: a case report. 2012; 1(1)

[31] Gronthos S, Mankani M, Brahim J, Robey PG, Shi S. Postnatal human dental pulp stem cells (DPSCs) in vitro and in vivo. Proc Natl Acad Sci USA 2000;97:13625-13630.

[32] Jamal M, Chogle S, Goodis H, Karam S. M. Dental stem cells and their potential role in regenerative medicine. J Med Sci. 2011 4(2), 53-61.

[33] Arthur A, Rychkov G, Shi S, Koblar SA, Gronthos S. Adult human dental pulp stem cells differentiate toward functionally active neurons under appropriate environmental cues. Stem Cells 2008;26: 1787–1795.

[34] Patel AN, Park E, Kuzman M, Benetti F, Silva F J, Allickson J G. Multipotent menstrual blood stromal stem cells: isolation, characterization, and differentiation. Cell transplantation. 2008;17(3), 303-311.

[35] Bender JG, Unverzagt KL, Walker DE, Lee W, Epps DEV, Smith DH, Stewart CC, Bik To L: Identification and comparison of CD34-positive cells and their subpopulations from normal peripheral blood and bone marrow using multicolor flow cytometry. Blood 1991;77:2591-2596

[36] Lim JY, Jeong CH, Jun JA, Kim SM, Ryu CH, Hou Y, Oh W, Chang JW, Jeun SS. Therapeutic effects of human umbilical cord blood-derived mesenchymal stem cells after intrathecal administration by lumbar puncture in a rat model of cerebral ischemia. Stem Cell Res Ther. 2011;2(5):38.

[37] Mehta T, Feroz A, Thakkar U, Vanikar A, Shah V, Trivedi H. Subarachnoid placement of stem cells in neurological disorders. Transplant Proc. 2008;40(4):1145-7.

[38] Fischer UM, Harting MT, Jimenez F, Monzon-Posadas WO, Xue H, Savitz SI, Laine GA, Cox CS Jr. Pulmonary passage is a major obstacle for intravenous stem cell delivery: the pulmonary first-pass effect. Stem Cells Dev. 2009;18(5):683-92)

[39] Kang SK, Lee DH, Bae YC, Kim HK, Baik SY, Jung JS. Improvement of neurological deficits by intracerebral transplantation of human adipose tissue-derived stromal cells after cerebral ischemia in rats. Exp Neurol 2003, 183:355-366

[40] Nakamura Y, Okudera T, Hashimoto T: Vascular architecture in white matter of neonates: Its relationship to periventricular leukomalacia. J Neuropathol Exp Neurol 1994;53:582-589

[41] Takashima S, Tanaka K: Development of cerebrovascular architecture and its relationship to periventricular leukomalacia. Arch Neurol 1978; 35:11-16

[42] Volpe JJ: Neurobiology of periventricular leukomalacia in the premature infant. Pediatr Res 2001;50:553-562

[43] Hagel C,Dimitrios S. Neuropathology of cerebral palsy. Cerebral Palsy: Principles And Management (2004): 49.

[44] Back SA, Rivkees SA. Emerging concepts in periventricular white matter injury. Semin Perinatol. 2004;28(6):405-14.

[45] Skaper SD, Giusti P, Facci L. Microglia and mast cells: two tracks on the road to neuroinflammation. FASEB J. 2012;26(8):3103-17.

[46] Mundkur N. Neuroplasticity in children. The Indian Journal of Pediatrics 2005; 72(10):855-857

[47] Goldman SA. Progenitor Cell–Based Treatment of the Pediatric Myelin Disorders.Arch Neurol. 2011;68(7):848-856.

[48] Alvarez P, Carrillo E, Vélez C, Hita-Contreras F, Martínez-Amat A, Rodríguez-Serrano F, Boulaiz H, Ortiz R, Melguizo C, Prados J, Aránega A. Regulatory systems in bone marrow for hematopoietic stem/progenitor cells mobilization and homing. Biomed Res Int. 2013;2013:312656.

[49] Gnecchi M, Zhang Z, Ni A, Dzau, VJ. Paracrine mechanisms in adult stem cell signaling and therapy. Circulation research, 2008; 103(11), 1204-1219.

[50] Daadi MM, Davis AS, Arac A, Li Z, Maag AL, Bhatnagar R, Jiang K, Sun G, Wu JC, Steinberg GK. Human neural stem cell grafts modify microglial response and enhance axonal sprouting in neonatal hypoxic-ischemic brain injury. Stroke. 2010;41(3): 516-23

[51] Sharma S, Yang B, Strong R, Xi X, Brenneman M, Grotta JC, Aronowski J, Savitz SI. Bone marrow mononuclear cells protect neurons and modulate microglia in cell culture models of ischemic stroke. J Neurosci Res. 2010;88(13):2869-76.

[52] Brenneman M, Sharma S, Harting M, Strong R, Cox CS Jr, Aronowski J, GrottaJC,Savitz SI. Autologous bone marrow mononuclear cells enhance recovery after acute ischemic stroke in young and middle-aged rats. J Cereb Blood Flow Metab. 2010;30(1): 140-9.

[53] Bae SH, Kong TH, Lee HS, Kim KS, Hong KS, Chopp M, Kang MS, Moon J.Long-lasting paracrine effects of human cord blood cells on damaged neocortex in an animal model of cerebral palsy. Cell Transplant. 2012;21(11):2497-515

[54] Rosenkranz K, Tenbusch M, May C, Marcus K, Meier C. Changes in Interleukin-1 alpha serum levels after transplantation of umbilical cord blood cells in a model of perinatal hypoxic-ischemic brain damage. Ann Anat. 2013;195(2):122-7.

[55] Wasielewski B, Jensen A, Roth-Härer A, Dermietzel R, Meier C. Neuroglial activation and Cx43 expression are reduced upon transplantation of human umbilical cord blood cells after perinatal hypoxic-ischemic injury. Brain Res. 2012;1487:39-53

[56] Meier C, Middelanis J, Wasielewski B, Neuhoff S, Roth-Haerer A, Gantert M, Dinse HR, Dermietzel R, Jensen A. Spastic paresis after perinatal brain damage in rats is reduced by human cord blood mononuclear cells. Pediatr Res. 2006;59(2):244-9.

[57] Wang XL, Zhao YS, Hu MY, Sun YQ, Chen YX, Bi XH. Umbilical cord blood cells regulate endogenous neural stem cell proliferation via hedgehog signaling in hypoxic ischemic neonatal rats. Brain Res. 2013;1518:26-35

[58] Geissler M, Dinse HR, Neuhoff S, Kreikemeier K, Meier C. Human umbilical cord blood cells restore brain damage induced changes in rat somatosensory cortex. PLoS One. 2011;6(6):e20194.

[59] Pimentel-Coelho PM, Magalhães ES, Lopes LM, deAzevedo LC, Santiago MF, Mendez-Otero R. Human cord blood transplantation in a neonatal rat model of hypoxic-ischemic brain damage: functional outcome related to neuroprotection in the striatum. Stem Cells Dev. 2010;19(3):351-8.

[60] Tanaka N, Kamei N, Nakamae T, Yamamoto R, Ishikawa M, Fujiwara H, Miyoshi H, Asahara T, Ochi M, Kudo Y. CD133+ cells from human umbilical cord blood reduce cortical damage and promote axonal growth in neonatal rat organ co-cultures exposed to hypoxia. Int J Dev Neurosci. 2010;28(7):581-7.

[61] Park KI, Himes BT, Stieg PE, Tessler A, Fischer I, Snyder EY. Neural stem cells may be uniquely suited for combined gene therapy and cell replacement: Evidence from engraftment of Neurotrophin-3-expressing stem cells in hypoxic-ischemic brain injury. Exp Neurol. 2006;199(1):179-90.

[62] Qu SQ, Luan Z, Yin GC, Guo WL, Hu XH, Wu NH, Yan FQ, Qian YM. Transplantation of human fetal neural stem cells into cerebral ventricle of the neonatal rat following hypoxic-ischemic injury: survival, migration and differentiation. Zhonghua Er Ke Za Zhi. 2005;43(8):576-9.

[63] Zheng T, Rossignol C, Leibovici A, Anderson KJ, Steindler DA, Weiss MD.Transplantation of multipotent astrocytic stem cells into a rat model of neonatal hypoxic-ischemic encephalopathy. Brain Res. 2006;1112(1):99-105.

[64] Titomanlio L, Bouslama M, Le Verche V, Dalous J, Kaindl AM, Tsenkina Y, Lacaud A, Peineau S, El Ghouzzi V, Lelièvre V, Gressens P. Implanted neurosphere-derived precursors promote recovery after neonatal excitotoxic brain injury. Stem Cells Dev. 2011;20(5):865-79.

[65] Chen A, Siow B, Blamire AM, Lako M, Clowry GJ. Transplantation of magnetically labeled mesenchymal stem cells in a model of perinatal brain injury. Stem Cell Res. 2010;5(3):255-66.

[66] Yasuhara T, Matsukawa N, Yu G, Xu L, Mays RW, Kovach J, Deans RJ, Hess DC, Carroll JE, Borlongan CV. Behavioral and histological characterization of intrahippocampal grafts of human bone marrow-derived multipotent progenitor cells in neonatal rats with hypoxic-ischemic injury. Cell Transplant. 2006;15(3):231-8.

[67] Webber DJ, van Blitterswijk M, Chandran S. Neuroprotective effect of oligodendrocyte precursor cell transplantation in a long-term model of periventricular leukomalacia. Am J Pathol. 2009;175(6):2332-42

[68] Luan Z, Liu W, Qu S, Du K, He S, Wang Z, Yang Y, Wang C, Gong X. Effects of neural progenitor cell transplantation in children with severe cerebral palsy. Cell Transplant. 2012; 21 Suppl 1:S91-8.

[69] Chen G, Wang Y, Xu Z, Fang F, Xu R, Wang Y, Hu X, Fan L, Liu H. Neural stem cell-like cells derived from autologous bone mesenchymal stem cells for the treatment of patients with cerebral palsy. J Transl Med. 2013;11:21.

[70] Min K, Song J, Kang JY, Ko J, Ryu JS, Kang MS, Jang SJ, Kim SH, Oh D, Kim MK, Kim SS, Kim M. Umbilical cord blood therapy potentiated with erythropoietin for children with cerebral palsy: a double-blind, randomized, placebo-controlled trial. Stem Cells. 2013;31(3):581-91.

[71] Papadopoulos KI, Low SS, Aw TC, Chantarojanasiri T. Safety and feasibility of autologous umbilical cord blood transfusion in 2 toddlers with cerebral palsy and the role of low dose granulocyte-colony stimulating factor injections. Restor Neurol Neurosci. 2011; 29(1): 17-22.

[72] Li M, Yu A, Zhang F, Dai G, Cheng H, Wang X, An Y. Treatment of one case of cerebral palsy combined with posterior visual pathway injury using autologous bone marrow mesenchymal stem cells. J Transl Med. 2012;10:100.

[73] Jensen A, Hamelmann E. First autologous cell therapy of cerebral palsy caused by hypoxic-ischemic brain damage in a child after cardiac arrest-individual treatment with cord blood. Case Rep Transplant. 2013;2013:951827.

[74] Wang L, Ji H, Zhou J, Xie J, Zhong Z, Li M, Bai W, Li N, Zhang Z, Wang X, Zhu D, Liu Y, Wu M. Therapeutic potential of umbilical cord mesenchymal stromal cells

transplantation for cerebral palsy: a case report. Case Rep Transplant. 2013;2013:146347

[75] Purandare C, Shitole DG, Belle V, Kedari A, Bora N, Joshi M. Therapeutic potential of autologous stem cell transplantation for cerebral palsy. Case Rep Transplant. 2012;2012:825289.

[76] Sharma A, Sane H, Paranjape A, Gokulchandran N, Kulkarni P, Nagrajan A, Badhe P. Positron emission tomography-computer tomography scan used as a monitoring tool following cellular therapy in cerebral palsy and mental retardation-a case report. Case Rep Neurol Med. 2013;2013:141983

[77] Sharma A, Kulkarni P, Sane H, Gokulchandran N, Badhe P, Lohia M, Mishra P. Positron Emission Tomography-Computed Tomography Scan Captures the Effects of Cellular Therapy in a Case of Cerebral Palsy. J Clin Case Rep; 2012, 2:13

[78] Sharma A, Gokulchandran N, Chopra G, Kulkarni P, Lohia M, Badhe P, V.C.Jacob. Administration of autologous bone marrow derived mononuclear cells in children with incurable neurological disorders and injury is safe and improves their quality of life. Cell Transplantation. 2012; 21 Supp 1: S1–S12.

[79] Pösel, Claudia, Karoline Moller, Wenke Frohlich, Isabell Schulz, Johannes Boltze, Daniel-Christoph Wagner. Density Gradient Centrifugation Compromises Bone Marrow Mononuclear Cell Yield. PloS one. 2012; 7(12):e50293.

[80] Debuse D, Brace H. Outcome measures of activity for children with cerebral palsy: a systematic review. Pediatr Phys Ther. 2011;23(3):221-31

[81] Bax M, Tydeman C, Flodmark O. Clinical and MRI Correlates of Cerebral Palsy: The European Cerebral Palsy Study. JAMA. 2006;296(13):1602-1608.

[82] Li DK, Li MJ, Traboulsee A, Zhao G, Riddehough A, Paty D. The use of MRI as an outcome measure in clinical trials. Adv Neurol. 2006;98:203-26.

[83] M. E. Raichle. Visualizing the mind. Scientific American. 1994; 270(4): 58–64

[84] Pauwels EKJ, Ribeiro MJ,. Stoot JHMB, McCready VR, Bourguignon M, Mazière B.FDG accumulation and tumor biology. Nuclear Medicine and Biology. 1998;25 (4): 317–322.

[85] Thie JA. Understanding the standardized uptake value, its methods, and implications for usage. Journal of Nuclear Medicine. 2004; 45(9):1431–1434.

[86] Sharma A, Gokulchandran N, Badhe P, Kulkarni P, Mishra P, Shetty A, Sane H. An Improved Case of Autism as Revealed by PET CT Scan in Patient Transplanted with Autologous Bone Marrow Derived Mononuclear Cells. J Stem Cell Res Ther 2013, 3:2

[87] Sharma A, Gokulchandran N, Shetty A, Sane H, Kulkarni P, Badhe P. Autologous Bone Marrow Mononuclear Cells may be Explored as a Novel. Potential Therapeutic Option for Autism. J Clin Case Rep 2013, 3:7

[88] Varrone A, Asenbaum S, Vander Borght T, Booij J, Nobili F, Någren K, Darcourt J, Kapucu OL, Tatsch K, Bartenstein P, Van Laere K; European Association of Nuclear Medicine Neuroimaging Committee. EANM procedure guidelines for PET brain imaging using [18F]FDG, version 2. Eur J Nucl Med Mol Imaging. 2009;36(12): 2103-10.

[89] Devine CE, Mawlawi O. Radiation safety with positron emission tomography and computed tomography. Semin Ultrasound CT MR. 2010;31(1):39-45.

[90] Ogawa S, Lee T, Nayak AS, Glynn P. Oxygenation-sensitive contrast in magnetic resonance image of rodent brain at high magnetic fields. Magn Reson Med. 1990;14:68–78.

[91] Cao Y, D'Olhaberriague L, Vikingstad EM, Levine SR, Welch KM. Pilot study of functional MRI to assess cerebral activation of motor function after post stroke hemiparesis. Stroke. 1998;29(1):112-22.

[92] Wright T, Nicholson J: Physiotherapy for the spastic child: An evaluation. Dev Med Child Neurol 15:146-163, 1973

[93] vanPraag H, Christie BR, Sejnowski TJ, Gage FH. Running enhances neurogenesis, learning, and long-term potentiation in mice. Proceeding of the National Academy of Sciencesof the U S A. 1999;96:13427-31

[94] Colcombe S, Kramer AF. Fitness effects on the cognitive function of older adults: A metaanalytic study. Psychol Sci. 2003;14:125-30

[95] Möbius-Winkler S, Hilberg T, Menzel K, Golla E, Burman A, Schuler G, Adams V. Time-dependent mobilization of circulating progenitor cells during strenuous exercise in healthy individuals. J Appl Physiol. 2009;107(6):1943-50.

[96] Sandri M, Adams V, Gielen S, Linke A, Lenk K, Kränkel N, Lenz D, Erbs S, Scheinert D, Mohr FW, Schuler G, Hambrecht R. Effects of exercise and ischemia on mobilization and functional activation of blood-derived progenitor cells in patients with ischemic syndromes: results of 3 randomized studies. Circulation. 2005;111(25):3391-9.

[97] Petersen AM, Pedersen BK. The anti-inflammatory effect of exercise. J Appl Physiol. 2005;98(4):1154-62.

[98] Gielen S, Adams V, Möbius-Winkler S, Linke A, Erbs S, Yu J, Kempf W, Schubert A, Schuler G, Hambrecht R. Anti-inflammatory effects of exercise training in the skeletal muscle of patients with chronic heart failure. J Am Coll Cardiol. 2003;42(5):861-8.

[99] Lin YS, Jan MS, Tsai TJ, Chen HI. Immunomodulatory effects of acute exercise bout in sedentary and trained rats. Med Sci Sports Exerc. 1995;27(1):73-8

[100] Sharma A, Sane H, Badhe P, Kulkarni P, Chopra G, Lohia M, Gokulchandran N. Autologous Bone Marrow Stem Cell Therapy shows functional improvement in hemorrhagic stroke- a case study. Indian Journal of Clinical Practice, 2012:23(2):100-105

Section 3

New Possibilities

Chapter 8

Brain Computer Interfaces for Cerebral Palsy

Pedro Ponce, Arturo Molina, David C. Balderas and
Dimitra Grammatikou

Additional information is available at the end of the chapter

http://dx.doi.org/10.5772/57084

1. Introduction

Cerebral Palsy (CP) is a group of disorders that affect movement and posture, causing activity limitation to the person who suffers from it. It is caused by a lesion that occurred in the developing brain, usually before birth but also during or after. Cerebral palsy manifests itself early in life, during infancy or preschool years with delayed or aberrant motor progress and it is non-progressive, which means that at the time of the diagnosis, the disturbance that incited the cephalic lesion is no longer active. At the moment there is no cure for cerebral palsy (Bax 2005).

Cerebral Palsy is a condition which affects approximately 2 out of every 1000 newborns. The total number of children with cerebral palsy has remained stable since 1970, but at the same time there has been a consistent rise in the risk of cerebral palsy associated with preterm infants (Thornhill 2009). Since it was first reported by Little in 1861, it has been widely documented and it has attracted research interest.

According to the World Health Organization (WHO), more than one billion people of the world's population lives with a disability and this number is rising as the population grows, the increase of chronic health conditions, and the life expectancy becomes higher (World report on disability, 2011).

There are 3 possibilities to restore function:

- Using remaining muscular pathways as substitute for paralyzed muscles. The use of eye or hand movement can give control over communication devices.
- Using EMG, above the level of lesion, as control for paralyzed muscles.
- Provide the brain with a new form of communication and control that uses no muscular paths.

Despite the motor limitations of the physically disabled, most of the time the brain activity remains intact. Brain-Computer Interfaces (BCI) are communication devices that translate signals from the brain or nervous system (e.g. Electroencephalogram (EEG)) into electrical signals For the control of devices, allowing people to regain some form of control and regain interaction with the environment (Rao 2010, Wolpaw 2002). Even though, BCIs where initially developed as assistive devices for people with severe neuromuscular disorders, such as brain or spinal cord injury, cerebral palsy, muscular dystrophies, multiple sclerosis, and numerous other conditions; the increasing interest in non-medical application looking for an improved technology for Human-Computer Interaction (HCI) such as exoskeletons, robot or wheelchair control, or augmented reality (Lotte 2013), has generated clinical, scientific and commercial interest in the use of BCI's for an augmentative communication and control technology.

2. Examples of success

Despite, that BCIs have shown their possibility as communication and control device through spelling devices (Donchin 2000), used as control of prosthesis (Tenore 2008), web browsing (Mugler 2010), for control in a virtual reality environment (Lotte 2009) and for entertainment (Rao 2010), there are still more possible applications and room for improvement, using combined technology (e.g. Hybrid BCIs), improving or creating new classification algorithms, and better recording technology. In this chapter we give a brief description of the recording technology, pattern selection, current classification algorithms used for BCI and the state of the art as well as future technology.

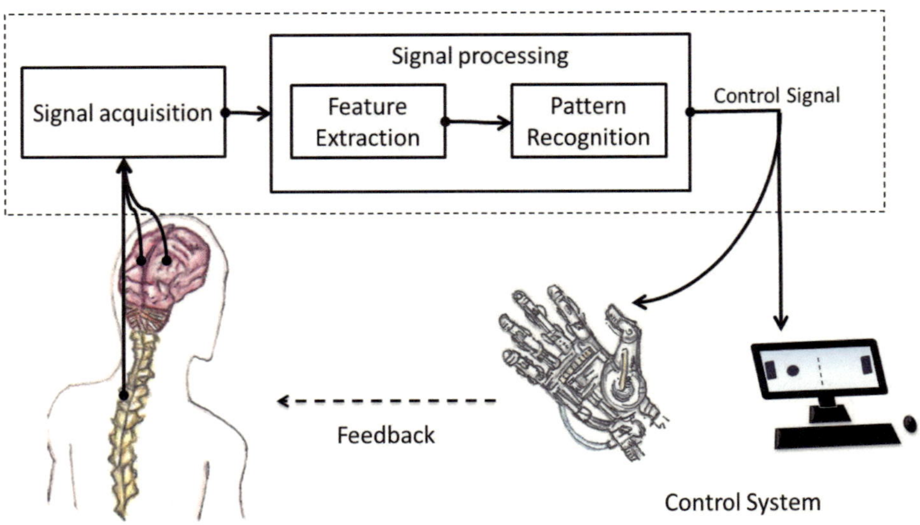

Figure 1. Basic components of a BCI. The image illustrates the map between the input and output through the translating algorithm. Signals are acquired by electrodes and then translated into a control signal for an external device (e.g. wheelchair, neuroprosthesis or exoskeleton) using a sequence of processing steps.

3. Evaluation and classification

3.1. Neuromotor examination of neonates and infants

The diagnosis of CP is made largely through clinical observations. The natal history is of vital importance for the identification of reasons for concern and the determination of the cases which merit closer monitoring. Failure to meet gross motor milestones is often the initial concern of parents. Significantly delayed motor milestones, persistence of primitive reflexes, and abnormal postural reactions are additional reasons for concern and referral to a neurologist or expert in neurodevelopment for evaluation. Clusters of symptoms or evolving abnormal movement patterns may be indications of CP and thus should be explored further with diagnostic instruments.

Instruments like the Hammersmith Neurological Examination (Dubowitz 1999), the Amiel Tison Neurological Assessment (Tison 2002) or the INFANIB (Infant Neurological International Battery for the Assessment of Neurological Integrity in Infancy) (Ellison 1994) have proven extremely valuable in the earlier identification of the difficulties that at-risk neonates and, as a result, a better targeted, early intervention.

These instruments offer a neurological or neuromotor exploration of the neonate and the infant, assessing the existence of primitive reflexes, the automatic system or any other involuntary movements, that appear in normal infants and should be integrated by the 9th month of life. Their persistence past that age is a reflection of abnormalities in terms of control in the central nervous system and may indicate cerebral palsy. The persistence of primitive reflexes causes changes in muscle tone and the position of limbs, which makes it interfere with the development of voluntary motor movements by causing changes in muscle tone and the position of the limbs. Failure to develop protective reflexes such as the parachute response or an asymmetrical response is also taken into consideration.

The instruments also take into consideration the age when the infant met the motor milestones (head control, sitting, voluntary grasp, ability to kick, rolling, crawling, standing and walking) and may include some items of attention, sensory function or self regulation, as well as muscle tone and posture are considered.

4. Nature and typology of the motor disorder

Cerebral Palsy is a symptom complex with various types and degrees of motor impairment. Depending on the area of the brain that has been affected, according to the SCPE (Elison P 2004, SCPE working group 2002) we may identify the predominant motor characteristics of the condition as of three types: Spastic, dyskinetic and ataxic.

Spastic CP results from defects or damage occurring in the brain's corticospinal pathways, also described as upper motor neuron damage. Spastic CP accounts for almost 84% of all cases of CP, with cognitive impairments seen in approximately 30% of the cases with CP (SCPE working group 2000, Palisano 1997). Although increased , as well as muscle tone is the predominant feature observed, hyperreflexia, clonus, extensor Babinski response, and persistent primitive reflexes are commonly seen.

Dyskinetic and ataxic CP are caused by damage to nerve cells outside of the pyramidal tracts in the basal ganglia or the cerebellum. Dyskinetic CP is then further divided into athetoid and dystonic. It accounts for 15% to 20% of all cases of CP, with dyskinetic accounting for 10% to 15% and ataxic approximately 5%. The resulting disability is global with abnormal tone regulation, postural control, and coordination (SCPE Working Group, Palisano 1997)

It is actually quite common to see many different combinations of types of CP, since this depends on the area of brain damage; the types overlap very frequently, which can make it very difficult to precisely label the resulting disability within the typical subtypes. As a result, when not one type dominates we make reference to a "mixed" category.

5. Functional motor abilities

What is of particular interest for the parents of the children affected is evaluating the functional consequences of the condition. The Gross Motor Function Classification System (GMFCS) (Eliasson 2006) was developed as an evaluation tool in order to offer a prognosis or to assess differences in motor functions after an intervention. It recognizes motor function as dependent on age due to the expected change of the developing child. It separates clusters of periods (0-2, 2-4, 4-6 and 6-12 years of age).

Degree of functionality	Gross Motor Function Classification System (GMFCS)	Manual Ability Classification System (MACS)
Level I	Child´s ability to walk is not affected	Child handles objects easily and successfully
Level II	Child´s ability to walk is slightly affected	Child handles objects with somewhat reduced quality
Level III	Child walks with assistive device	Child handles objects with difficulty
Level IV	Limited self-mobility with assistive device	Child handles only a few, adapted objects
Level V	No self-mobility	Child cannot handle objects

Table 1. Classification of gross motor function and manual ability in children with cerebral palsy.

In this chapter we have cited the classification of a child´s gross motor function between 6 and 12 years of age, which is divided into five levels, based on functional mobility or activity limitation. Particular emphasis is made on the function of sitting and walking. Children in level I are the most independent (motor function) and children in Level V are the least according to the **Gross Motor Function Classification System for Cerebral Palsy**.

The Manual Ability Classification System for Children with Cerebral Palsy (MACS) (Bottcher 2010) is widely used to evaluate and classify how children with cerebral palsy use their hands to handle objects in daily activities. Like the Gross Motor Function Classification System, MACS describes five levels. The levels are based on the children's self-initiated ability to handle objects and their need for assistance or adaptation to perform manual activities in everyday life.

6. Accompanying impairments

A child with cerebral palsy often has other conditions related to developmental brain abnormalities, such as intellectual disabilities. Almost 50% of children with CP have an average intelligence, 20% have an intelligence slightly lower than average (borderline intelligence). The rest 30% its not mentioned if its more inteligent or not than average. Most patients that have spastic tetraparetic, discinetic and ataxic have a severe mental discapacity (SCPE Working Group 2002)

There have been studies that prove that children with CP with average intelligence have attentional deficits or problems with the executive functions, which may partially account for the behavioral problems that sometimes present. (Guzzetta 2001) They might have deficits in visioperceptive functioning. The child has difficulties recognizing the spatial relations between objects, as well as between objects and his own body. This results frequently in a constructive dyspraxia. The saccadic movement of the eye to focus on an object that appears periferically at the previous point of focus are slow and dyspraxic, which constitutes an added difficulty in order to achieve the perceptive integration. The proprioceptive-visual integration of the parietal lobe is necessary in order to orient the movements and postures of the upper limbs to reach for and manipulate the surrounding objects and starting the proceeding automatic movement that experience and repetition offers. These deficits are completely independent from the vision problems that may coexist (Guzzetta 2001).

Language problems are also common and their severity depends on the timing that the lesion took place, in the prelinguistic period or later, when the linguistic function has already started to form.

7. Implications for everyday life

Although there are many compelling reasons to give the diagnosis as early as possible (parents frustration of handling a child with abnormal tone such as feeding, sleep, and temperament problems, plan in advance for long-term treatments and management options that may be needed by the child, possible increased insurance benefits and in some cases federal assistance, benefits that come from an early intervention) the diagnosis should not be formally made until the second year of age. For the SCPE in Europe minimum age of 4 years old is required to make a diagnosis so that transitory alterations of neurodevelopment or degenerative diseases may not be confused with CP (SCPE Working Group 2002).

The diagnosis has an impact on the life of the family and, of course, the child. The major issue of concern is, for most parents, walking. Once confronted with the diagnosis, the first question that parents ask their child´s health care specialist if the child is going to walk. Children will CP will experience some degree of difficulty with movement. This can range from problems like clumsiness that does not disrupt everyday life activities all the way to difficulties with walking. The child may move slowly, may need to use a walking aid or a wheelchair.

Simple activities like dressing, bathing, eating can be a real challenge to the child with CP and their family. The activities can take longer, especially if the child needs more assistance, physical help or specialized equipment.

Language problems are common among children with CP. Children may have difficulties with both verbal and non verbal aspects of language. The expression and understanding of the formal aspects of language can be affected (for example articulation or denomination) which may eventually lead to problems with reading and writing or even interfere with the child´s ability to communicate verbally. The other aspect of language that can be impaired is pragmatics, which refers to the ability to place words in the context of one´s own mind and the interlocutor´s, which creates problems in the child´s social adjustment.

8. Attentional processes

The aim is for every child with CP to achieve their potential. Depending on the child´s individual characteristics decisions must be taken that will determine whether he would benefit more from mainstream placement in a school or from a placement in a more specialized environment that could tend to his needs.

In order for the interface to be able to read the brain signal, the child needs to be focused. Not always is it possible for all children to emit a signal strong enough so that it can be captured by the interface. The emission of a strong signal depends on the attention of the child which can be negatively affected by a variety of factors which have no relation with the interface but which affect its ability to read the brain signals.

The attention of the child can be hindered by three main factors which are at constant interplay and affect the prefrontal cortex and the ability of the child to focus on a particular task. The three main factors are:

a. Cognitive

b. Emotional

c. Behavioral

Although the cognitive function of children with CP has not been systematically studied, and more research is needed, there is evidence suggesting that children with CP and normal

http://dx.doi.org/10.5772/57084

intelligence present impairments in executive functions. Executive functions are the brain functions that regulate and control impulse, anticipate consequences, put attention, regulate emotion, allow flexibility, plan and monitor results. Executive functions are highly fragile because they are the last cognitive area to mature. They involve the prefrontal cortex and they rely on an extensive interconnectivity with other parts of the brain. Damage to that area results in slower information processing, and a decrementation in sustained attention performance, which is necessary for the reading of the signal by the interface (Guzzetta 2001).

In the case of an intellectual disability, which as we have seen affects almost half of the children with CP, we cannot speak of attention problems. The degree of cognitive impairment is such that the attention processes cannot reach the required level so that it can be captured by the interface.

The attentional processes are also going to be affected by the emotional problems that the child may be experiencing (Parkes 2008). This is also an issue that has not been researched but there is enough evidence to suggest that children with CP, like children with some sort of a disability in general, are more likely to suffer from depression, anxiety and low self esteem. This is associated with the severity and visibility of the condition, which affects the child s ability to control his own body and the way his peers may perceive him as being different from them. (McDermott 1996) The lack of social support, the anxiety of the parents, the child´s inability to use words to express his emotions are all factors that put the child at increased risk to experience emotional problems. Emotional problems hinder the ability of a child to focus and pay enough attention so as to send a strong signal to the interface.

Bottom of Form

Children with CP have behavioral problems like being defiant and disobedient. The behavior problems reported by parents were 5 times more likely in children with cerebral palsy compared with children having no known health problems. Behavioral problem are associated with some kind of combination of the impairment, the environment and interpersonal relationships. Damage to the prefrontal cortex affects, as we have seen, cognitive flexibility, the abilities for strategic planning, tolerance to frustration, behavioral inhibition (hyperactivity-impulsivity) as well as the associated impairment of inattention. The child that has trouble maintaining his attention on the signal is more likely to refuse to try or abandon the task.

It is important to mention epilepsy as one of the factors that cause behavior and attention difficulties in children with cerebral palsy.

Epilepsy affects 7 to 50% children with cerebral palsy. Epilepsy, in itself, takes away part of the vitality of the brain, with the frequent crises affecting the cognitive abilities of the child. Behavior and attention difficulties are highly common in children with cerebral palsy who have epilepsy. Furthermore, the crises are frequently a motive for the child to stop receiving education. Medical treatment for epilepsy can be helpful, keeping in mind that although antiepileptic drugs may impair the cognitive functions of the child, with the careful monitoring of the physician and the new medical intervention, this side effect would be very infrequent.

9. Communication system

There are some limits that can be solved using the brain activity. This activity allows the communication between the processor and the person.

9.1. Brain signals

Through the recording and processing of direct brain electrical activity via signal processing and machine learning algorithms, BCIs enables communication and control to assistive devices. Although the aim of a BCI is to identify and translate brain electrical signals into commands, it is not a thought-reading device or systems able to literally translate arbitrary cognitive activities. BCIs are design for translation of well characterized a priori defined brain activity patterns through the use of machine learning techniques and patterns recognition methods into commands.

Considered as a control system, a BCI has an input (e.g. EEG), an output (e.g. control signal), and components that translate input into output, a protocol that determines the timing operation and in some cases some feedback is provided to the user (Figure 1.).

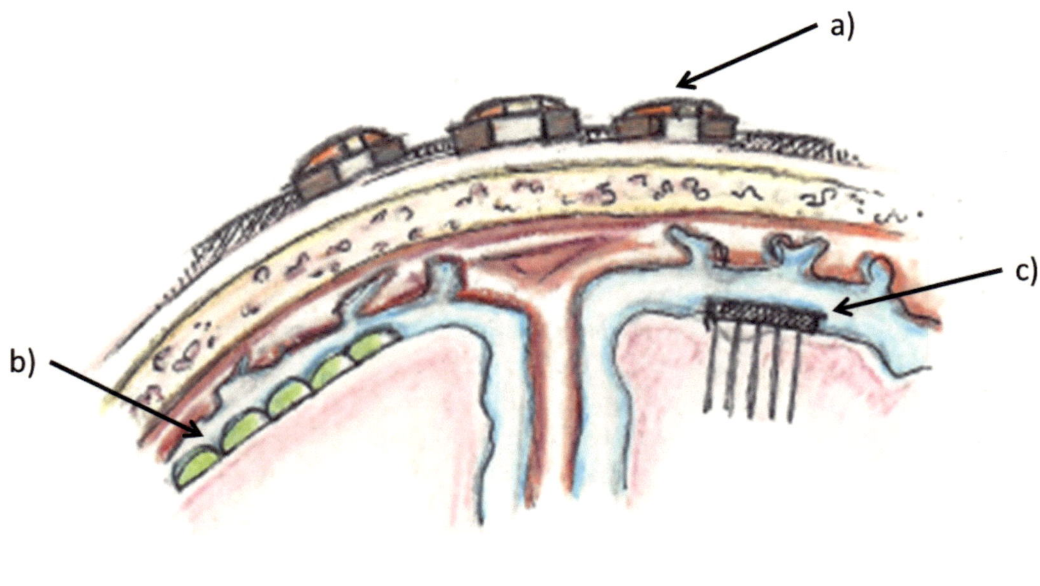

Figure 2. Exemplification on EEG (a), ECoG (b) and Single-neuron recording (c) electrode placement over the head

9.2. Functional neuroimaging

Invasively or noninvasively brain activity is recorded either from recording electrical activity through electrodes (EEG, Electrocorticography (ECoG) or from single-neuron recordings within the brain), recording magnetic fields using magnetoencephalography (MEG)), or recording metabolic activity reflected in changes in blood flow (positron emission tomography (PET), functional magnetic resonance imaging (fMRI) and functional Near Infrared (fNIR)). Despite the fact that MEG, PET, fMRI and fNIR have shown success for BCI applications these techniques are still technically demanding and expensive technologies that require sophisticated equipment that can be operated only in special facilities. Furthermore, PET, fMRI and fNIR techniques depend on metabolic processes, such as blood flow, having long latencies and thus less suitable for the control of BCIs.

On the other hand, the non-invasive EEG and the invasive ECoG and single neuron recordings (Figure 2.), are methods that have relative low costs, are simpler to use and have higher temporal resolutions, making them more practical to the use with BCIs.

Invasive techniques such single-neuron recording and ECoG take recordings over the cortex; while single-neuron recording records the activity within the cortex, ECoG records the activity over the cortical surface of the brain. Single-neuron recordings and ECoG does not record single neuron activity but records activities over small regions of the brain giving them a high spatial resolution, and as it is implanted directly over the cortex, they have a high bandwidth, high SNR and high amplitude. Since ECoG electrodes do not penetrate the cortex, recorded signals are also not subjected as heavily to immune response, possess lower risk to implant as well. Furthermore, maintaining long term reliable recording with implantable electrodes is difficult.

Although, ECoG has a higher spatial resolution compared to EEG (i.e. 1.25 - 1.4mm vs centimeters) higher frequency bandwidth ([19, 11] (i.e. 0–500Hz vs. 0–40Hz), have higher signal amplitude (50 – 100μV maximum vs. 10 – 20μV maximum), and being less susceptible to artifacts (i.e. EMG, EOG or electrical devices), EEG has become the most common source for brain activity due to its none invasiveness (requiring no craniotomy (surgical incision of the skull)), being more practical for everyday situations. EEG measures the potential over the scalp, reflecting the collective activity over large population of neurons located underneath the sensor position.

9.3. Recording and processing

Brain signal recordings, like EEG or ECoG, are obtained with electrodes attach from the surface of the skull or to the surface of the brain measuring difference over the potential that reflect the activity within the brain. The electrodes are connected to biosignal amplifier where they are amplified and go through an analog-digital conversion. These signals are sent to the signal processing system that is in charge to perform the feature extraction and classification. Finally, a signal will be send to the control system as final output. The BCI can be design to present feedback that is beneficial to learn the BCI control faster.

The electrodes measure a difference in potential (i.e. the voltage) between two electrodes. The difference in potential reflects neural activity below the electrode. There are different EEG electrode montages. Usual EEG recordings use unipolar montage rather than bipolar electrodes, meaning that they use a common reference for all electrodes. A ground is added to keep the voltage levels close to the amplifier ground voltage level. The reference and ground can be positioned everywhere within the array of electrodes, but they are normally placed either over the ear or the mastoids (the temporal bone behind the ear). There also exist a bipolar and Laplacian montage that each electrode represents the difference between the electrode and its surrounding electrodes (see Figure 3).

EEG recordings electrodes usually use small metal plates made out of gold or Ag/AgCl. Alternative materials such as Tin have been used, but they present drifting noise below 1HZ, making them unsuitable for some applications, such as Slow Cortical Potentials. The electrodes could be either passive or active (i.e. pre-amplified with gain 1-10) disks that are connected through a cable to the biosignal amplifier. Active electrodes are less susceptible to environmental noise, and can work with higher skin impedance than passive electrodes. There exist also dry and wet electrodes. As the dry electrodes normally use an array of pins to go through the hair have contact with the skin, the wet electrodes use a gel that reduces the impedance and make a better connection to the skin. Even though dry electrodes have the advantage that require less preparation and cleaning time (not requiring conductive gel) and They are proven to be an alternative for EEG recordings (Zander 2002), More in-depth research is necessary for their successful dailybased application. daily-base application.

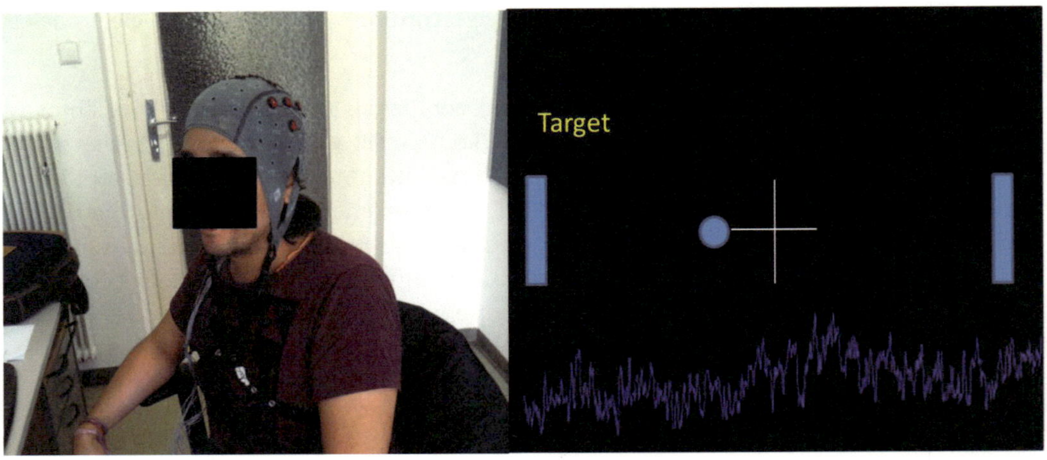

Figure 3. Examples of non-invasive BCI, with visual stimuli and virtual control of ball movement

9.3.1. Electrode distribution

The standard EEG electrodes naming and position on the scalp are according to the international 10-20 electrode system (Jasper 1958). The system ensures that different laboratories share the same names over electrode positioning. It is based on arcs dividing the scalp in an array, using the Nasion and Inion as longitudinal reference points (i.e. front and back respectively) and the left and right Pre-Auricular points as lateral reference points (Figure 4a.). The intersection between the longitudinal line and the lateral is called the Vertex, and at this point it is located over the center. The 10-20 system identify each point using each lobe (Frontal-F, Temporal-T, Central-C, Parietal-P and Occipital-O) and each hemisphere (Left-Odd, Right-

http://dx.doi.org/10.5772/57084

Even numbers and z or zero over the midline) as a marking. An extension to this configuration is using 70 electrodes (Figure 4b.), subdividing in between the 10-20 arcs (using the combinations of the letters for reference). In addition, the letters A and Fp are used to identify the earlobes and frontal polar sites respectively[1]. The electrode can either be placed directly over the scalp, Which requires practice and is time consuming. A second technique is using caps, Which already have the marking of the electrodes and their position, some even have the electrodes pre-mounted making them easier to work with. These caps come in different sizes and can be adjusted to different persons.

Currently electrode caps are mainly intended to be used in a laboratory environment, therefore being expensive, the electrode and cap placement is difficult and need special preparation, making them not practical for an everyday use. Current commercial caps, such as the Emotive Epoc, they present an alternative for an everyday use being more economically accessible, they have the drawback that they use wet electrodes that dry quickly, and they are difficult to position and remain stable, which is problematic as it gives a high variance over recordings, as well as the same recording, and they are not possible to use in a different configuration. For the remaining of this section we will only focus on laboratory caps and EEG recording, as these are the most commonly used for BCI.

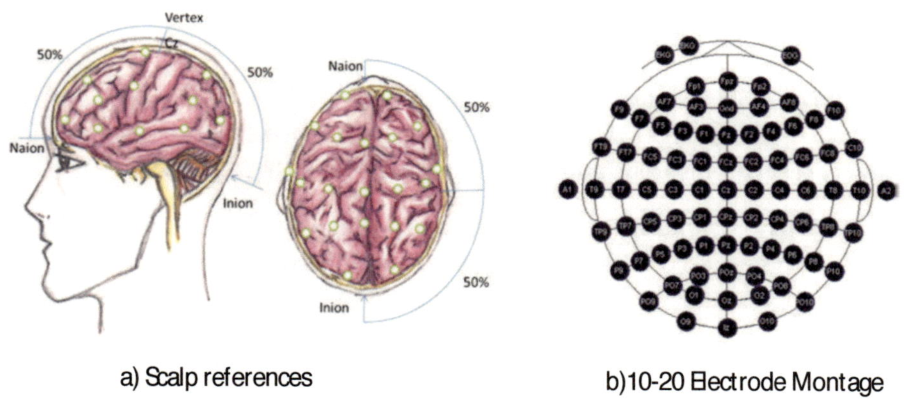

Figure 4. EEG Electrode Montage (Jasper 1958)

9.3.2. *Artifacts*

Since EEG is further away from the neurons it has low spatial resolution and very noisy overview of ongoing brain activity. There are mainly two sources of noise while performing EEG recordings: Environment and Physiological artifacts.

Environment artifacts: Electrical power lines and/or surrounding electrical equipment become a problem while recording EEG. Their frequencies can overlap with the EEG feature the

1 There is also the AF marking the subdivision in between the Frontal and Frontal polar site

classification algorithm is working on. For laboratory settings these artifacts are normally solved by using some isolation from environmental signals, avoiding the interference with EEG recordings. A notch pass filter over the frequency over the power line (50 or 60 Hz for America or Europe recordings respectively) may also be applied suppressing signals in a narrow band.

Physiological artifacts: Although, muscle activation, eye movement or eye blinking can serve as communication signals for HCI, they can mask EEG frequencies and mislead researchers by mimicking EEG-based control and/or hide EEG features. Some EEG recording incorporate these signals as either control signals, or they used to filter the desired signal. A different way of control is to remove the EEG recordings that have been contaminated by artifacts, leaving only the trials that were not contaminated for the training of the classifier.

9.4. BCI signal processing

To design a BCI, we need to decide on the type of signal, the location, the desired feature and the appropriate classification technique.. In this section a description of the different types of signals, the different types of feature extraction that has been used, and finally a brief description of the different machine learning algorithms available is presented.

9.5. Evoked potentials and oscillatory activity patterns

The two major types of EEG signals used in BCI are Evoked Potentials (EPs) and changes in the spontaneous oscillatory EEG activity, also known as event-related desynchronization (ERD), and event-related synchronization (ERS) (Pfurtscheller 1999(2)).

EPs are electrical potential shifts that are time-locked to perceptual events, such as a rare visual or audio stimulus. Time-locked implying here that the time between the event and the time potential shift is approximately constant. Due to its low Signal-to-Noise Ratio (SNR) are typically analyzed by averaging EEG data over time beginning of the perceptual event for duration over 1s. There are different types of EPs based on the source of stimulus (e.g. visual, auditory or tactile)

On the other hand, oscillatory activity can be voluntary induced by the user (e.g. imagination of kinesthetic body movement, aka motor imagery, Neuper 2005). Such imagery usually generates a decrease or increase in power in a particular frequency band (ERD or ERS (ERDS) respectively). ERS are normally associated with an ERD appearing either after the termination of the movement or simultaneously to the ERD, but in other areas of the cortex. Although, Oscillatory patterns detection is less robust and reliable compare to EP, which as synchronous signal (i.e. knowing its time and shape) requires little adaptation and its detection is robust, as an asynchronous BCI it allows the user to send information at their own pace, unlike synchronous BCIs that require to follow the cues or prompts from the system.

Figure 5 illustrates the use of EP and ERDS for achieving brain-computer interaction in physical and virtual environment.

http://dx.doi.org/10.5772/57084

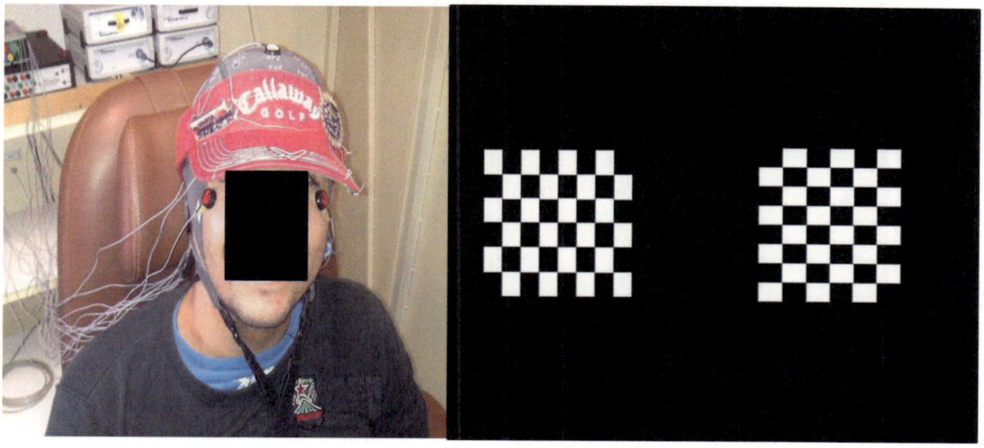

Figure 5. Examples of evoke potentials with either lights or screen squares flickering at different frequencies

9.6. BCI control commands

Within these two ways of brain signal extraction there are four main strategies to consider for input at a BCI system. extraction there exist mainly 4 common strategies are considered for input of a BCI system, a) Motor imaginary. b) Slow Cortical Potentials, c) the P300 wave of visual evoke potentials, and d) Steady State Visual Evoked Potential (SSVEP)

9.6.1. Motor imaginary

It is believed that the mechanisms of brain operations are characterized by groups of neurons synchronizing themselves to a certain physiological frequency (Engel 2001). These oscillations has been divided into different frequency bands and are referred as brain rhythms (Delta [0.1 – 4Hz], Theta [4 – 7Hz], Alpha [8 – 12Hz], Mu [8 – 13Hz][2], Beta [12 – 30Hz], Gamma wave [30 – 100Hz] Figure 5a.). Movements are normally accompanied by changes of the Mu and Beta rhythm over the motor/sensory cortical areas (see Figure 5b.). For example movements of the hand are associated with decrease of power (or desynchronization) over the Mu rhythms and associated with a decrease over the Beta rhythm, particularly contralateral to the movement. The same effect occur with motor imagery (Neuper 2005), making the Mu/Beta rhythm as base for a BCI. The most common approach used for classification is to calculate the bandpower in a specific frequency band and then use discrimination via some machine learning technique (e.g. Fisher linear discriminant analysis).

9.6.2. Slow cortical potentials

Slow cortical potentials (SCPs) are slow voltage changes generated over the cortex. These changes in potential occur over 0.5–10s. These potentials can be divided in Negative SCPs,

2 Although the Alpha and Mu rhythm occur over the same frequency, one is located over the resting visual cortex at the back of the scalp, while the second is found over the motor cortex

typically associated with movement and other typical cortical activation, and Positive SCPs that is associated with a reduce cortical activation, the viability of the use of SCPs that after a period of learning user has gain control selecting words or pictograms from a computerized language (Birbaumer 2000) or used with patients Suffering from a locked-in condition such as amyotrophic lateral sclerosis (ALS) (Kübler 1999). The drawback of using SCPs is that It requires a long training process that allows the user to gain control, a normal training can go for several weeks or even months. The normal training for a SCP based BCI users first learn to move a cursor vertically on a monitor selecting targets at the top or the bottom of the screen. Next, a split keyboard into two where an area is selected, the selected characters are once more split into two and once more selected, and this is done until the final choice is made.

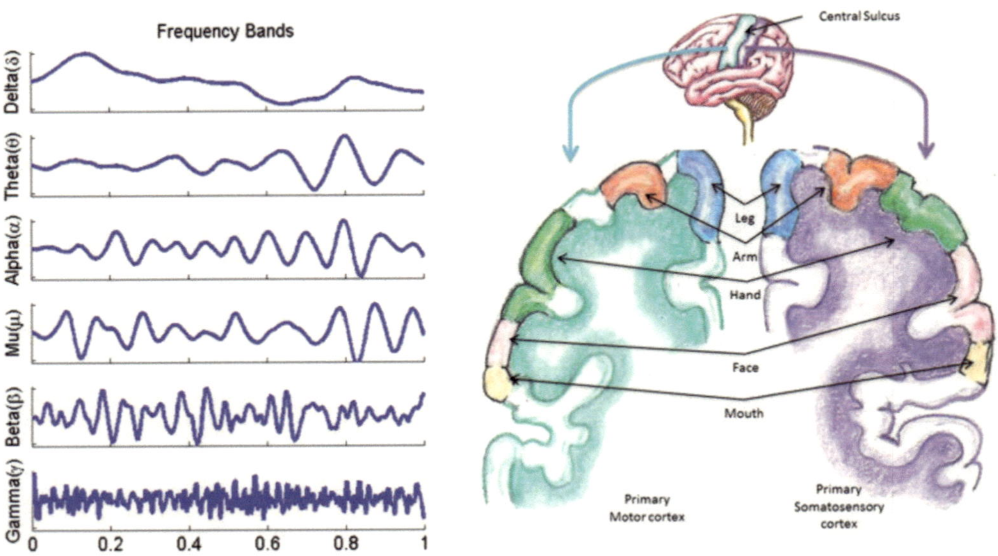

a) EEG Rhythms

b)Motor and somatosensory

Figure 6. Example of different types of normal EEG rhythms (Lotte 2009) and Primary Motor and somatosensory cortical homunculus

9.6.3. The P300 wave of visual evoke potentials

The P300 (P3) wave of visual evoke potentials (henceforth referred only as P300) is a positive wave that appears 300ms after a stimulus is presented (Figure 7a.). It was first described by Sutton (Sutton 1965). The most known paradigm for P300 is the one described by Farwell-Donchin 1988, where characters and numbers are represented in a six by six grid (Figure 7b.)

http://dx.doi.org/10.5772/57084

where the rows and columns are repeatedly flashed and when the character containing the chosen character is flashed a P300 is evoked. These characters (number or letters) can be replaced with different symbols and used not only as spelling device, but for navigation or for control tasks, using symbols as arrows or object selection. P300 can be used as a lie detector, providing certain stimuli (e.g. picture, phrase or word related to the lie) that P300 is generated if the subject has knowledge of the stimuli presented (Farwell 2001, Farwell 2012)

As a normal procedure, the P300 of different repetitions are averaged, to reduce the effect of artifacts, or the presence of different mental activity that could masked the ongoing P300. Different features and classification techniques (e.g. Linear Discriminant Analysis or Supported vector machines, Krusienski 2006) has been used for P300 based systems, these techniques will be described in a following section.

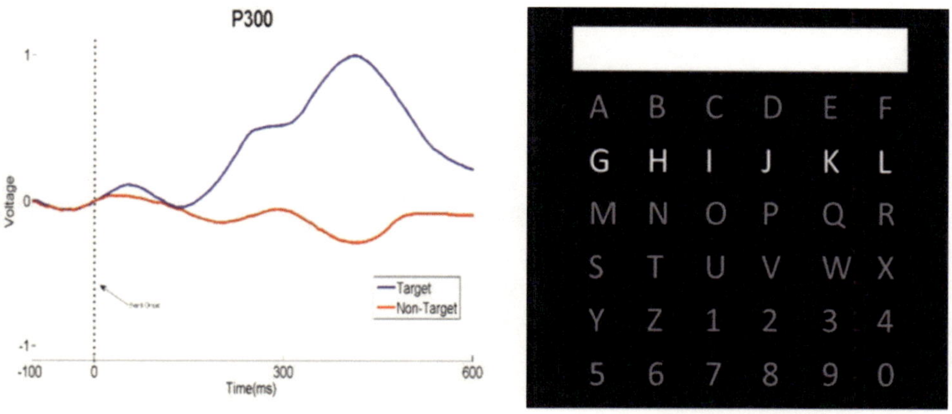

a) P300 Wave b) P300 paradigm

Figure 7. P300 wave and the classical P300 spelling paradigm described by Farwell-Donchin 1988. Figure 6a show a change of potential occur approximately 300ms after the stimulus is presented (Picture adapted from Scherer 2013). In Figure 6b shows the classical spelling paradigm, where a P300 potential is generated if either the row or column flashes over the letter desired.

9.6.4. Steady State Visual Evoked Potential

Steady State Visual Evoked Potential (SSVEP) are brain responses to visual stimulus (e.g. flickering LEDs or phase-reversing checkerboards), flashing at constant frequencies between approximately 6 – 100Hz. SSVEP is a frequency-locked signal that manifests itself as an increase of the EEG amplitude of the stimulated frequency over the occipital lobe. Classification of SSVEP is done either using FFT-spectrum analysis or by the use of canonical correlation analysis (CCA) or finally by using of the minimal energy approach.

SSVEP have shown to be independent to eye movement, making them a good alternative for people with well preserve eye acuity but are incapable to moving their eyes (Brendan 2008). Some drawbacks while using SSVEP is that if a computer screen is used only frequencies that entire division over its base refresh rate (e.g. a 60Hz screen only 30, 20, 15, 12, 10 or lower are possible). A second drawback is that SSVEP are usually developed with a short number of flickering channels trying to avoid distraction and hence lower performance.

9.7. Feature extraction

Even though the amount of electrodes, the number of tasks performed, the high sample frequency required, the classes and the different patterns make that the amount of data recorded large, the normal training data set is short. Identifying, Selecting and extracting the relevant properties or features of the signals that better describe the EEG signals are essential steps in the design of a BCI. The correct selection of the features is crucial, if the features extracted from EEG are not relevant and do not accurately describe the EEG signals employed, the classification algorithm will have trouble selecting the class or label the user intended. The feature extraction could be divided in two main groups: temporal and frequential methods, a third group can be added as hybrid between temporal and frequency techniques.

9.7.1. Temporal methods

Features that present a time dependent variation can be treated using a temporal method. The changes can be as the ones that occur on P300 wave, which depend of the flashing of the selected command to 300ms later to be generated. The main temporal methods are the parametric models (e.g. AR or AAR) modeled the signal using a weighted sum of values, the Hjorth parameters that describe the dynamics of the signal by the use of three measures (activity, mobility and complexity) and finally the signal amplitude method that concatenates the electrodes amplitude into a feature vector that is used as input into the classification algorithm.

9.7.2. Frequency methods

The different oscillations or rhythms that characterize the EEG signals present variations while performing a mental task (e.g. motor imagery) or with a steady state evoked potential that a change in the oscillation is highly related to the stimulus frequency. Frequency methods are commonly used for the ease of application and computational speed. The most commonly used methods are power spectral densities and band powers. The third method uses a feature that can be located both in time and frequency domain. This method uses the Short Time Fourier Transform or the Wavelet transform to have a time-frequency representation of the signal.

9.8. Classification techniques

After the features have been selected the next step is to translate them into a command. This translation can use regression/classification methods. There are different classification methods and they can be divided using their classifier properties into: Linear classifiers, neural networks, non-linear Bayesian classifiers, nearest neighbor classifiers and combinations of classifiers (Lotte 2007).

http://dx.doi.org/10.5772/57084

9.8.1. Linear classifiers

Linear classifiers are discriminant algorithms that use linear functions to separate between classes. The most common used for BCI are Linear Discriminant Analysis (also known as Fisher's LDA) and supported vector machines. These two methods separate the data using hyperplanes, for two-classes they are divided depending on the side of the hyperplane they are located (see Figure 8.). For LDA and SVM the popular method to solve a multiclass situation (N-number of classes) is selecting a class and separating it from the rest, this technique is referred "One Versus the Rest" (Schlögl 2005). This technique is very computational efficient and suitable for online classification. One drawback of LDA is when it deals with complex nonlinear EEG data (Garcia 2003). Even though SVM is originally linearly, it can be expanded using the "kernel trick". The trick consists of mapping the data into another space, using a kernel function. For BCI usually the Gaussian or radial basis function $K(x,y)=exp[-||x-y||/(2\sigma 2)]$. This trick gives a better generalization, but has lower speed execution (Lotte 2007).

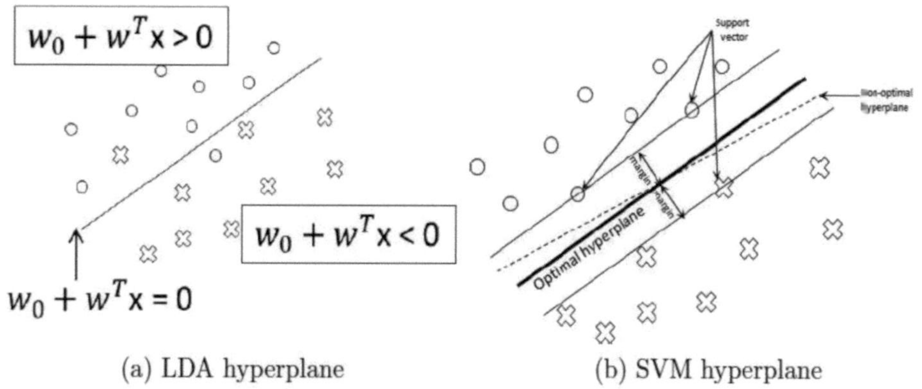

Figure 8. LDA and SVM hyperplanes that separate between two classes (circles and crosses).

LDA has been used for motor imagery (Pfurtscheller 1999), for a multiclass asynchronous motor imagery (Scherer 2004), as well as for P300 (Congedo 2006).

9.8.2. Neural networks

The second most used for classification method for BCI is using Neural Networks (NN). NN are non-linear classifiers that use assembly of neurons to produce the boundaries. The most used technique is the Multilayer Perceptron (Bishop 1995), that uses an input layer where the features are inserted, some hidden layers for processing and finally an output that defines the class (Figure 9.). Even though NNs can adapt to any number of classes and composed with enough neurons they can approximate any function, they are susceptible to over training and noise (Bishop 1995).

A conventional feed-forward artificial neural network (ANN´s) is a system constructed by a finite number of basic elements called neurons, which are grouped in layers. Every neuron is highly interconnected in the whole topology; the structure has a number of inputs and outputs that depends on the system that will be approximated.

A neuron is the basic element in an artificial neural network that simulates biological neurons which receives electrical impulses which are received through its dendrites, from other neuron's axons. Those electrical impulses are added in order to have a final potential. This potential must exceed a certain level to have the neuron generate an electrical impulse on its axon. If the level required is not met, then the axon of that neuron doesn't fire its axon. Neurons can be divided as: dendrites which are channels of input signals, core cell that processes all these signals and axons that transmit output signals of the processed information came from dendrites.

The ANN´s are applied to approximate normally a non-linear system as universal approximations. The first step to design an ANN´s is to train the neural network in order to fix the interconnection namely weights which are between the neurons. The training process can take a lot of time in the case of the back-propagation algorithm. After training the ANN´s the response could show a high-quality behavior, when a new input signals is presented to the system.

In other words the ANN´s could generalize any input signal. These ANN´s mimic the human brain, on the basic process of learning and generalization. Normally the process of training the ANN´s is slow and defining the correct topology could be complicated. The main advantages of artificial neural networks are:

- Ability to generalize and learn.
- Acquire knowledge from internal and external parameters.
- Ability to learn from examples and adapt to situations based on its findings.
- Generalization of knowledge to Production of adequate responses to unknown situations.
- Artificial neural networks solve complex problems that are difficult to manage by approximation.
- Produce linear or non-linear relationships
- Fault Tolerance

An extension to this technique is the Restricted Boltzmann Machines (RBM) that have a bidirectional connection between the layers (see Figure 9b.), this quality allows the RBMs to be train as normal NN and retrained using back propagation (Hinton 1986). Success of NN can be seen in (Kalcer 1993, Pfurtscheller 1996,
Hsu 2012) while for RBM in (Balderas 2011).

9.8.3. Non-linear Bayesian classifiers

There are mainly types Bayesian classifiers used for BCI systems: Bayes quadratic and Hidden Markov Models (HMM). Both these classifiers produce nonlinear decision boundaries. Furthermore, they are generative, which allows them to reject uncertain samples more efficiently than discriminative classifiers (Lotte 2007). While Bayesian assign the class to the feature vector with the highest probability, HMM is probabilistic automaton that can provide the probability of observing a given sequence of feature vectors (Rabiner 1989, Lotte 2007).

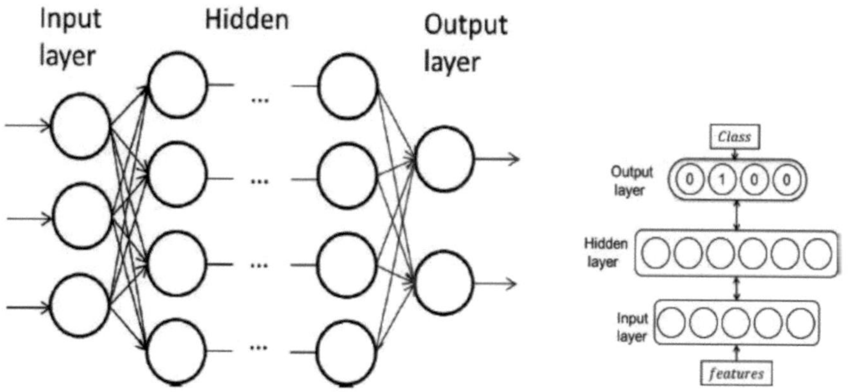

(a) Arquitecture of multilayer perceptron (b) General and Restricted Boltzmann Machines

Figure 9. Neural Networks architectures having Multilayer Perceptron and RBM

9.8.4. Nearest Neighbor classifiers

Nearest Neighbor classifiers are also used in BCIs with the k Nearest Neighbor (kNN) and Mahalanobis Distance (MDist) as preferred Nearest Neighbor classifier methods. kNN assign to an unseen point the dominant class among its kNN within the training set. kNN algorithms are sensitive to the curse of dimensionality making them fail in several BCI experiments, however they may perform efficiently with low-dimensional feature vectors (Lotte 2007). Mahalanobis Dististance based classifiers use Mahalanobis distance to assign a class to a feature vector to the nearest prototype. Mahalanobis Distance has been used to detect motor imagery of the hand giving accuracies over the 80% (Ming 2009).

9.8.5. Combinations of classifiers

Combinations of classifiers are proposed trying to reduce the variance and thus increase classification accuracy. Voting, Boosting, Stacking and Random subspaces. Voting consists of assign different classifiers the input feature vector and select the class with the higher majority of votes (hence the name). Boosting uses several classifiers in cascade where the errors committed by previous classifier are focus by each classifier. Stacking uses several classifiers (level-0 classifiers) running through the input vector. The output of the different classifiers is then use as input vector for a meta-classifier (or level-1 classifier) which is responsible for the final decision. Lastly Random subspaces uses subsets of the original feature vector as training set for different classifiers and the final decision is made by majority voting.

9.9. State of the art

The design of a BCI comes with two major challenges, the non-stationary and inherent variability of the EEG signals. Data from the same experimental paradigm but recorded at different instances are likely to exhibit differences due to; for instance; shift of the electrodes positions between sessions or changes in the sensor mechanical properties of the electrodes (e.g. change in the impedances). Adding to this problem the noisy nonlinear superposition of the measured EEG activity can mask underlying neural patterns and hamper their detection. The user current mentally state (e.g. due to tiredness, workload or stress) may impact in the ability to focus and generate specific mental events. Due to these factors, statistical signal processing and machine learning techniques play a crucial role in recognizing EEG patterns and translating them into control signals.

9.10. Co-adaptive training

A normal training of a BCI uses information from a first or previous sessions EEG recordings are used to pre-train the pattern recognition algorithms for classification or regression. On a posterior session user uses the trained algorithm for control. One of the drawbacks is that the variability of brain activity requires that the system is robust enough to handle the changes. Adding to this, the high adaptability of the brain gives the problem on how much has to be relegated for the system and how much left for the brain. It has been shown that using invasive over single neuron or a population of neurons the can rapidly learn to generate an appropriate pattern for a fix task. The same adaptation using EEG could take months to have a similar level of performance (Kübler 1999). Adding to this normal neuromuscular activity depends of feedback to have a successful control. A strategy to improve the control over the BCI system has to have a control that uses feedback. A good strategy is to use co-adaptive training, with a self-optimizing pattern detector and user adaptation, using new data to update the system. So new data is collected in different session and used to update the classifier to user's most recent brain patterns. Feedback can be provided during the new session to helping to generate more distinct EEG patterns, which increases detection performance. An Online adaptation can be included to provide a faster update of the training parameters and have a faster co-adaptation.

9.11. Hybrid BCI

There exists almost no reason why different technology could be combined, combining different patterns (e.g. EP and motor imaging) or different recording technology, combining

http://dx.doi.org/10.5772/57084

invasive and non-invasive, or using different electrophysiological signals with the combination of BCI technologies.

9.12. HCI recording

A case study was made using a commercial HCI, the amplifier Emotiv EPOC, which can record EEG signals as well as movement from the head with an incorporated gyroscope. The interface was created using the amplifier gyroscope signals as control for the displacement and the direction for an electric wheelchair (see Figure 10.).

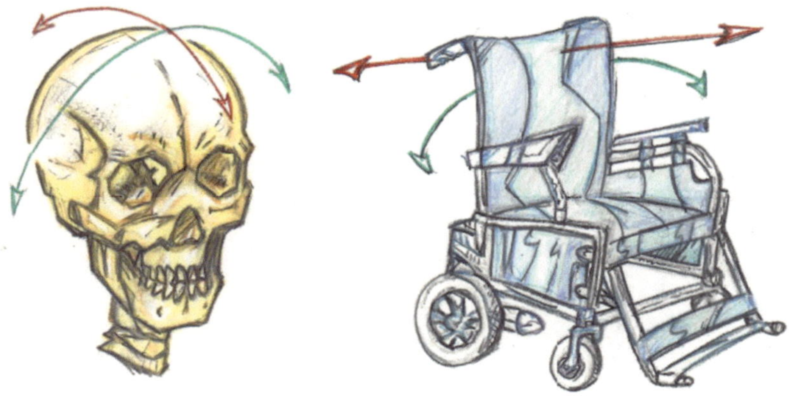

Figure 10. Example of the control with the movements of the head, translated with a gyroscope into the control of an electric wheelchair.

The gyroscope counts with two rotation axes that were used for displacement and turn. Also the velocity of displacement and turn was control depending on the amount of rotation the gyroscopes detect from the movement of the head.

The wheelchair counts with the displacement control of the two back wheels, giving it advance and turn control. This control was adapted to be controlled directly from a DAQ that has a direct interface with Labview.

9.13. Methodology

The case study was divided in three areas: Signal acquisition, Signal processing, and control signal (Figure 11.). For the signal acquisition we use the amplifier driver connected with Simulink (Matlab). Both signals were filtered and amplified using the rotation left-right for the turn and the rotation frontal-backward for forward or backward motion. The online process was done in the same interface that was used for recording in Simulink (Figure 12a.), which finally send the signals to Labview using a UDP protocol. Labview was finally in charge of the control signal (Figure 12b.) that had control over the wheelchair wheels.

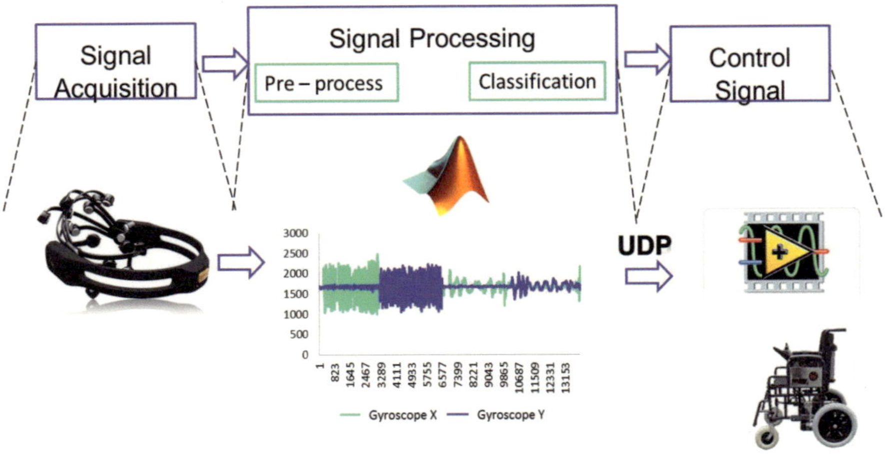

Figure 11. Control Process

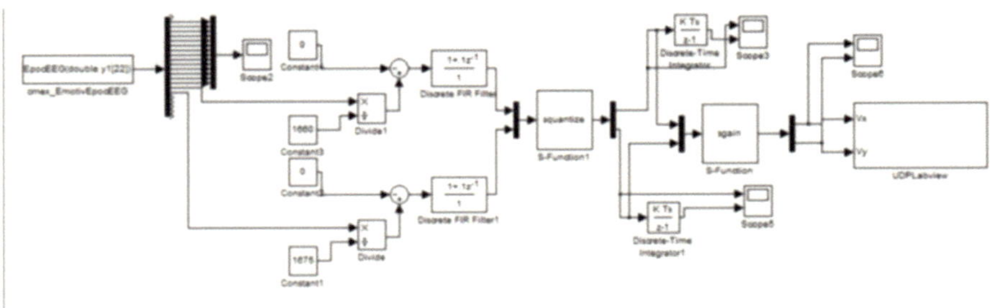

(a) Simulink block diagram

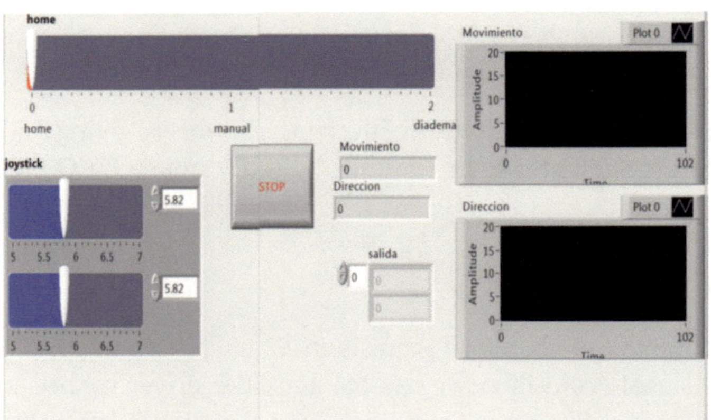

(b) Interface Labview

Figure 12. Matlab-Simulink and Labview interfaces

The control was first tested on a free environment and later on a simple maze (Figure 13.). Testing the manageability to make turns and understand the commands.

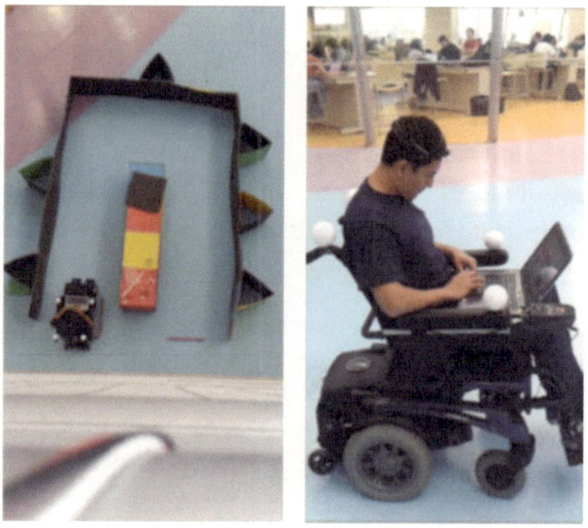
(a) Test 1 (b) Test 2

Figure 13. Interfaces de Matlab-Simulink y Labview

10. Conclusion

Even though the control with the Emotiv EPOC was limited, the viability of developing an HCI was shown using the gyroscope signals as control signals for a wheelchair.

Author details

Pedro Ponce, Arturo Molina, David C. Balderas and Dimitra Grammatikou

Institu to Tecnológico y de Estudios Superiores de Monterrey -Campus ciudad de México, Departamento de posgrado en ciencias de la ingeniería, Mexico City

References

[1] Brendan Z. Allison, Dennis J. McFarland, Gerwin Schalk, Shi Dong Zheng, Melody Moore Jackson, and Jonathan R. Wolpaw. Towards an independent brain–computer

interface using steady state visual evoked potentials. Clinical Neurophysiology, 119(2):399 – 408, 2008.

[2] David Balderas, Thorsten Zander, Fabian Bachl, Christa Neuper, and Reinhold Scherer. Restricted boltzmann machines as useful tool for detecting oscillatory eeg components. In Proc. of the 5th International Brain-Computer Interface Conference, Graz, Austria., pages 68–71, 2011.

[3] N. Birbaumer, A. Kubler, N. Ghanayim, T. Hinterberger, J. Perelmouter, J. Kaiser, I. Iversen, B. Kotchoubey, N. Neumann, and H. Flor. The thought translation device (ttd) for completely paralyzed patients. Rehabilitation Engineering, IEEE Transactions on, 8(2):190–193, 2000.

[4] Christopher M. Bishop. Neural Networks for Pattern Recognition. Oxford University Press, Inc., New York, NY, USA, 1995.

[5] M Congedo, F Lotte, and A L′ ecuyer. Classification of movement intention by spatially filtered electromagnetic inverse solutions. Physics in Medicine and Biology, 51(8):1971, 2006.

[6] Emanuel Donchin, Kevin M. Spencer, and Ranjith Wijesinghe. The mental prosthesis: Assessing the speed of a p300-based brain–computer interface. IEEE Transactions on Rehabilitation Engineering, 8, 2000.

[7] Andreas K. Engel, Pascal Fries, and Wolf Singer. Dynamic predictions: Oscillations and synchrony in top-down processing. Nature Reviews Neuroscience, 2:704–716, 10 2001.

[8] L. A. Farwell and E. Donchin. Talking off the top of your head: toward a mental prosthesis utilizing event-related brain potentials. Electroencephalography and Clinical Neurophysiology, 70:510–523, 1988.

[9] Lawrence A. Farwell, DrewC. Richardson, and GrahamM. Richardson. Brain fingerprinting field studies comparing p300-mermer and p300 brainwave responses in the detection of concealed information. Cognitive Neurodynamics, pages 1–37, 2012.

[10] Lawrence A. Farwell and Sharon S. Smith. Using brain mermer testing to detect knowledge despite efforts to conceal. Journal of Forensic Sciences, pages 135–143, 2001.

[11] Charles M. Gaona, Mohit Sharma, Zachary V. Freudenburg, Jonathan D. Breshears, David T. Bundy, Jarod Roland, Dennis L. Barbour, Gerwin Schalk, and Eric C. Leuthardt. Nonuniform high-gamma (60 –500 hz) power changes dissociate cognitive task and anatomy in human cortex. The Journal of Neuroscience, 31(6):2091–2100, February 2011.

[12] G.N. Garcia, T. Ebrahimi, and J. Vesin. Support vector eeg classification in the fourier and time-frequency correlation domains. In Neural Engineering, 2003. Conference Proceedings. First International IEEE EMBS Conference on, pages 591–594, 2003.

[13] Geoffrey E. Hinton and Terrence J. Sejmowski. Learning and Relearning in Boltzmann Machines , volume 1, chapter 7, pages 282–317. MIT Press, Cambridge, 1986.

[14] WEI-YEN HSU. Application of competitive hopfield neural network to braincomputer interface systems. International Journal of Neural Systems, 22(01):51–62, 2012. PMID: 22262524.

[15] H.H. Jasper. The ten-twenty electrode system of the international federation. Electroencephalography and clinical neurophysiology, 10(2):371–375, 1958.

[16] J. Kalcher, D. Flotzinger, and G. Pfurtscheller. Graz brain-computer interface: an eeg-based cursor control system. In Engineering in Medicine and Biology Society, 1993. Proceedings of the 15th Annual International Conference of the IEEE, pages 1264–1265, 1993.

[17] Andrea Kübler, Boris Kotchoubey, Thilo Hinterberger, Nimr Ghanayim, Juri Perelmouter, Margarete Schauer, Christoph Fritsch, Edward Taub, and N. Birbaumer. The thought translation device: a neurophysiological approach to communication in total motor paralysis. Experimental Brain Research, 124(2):223–232, 1999.

[18] Dean J Krusienski, Eric W Sellers, Fran¸ cois Cabestaing, Sabri Bayoudh, Dennis J McFarland, Theresa M Vaughan, and Jonathan R Wolpaw. A comparison of classification techniques for the p300 speller. Journal of Neural Engineering, 3(4):299, 2006.

[19] Eric C Leuthardt, Gerwin Schalk, Jonathan R Wolpaw, Jeffrey G Ojemann, and Daniel W Moran. A brain–computer interface using electrocorticographic signals in humans. Journal of Neural Engineering, 1(2):63, 2004.

[20] Fabian Lotte. Study of electroencephalographic signal proccessing and classification techniques towards the use of brain-computer interfaces in virtual reality applications. PhD thesis, Intitute National des Sciences Appliquees de Rennes, January 2009.

[21] Fabian Lotte, M Congedo, A L′ ecuyer, F Lamarche, and B Arnaldi. A review of classification algorithms for eeg-based brain–computer interfaces. Journal of Neural Engineering, 4:R1–R13, 2007.

[22] Fabien Lotte, Josef Faller, Christoph Guger, Yann Renard, Gert Pfurtscheller, Anatole L′ ecuyer, and Robert Leeb. Combining bci with virtual reality: Towards new applications and improved bci. In Brendan Z. Allison, Stephen Dunne, Robert Leeb, Jos′ e Del R. Mill′ an, and Anton Nijholt, editors, Towards Practical Brain-Computer Interfaces, Biological and Medical Physics, Biomedical Engineering, pages 197–220. Springer Berlin Heidelberg, 2013.

[23] Dong Ming, Yuhuan Zhu, Hongzhi Qi, Baikun Wan, Yong Hu, and K. D K Luk. Study on eeg-based mouse system by using brain-computer interface. In Virtual Environments, Human-Computer Interfaces and Measurements Systems, 2009. VECIMS '09. IEEE International Conference on, pages 236–239, 2009.

[24] E.M. Mugler, C.A. Ruf, S. Halder, M. Bensch, and A. Kubler. Design and implementation of a p300-based brain-computer interface for controlling an internet browser.

Neural Systems and Rehabilitation Engineering, IEEE Transactions on, 18(6):599–609, 2010.

[25] Christa Neuper, Reinhold Scherer, Miriam Reiner, and Gert Pfurtscheller. Imagery of motor actions: Differential effects of kinesthetic and visual–motor mode of imagery in single-trial {EEG}. Cognitive Brain Research, 25(3):668 – 677, 2005.

[26] World Health Organization. World report on disability, 2011.

[27] Gert Pfurtscheller, J. Kalcher, Christa Neuper, D. Flotzinger, and M. Pregenzer. On-line eeg classification during externally-paced hand movements using a neural networkbased classifier. Electroencephalography and Clinical Neurophysiology, 99(05): 416–425, 1996.

[28] Gert Pfurtscheller and Fernando Lopes da Silva. Eeg event-related desynchronization(erd) and event-related synchronization(ers). Electroencephalography:Basic Principles and Clinical Neurophysiology, 4:958–967, 1999.

[29] Gert Pfurtscheller and Fernando Lopes da Silva. Event-related EEG/MEG synchronization and desynchronization: basic principles. Clinical Neurophysiology, 110:1842–1857, 1999.

[30] L. Rabiner. A tutorial on hidden markov models and selected applications in speech recognition. Proceedings of the IEEE, 77(2):257–286, 1989.

[31] Rajesh P.N Rao. and Reinhold Scherer. Brain–computer interfacing: more than the sum of its parts. IEEE Signal Processing Magazine, 27(4):152–150, July 2010.

[32] R. Scherer, G.R. Muller, C. Neuper, B. Graimann, and G. Pfurtscheller. An asynchronously controlled eeg-based virtual keyboard: improvement of the spelling rate. Biomedical Engineering, IEEE Transactions on, 51(6):979–984, 2004.

[33] Reinhold Scherer, Josef Faller, David Balderas, Elisabeth V. C. Friedrich, Markus Pr° oll, Brendan Allison, and Gernot M° uller-Putz. Brain–computer interfacing: more than the sum of its parts. Soft Computing, 17(2):317–331, February 2013.

[34] Alois Schlögl, Felix Lee, Horst Bischof, and Gert Pfurtscheller. Characterization of four-class motor imagery eeg data for the bci-competition 2005. Journal of Neural Engineering, 2(4):L14, 2005.

[35] S Sutton, M Braren, J Zubin, and ER John. Evoked-potential correlates of stimulus uncertainty. Science (New York, NY), 150(3700):1187–1188, 1965.

[36] Francesco Tenore, Ander Ramos, Amir Fahmy, Soumyadipta Acharya, Ralph EtienneCummings, and Nitish V. Thakor. Towards the control of individual fingers of a prosthetic hand using surface emg signals. In Proceedings of the 29th Annual International, pages 6145–6148, 2008.

[37] Johnathan R. Wolpaw, Niels Birbaumer, Denis J. McFarland, Gert Pfurtscheller, and Theresa M. Vaughan. Brain-computer interfaces for communication and control. Clinical Neurophysiology, 113:767–791, 2002.

[38] Thorsten Oliver Zander, Moritz Lehne, Klas Ihme, Sabine Jatzev, Joao Correia, Christian Kothe, Bernd Picht, and Femke Nijboer. A dry eeg-system for scientific research and brain-computer interfaces. Fronti

[39] Bax M, Goldstein M, Rosenbaum P, Leviton A, Paneth N. (2005) Proposed definition and classification of cerebral palsy, April 2005. Dev Med Child Neurol 47: 571–576.

[40] Cerebral Palsy: Classification and Epidemiology, Amy Thornhill Pakula, Kim Van Naarden Braun,Marshalyn Yeargin-Allsopp, Phys Med Rehabil Clin N Am 20 2009) 425–452 . Published by Elsevier Inc.

[41] Dubowitz LM, Dubowitz V, Mercuri E. The neurological assessment of the preterm and full-term newborn infant. 2nd ed. London: Mac Keith Press; 1999.

[42] Amiel –Tison C. Update of the Amiel-Tison neurologic assessment for the term neonate or at 40 weeks corrected age. Pediatr Neurol 2002:27.196-202

[43] Ellison P: The INFANIB: A reliable method for the neuromotor assessment of Infants, Therapy Skill Builders, 1994

[44] SCPE Working Group. Prevalence and characteristics of children with cerebral palsy in Europe. Dev Med Child Neurol 2002; 44: 633-40

[45] SCPE Working Group. Surveillance of cerebral palsy in Europe: a collaboration of cerebral palsy surveys and registers. Dev Med Child Neurol 2000; 42: 816-24

[46] Robert Palisano, Peter Rosenbaum, Stephen Walter, Dianne Russell, Ellen Wood, Barbara Galuppi, Gross Motor Function Classification System for Cerebral Palsy, Dev Med Child Neurol 1997;39:214-223

[47] Eliasson, A.-C., Krumlinde-Sundholm, L., Rösblad, B., Beckung, E., Arner, M., Öhrvall, A.-M. and Rosenbaum, P. (2006), The Manual Ability Classification System (MACS) for children with cerebral palsy: scale development and evidence of validity and reliability. Developmental Medicine & Child Neurology, 48: 549–554.

[48] Bottcher T, Flachs EM, Uldall P. Attentional and executive Impairments in children with spastic cerebral palsy. Developmental Medicine & Child Neurology. 2010; 52: E-42-47.

[49] Guzzetta A, Fazzi B, Mercuri E, Bertuccelli B, Canapicchi R, Van Hof Duin J et al. Visual Function in children with hemiplegia in the first years of life. Dev Med Child Neurol 2001;43:321-9

[50] Cognitive Deficits in Depression: Possible implications for functional neuropathology, Marie-Paule Austin, Philip Mitchel and Guy M. Goodwin, British Journal of Psychiatry (2001), 178, 200-206

[51] Jackie Parkes, Chris McCusker, Common psychological problems in cerebral palsy Paediatrics and Child Health Volume 18, Issue 9 , Pages 427-431, September 2008).

[52] Suzanne McDermott, Ann L. Coker, Subramani Mani, Shanthi Krishnaswami, Richard J. Nagle, Laura L. Barnett-Queen and Donald F. Wuori, A Population-Based Analysis of Behavior Problems in Children with Cerebral Palsy Journal of Pediatric Psychology, Volume 21, Issue 3, Pp. 447-463) 1996.

Chapter 9

Virtual Reality in Rehabilitation of Children with Cerebral Palsy

Mintaze Kerem Gunel, Ozgun Kaya Kara,
Cemil Ozal and Duygu Turker

Additional information is available at the end of the chapter

http://dx.doi.org/10.5772/57486

1. Introduction

Cerebral Palsy (CP) is a continuous but non-progressive motion/posture, motor function disorder resulting from a lesion in the developing brain primarily damaging the areas responsible for the postural control, affecting 2.1/1000 live births [1, 2]. That is the most common cause of physical disability in the early childhood period and a serious disorder also affecting the family, and the child's education and social life [3]. Recent researches, finding an accepted international definition of CP have been somewhat a challenge, but the proposed prevailing international definition [3-5]. In the last decade, the International Classification of Functioning, Disability and Health "ICF" developed by The World Health Organization constituted a new defining to understand CP and to think about intervention possibilities for clinicians, researchers as well as families [6, 7].

ICF is the transition from an "outcome of disease" classification to a "components of health" classification. This system, which aims to develop a common language and framework among health professionals considering above-mentioned multidisciplinary approach and communication, has rapidly become a focus worldwide. ICF is a system that classifies health and health-related fields. These fields are divided into two parts as the body, and the individual and social perspective: "body functions" and "body structures" form the first group, and the "activity and participation" field forms the second group. In recent literature, assessment tools, intervention techniques and researches outcomes are investigated in according to the dimensions (ICF): body function (e.g. physiological, psychological); structures (e.g. anatomical); activities (tasks); and participation (life roles). Thus, this "bio psychosocial" model of disability encourages a more holistic approach to rehabilitation (Figure I) [6]. "ICF Classification" for children with CP is a beneficial system that can be used in the formulation of problems in different areas, to build a bridge between professionals themselves and with the family. ICF can help the child's physiotherapist in deciding about the process and in determining special functional objectives [4]. ICF-CY (Child and Youth version: ICF-CY) is the updated version that specially attended to aspects of learning, behavior and development for use in children and adolescents from 2007 [8].

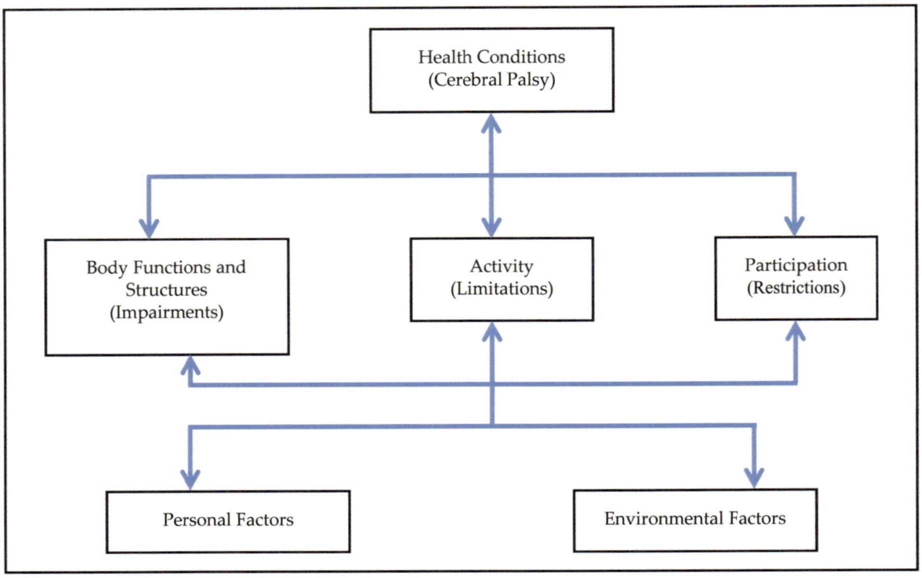

Figure 1. ICF bio psychosocial model

When viewed from ICF aspect, CP affect on a child's "functioning" (including of body structures [e.g. limbs, eyes], body functions [e.g. sensory, neuromusculoskeletal and mental functions], activities [e.g. walking, writing, learning] and participation [e.g. playing sport, going to concert] and besides that personal [e.g. motivation, anxiety, toleration, age, gender] and environmental factors [e.g. architectural accessibility, policies, physical accessibility of cultural, athletic or recreational centers] influence independence of CP children in their daily life and leisure activities. Moreover, CP may cause "disabilities", such as impairment, activity limitations and participation restrictions [9, 10]. Beckung et al. have investigated activity limitations and participation restrictions with gross and fine motor functions under the mobility, education and social relationship sub-headings proposed by ICF in children with CP. They indicated that effect of a child's impairment or activity limitation on participation might vary depending on environmental factors [11].

At first, most clinicians and researchers thought that CP is primarily a movement disorder but recently it is an umbrella term used to define a group of permanent conditions, indicating that there is heterogeneity in these conditions such as visual, cognition, perceptual and/or behavior, sensation problems and learning disabilities [12]. Neural disorders such as spasticity, co activation of agonist-antagonist muscles, muscle weakness, lack of selective motor control and restriction of normal joint movement and knee pain affect gross and fine motor functions in CP [9]. Many children with CP may never attain the abilities of their peers who are typically developing or lose some of them with growing [13]. There are many potential problems children with CP and these influence complexity of therapy planning and execution.

In CP treatment, the rehabilitation modalities are considerably wide including conventional interventions and new rehabilitation techniques. Medical and surgical approaches, physiotherapy, ergotherapy, speech therapy, orthoses and other supportive devices, recreational activities, school and education adaptation and psychological support can be considered in these modalities. The main aims of the rehabilitation in CP are to help the child achieve the highest possible physical, cognitive, psychological and social independence level within his/her physiological and anatomical deficiencies and environmental limitation, to reduce the effects of physical disorder to minimum, to improve the independence in daily life activities and social life, to increase the life quality of the child and thus support the quality of parents and siblings life [14].

Physiotherapy modalities in CP aim not only to improve the movement ability of the child, but also to reach normal level in all development stages. Rehabilitation in the children with CP depends on the clinical type, accompanying disorders, chronological age and the socio-economic factors. Especially visual, hearing, cognitive disorders, attacks, learning disorder, emotional state problems are among the problems that affect the success of the rehabilitation [15]. Thus, a multidisciplinary team is required in the treatment of these heterogeneous groups of problems. This team must include specialist physicians (pediatricians, pediatric neurologist, orthopedist, neurosurgeon, neonatologist, child psychiatrist, dentist and all related physicians), physiotherapist, ergotherapist, psychologist, child development specialist, dietitian, social service expert and caregivers for the child.

The children with CP are referred routinely to physiotherapy in the very early ages. But the main question is whether there is a scientific evidence to confirm the use of physiotherapy so often. Recently the most debated subject is if the treatment modalities performed in CP have an effect on the neurological process. The studies and clinical literature about the effects and density of the treatment are still unclear with lack of evidence and discussions on the field are predicted to continue in the future. Over the past 20 years, the early interventions for the children with disabilities have focused on evidence-based practice including the child and family centered approach. Therefore the examination of the efficiency of these applications is of great importance for clinicians and researchers [16]. It is accepted that physiotherapy applications play an important role in the CP rehabilitation, but which method, how much intensity and how long should be applied? These questions cannot be answered easily. The most criticism of denominated therapy modalities is that the lack of scientific baseline and evidence of activity. The evidences that support any efficiency of modality or that indicate the superiority of any other modality, are limited, and physiotherapists have been increasingly searching to evidence- based applications. To evaluate the efficiency of the therapy modality, which is performed for any motor problem or physical deficiency, is difficult due to several reasons. The main reason is the absence of standardized specific treatment. In other words there is no dosage application under the specific, stable procedures in many cases. Researches regarding interventions in CP indicate that 30%-40% of interventions have no reported evidence-based, other 20% of interventions informed ineffective, unnecessary or harmful. The current review is showed botulinum toxin (BoNT), selective dorsal rhizotomy, casting, constraint-induced movement therapy, bimanual training, context-focused therapy, goal-directed treatment, and occupational therapy following BoNT are effective interventions for body structures and functions level or the activity levels on the ICF but unfortunately, there were no evidence-based effective interventions for improving participation, environment or personal factors levels of the ICF. A high incidence of CP intervention studies, approximately 70% are low level evidence-based and required to increase their research quality to prove effectiveness of treatment. These modalities are assistive technologies, animal assisted therapy, strengthening, hippotherapy, hydrotherapy, early intervention, cognitive behavior therapy, communication training, orthoses, oral-motor therapy, play therapy, stretching, treadmill training and parent training. In addition, ineffective interventions are determined neurodevelopmental therapy (NDT), craniosacral therapy, hyperbaric oxygen, hip bracing and sensory integration [10]. There is a contradiction for NDT that is commonly used interventions all over the world. Todays' Bobath therapists focus on motor learning principles, family-centered practice, orthoses, BoNT, assistive devices, functional training, constraint movement treatment, strengthening and bimanual treatment within right body alignment and aim to improve posture. NDT may influence functional motor gains (low-evidence) but there are great requirement rigorous future researches to demonstrate effectiveness of NDT.

Also, physiotherapist focuses on the functional movement and gross motor skills in the treatment of motor disorder of the child with CP. Positioning, sitting, gait with or without orthoses, the use of wheelchair, transfers are some of the areas on which the physiotherapist work. Physiotherapist plans the physiotherapy and home programs, provides the school arrangements, and makes decision about orthoses and supportive devices. Physiotherapists teach the families how to feed, bath, cloth and hold their children during daily living activities; and also give advices about assistive devices [17]. Thus, the physiotherapists aim to maximize the child's performance by focusing the needs of the child [16].

The frequency of physiotherapy is not definitive, but some families and professionals think that physiotherapy is more useful when its frequency is increased. Recently, physiotherapist focus to solve the needs of the family and child and decide which therapy and frequency should be applied to the child [16]. Dosage or duration of treatment can be arranged coherently, but the procedures are based on the skill levels and specific aims of physiotherapists, therefore they can be varied. Even if the arrangements of treatment can be standardized (a condition of the treatment), the family of the child can never be standardized (another condition of treatment). All problems that accompany a research included low incidence and high heterogeneous condition, become complex with the process of growth and maturation.

All in all for the physiotherapists it is important to separate the evidence-based applications and clinical applications [10]. Therapists try to find a balance between the attractive and effective activities in the treatment process of children with CP [18]. CP is a heterogeneous group, so more general principles are used for treatment and rehabilitation [19]. During the last 10 years, popularity of performance based or "top-down" approaches based on motor

http://dx.doi.org/10.5772/57486

learning theory in which interventions focus directly specific task training in activities of interest and are not concerned with underlying impairments body structures and function, are gradually increasing such as goal directed therapy, constraint induced movement therapy [10]. While technologies come into our life, physiotherapy approaches changed and developed. Treatment of motor impairments with new complementary technologies such as robot-assisted therapy, locomotor therapy or computer based rehabilitation systems would improve motor development in children with CP, especially intense growth and adolescent period also after the multilevel surgical intervention, that effect muscle strength and body alignment. The success of the rehabilitation process depends on several factors: the intensity of therapy, repetition, and goal-directed or task-oriented therapy program are considered essential in achieving motor outcomes. During the past decades new technologies have been developed to improve sensory motor learning in children with CP. Motivation and active participation of children in intervention program play a fundamental role in the sensory-motor learning process and these are the key factors of successful outcomes [20]. Over the last three decades, there has been an increase in the number of individuals engaging in interactive computer plays [21]. Therefore, current studies focus on Virtual Reality (VR). VR as an intervention for sensory motor rehabilitation is promising tool in order to improve lower and upper limb function and also postural control in children with CP.

Virtual reality is a technology that provides a sense of presence in a real environment with the help of 3D pictures and animations formed in a computer environment and enable the person to interact with the objects in that environment. In other words, VR described as an improved form of human-computer interaction that allows the user to be part of and interact with a computer-generated environment [22]. A virtual environment (VE) is created by various computer technologies. The key specialty that separates VE from other forms of visual imaging, like video games or television, is real-time interaction. However, the interaction can be achieved in various ways. VE shows virtual or artificially produced sensory information, and allows the user to feel experiences similar to the events and activities in real life [23]. Interactive simulations that enable the participant to create an interaction between body movements and 3D area are therefore constituted [24]. The person sees and feels objects and events similar to those in the real world, can manipulate and move the virtual objects, and can do other things in the virtual environment he/she is in. Thus, "an imaginary presence feeling" occurs in the virtual world. In short, VR is the rebuilding of reality [25]. Some studies named these systems "Interactive computer play-ICP" is defined as any kind of any computer game or VR technology where the individual can interact and play with virtual objects in a computer generated environment. Fehlings et al indicated that there is a significant similarity between terms ICP and VR [26].

The use of VR applications started in the 1950s with a theatre machine called Sensorama. This machine, developed by the cinematographer Morton Heilig, was constructed to address all senses. Sensorama combined projected film, audio, vibration, wind, and odors, all to make the user feel as if they were actually being in the film rather than simply watching it. For example, one experience provided was driving a motorbike on the streets of New York. In addition VR was developed by the USA air and airplane industry during the 2nd World War. The Head

Mounted Display (HAD) that appeared in 1965 has been a milestone for VR applications. Ivan Sutherland developed (HMDs), which allowed users to be immersed inside a virtual environment with computer-generated scenes. To compare to the technology of 21th century both interface and realism were primitive, and the HMD was so heavy it had to be suspended from the ceiling. The potential of VR was recognized by researchers from many different areas and especially military [27]. The following decades VR is growing rapidly both regarding technological advancements and in areas of implementations. In order to accomplish the feeling of a strong presence, various stimulation modalities are provided to the users (audiovisual feedback). Since then VR has been successfully integrated into several areas of medicine and psychology as for example: training and education in surgical procedures; education of medical students; assessment and treatment of mental health problems including phobias and post-traumatic stress disorders; pain management through distraction; and in motor rehabilitation, where examples of explored areas are upper limb rehabilitation in persons with acquired brain injury; fall risk reduction in Parkinson's disease, particularly stroke rehabilitation and in pediatric rehabilitation field [28].

VR use in physiotherapy and rehabilitation has increased significantly in the last 6-7 years. Depending on the characteristics of the software used, VR-based therapies provide significant experiences to the user within the targets of the therapy. VR applications became the spark among new treatment modalities used for individuals with CP as computer technology became intriguing and motivating for children and young people and interest on the subject gradually increased. VR provides an opportunity for active learning, encourages the participant, and ensures motivation. It enables performing difficult movements in a secure environment and objectively shows the behaviors that are a result of these motions. An ever-increasing number of studies report that VR implementation in children with CP positively affects brain reorganization, plasticity, motor capacity, visual perceptive skills, social participation and personal factors [29].

2. Theoretical Base of Virtual Reality therapy

VR therapy includes the basic principles of the motor learning theories [19]. It provides this by enabling the user to continuously see the movements in 3D from the computer screen. VR provides repetitive practical and positive feedback in order to increase the functional independence in daily tasks. Holden et al reported the possibility of learning motor abilities in the virtual environment by individuals with a disorder. Movements learned in VE can be transferred to real life with equivalent motor tasks [30]. The first studies on this subject reported that the VR method was usable, a lot of fun but not successful for treatment. However, VR has been shown to provide motor recovery in the upper extremity in adult stroke rehabilitation [31].

According to the motor control and motor learning theories, motivation, repetition, and purposeful and special target-directed training should be used in the treatment of children with CP. The addition of games and social activities during rehabilitation is also important for

http://dx.doi.org/10.5772/57486

ensuring the development of the child. A good treatment should enable the therapy to be transferred to daily living activities and tasks. Treatment techniques based on motor learning theories use intensive practice of functional activities and show good results [32].

Active participation instead of passive practice is recommended for motor learning and cortical reorganization. Passive exercises have been shown not to enable maximum improvement in the affected upper extremity in patients with stroke. It is also necessary to provide and strengthen new motor skills, provide functional and duty- focused practices and increase motivation for re-learning and recovery after a stroke. Although motor learning is quite different in children with CP compared to patients with stroke and spinal cord injury, focusing on the activity and task is one of the most important aspects of the treatment in children [19, 33].

One of the major purposes of rehabilitation in child and adult patients is to restore the basic abilities. Recovery after neural damage usually depends on various factors such as the nature and amount of the rehabilitation. Conventional rehabilitation programs are shorter and less intensive to ensure optimal therapeutic results. They cannot adequately increase the motivation of the patient or support activity participation. Many studies have shown that the motivation of patients plays a critical role in treatment results. The virtual environment can also provide more intriguing and competitive conditions, by increasing the motivation of the patient and ensuring active participation, so that less time is used for regaining motor skills [34].

The results of the use of VR as a treatment approach in adults and children are promising. There are many factors lying behind the use of VR in rehabilitation; it provides a variety of environments, and an environment that is similar to the natural one can be designed to test performance and provide independent training. Designed scenarios can be used to train functional behavior in real life and improve functional performance [29]. The "mirror neuron system" is considered to be a mediator for relearning in cases of disturbed cortico-motor function [35]. The mirror neurons activate not only when performing a motor activity but also when observing, imagining, or listening to the same motor activity [7]. There are many important views on the development and structure of the mirror neuron system in children [36]. The development of this system in children supports motor learning and social function in daily life. Cortical reorganization implementations have been developed with mirror neuron system features. The results revealed the usefulness of transferring mirror neuron information to the treatment in the clinic. The premotor cortex has critical importance in motor learning and motor control. This cortex is divided into dorsal and ventral parts. The ventral premotor cortex (PMv), hand area of the primary motor cortex, anterior intraparietal area, and the supplementary motor area are associated with Brodmann 3, 1, 2. The dorsal premotor cortex (PMd) has more projection than PMv with the lateral intraparietal area, primary motor cortex, supplementary motor area, cingulate gyrus and Brodmann 5 [37, 38]. (Figure II). The position of movement towards the target has been coded in PMd cells. PMd neurons are active in the preparation phase of the movement and play a critical role in motor planning. They are responsible for planning and learning the movement and the constitution of the postural responses for the future. The ventral premotor cortex (PMv) is important in the sensorimotor processing of movement and is considered to be associated with the cognitive aspect of target-directed activities. It is a part of the mirror neuron mechanism. Brief pictures for motor movements are coded with two-way activation of PMv mirror neurons during the implementation and observation of the movement. We are therefore able to understand the movements of others in advance. Learning a motor skill is a cognitive and motor process. Motor learning is briefly gaining the movement skills for a complex target with practice. PMd is active during the early phase of motor learning, and is associated with spatial mapping while PMv is critical for motor learning in the sensorimotor transfer of vision-based motions. PMv neurons are involved in monitoring performance and deciding on the choice of motion in practical terms. The mirror neurons inside PMv play an important role in observational and mimic learning [7, 39].

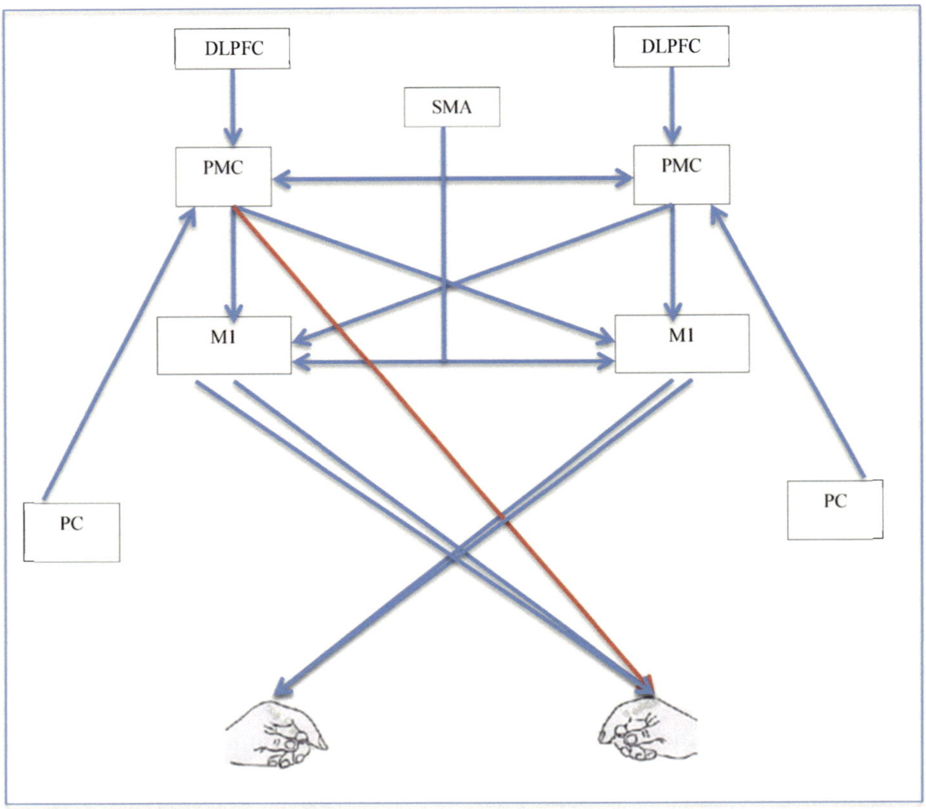

Figure 2. Kantak et al with kind permission "Premotor cortex (PMC) forms a part of the neural network involved in integration of sensory and cognitive information into goal-directed actions. PMC receives sensory information from the parietal cortex (PC), cognitive information from the dorsolateral prefrontal cortex (DLPFC) and supplementary motor cortex (SMA), and projects to the primary motor cortex (M1). In addition, it also has direct projections to the spinal cord via the corticospinal tract. These connections within the neural networks are plastic and are modified in response to injury, learning, and training/therapy [7].

The mirror neuron mechanism constitutes a physiological basis for motor memory and motor learning. The mirror neurons map the observed target activity in a pictorial and

http://dx.doi.org/10.5772/57486

kinematic manner and activate with the mobility recognition mechanism. It is believed that the primary motor cortex is facilitated with VR implementations [7]. In conclusion, activation of the mirror neuron system stimulates cortical reorganization and contributes to functional improvement [38].

3. The benefits of using Virtual Reality in rehabilitation

The first results in the literature showed that VR was a robust treatment method that was functional, target-directed and motivating [30, 40]. Studies, especially in the field of pediatric rehabilitation, have taken the various aspects into account (such as development of life skills, mobility, cognitive abilities, entertainment, motivation). VR provides specific and intensive treatment for children. However, evidence supporting VR implementation for the rehabilitation of children with neurological disorders is still limited [28, 31].

New technologies like VR play an important role in functional training and performance. VR allows intensive and motivational training. It enables the use of many interactive environments and multiple sensory feedbacks [41]. The use of this technology in disabled individuals ensures communication with others, improves social relations, and increases independence. VR meets important criteria for motor learning and motor control. VR applications also enable the therapist to train the child at home [30]. High levels of motivation, participation, and cooperation are essential components of a game system. These characteristics of the training support behavioral changes and neural plasticity. In conclusion, multi-sensory feedback explains the improvement in learning and performance. VR implementation in children with CP positively affects brain reorganization, plasticity, motor capacity, visual perceptive skills, social participation and personal factors (Table I) [42, 43].

The benefits of using Virtual Reality in rehabilitation
Increases motivation
Demonstrate target-directed functions more realistically
Provide an experience for the child according to his/her own motor learning capacity
Support motor learning
Support cortical reorganization
Provide interactive treatment

Table 1. The benefits of using Virtual Reality in rehabilitation

4. Virtual Reality systems used in pediatric rehabilitation

The popularity of computer technologies has increased between both children and adults. In addition video games are an important part of leisure activities for the young. Actually, one of every 4 children in the USA now has his/her own video game console at home. In last 10 years, active video game consoles give an opportunity to transform sedentary screen time into a period of physical activity. Examples are Sony PlayStation, Nintendo Wii, etc. [21]. Active video consoles based on VR concept and allow interactive physical activity. Nowadays, the computers systems focus on touch technology that are rapidly improved and become significant part of our personal, social and occupational life. Touch technology is frequently used in most area such as airports, cell phones, tablets due to easy manipulation of touch interface, flexibility and convenience [44]. The numbers of touch screen devices are gradually increased from 665 million in 2011 to 1350 million by 2014 [45]. According to our clinical experience many therapist use these touch technologies and active video consoles in CP rehabilitation to motivate children and take advantage of variable applications. Thus, children and young people with disabled or not are now more familiar with such technologies. They can be used as free-time activity and a socializing method.

VR is used in rehabilitation for the development of an interactive game environment so that the special aims of the treatment can be achieved. The first purpose of VR as a treatment modality is to develop the confidence and adequacy in motor-based and game-based activities that are impossible for the patient to accomplish in the real world [29, 31].

There are several methods where the users are in interaction with virtual reality technologies. The phrase 'interactive computer game' has been created to understand these differences [23]. An interactive computer game is any kind of computer game using virtual reality where the child can play and interact with virtual objects on the computer or the created environment. There are different types of VR systems that separate according to immersion degree and how the users interact with the system [25].

The virtual environment can be divided into 2 subgroups;

1 Immersion VR; the virtual environment is shown with a screen mounted on the head. Immersion means how much the user feels virtual environment like real. All immersive systems, users wear a head mounted display that brings them into a 3D virtual environment. Movement through VE is controlled by head movement.

2 Desktop VR; the designed images are seen on the computer screen, or on the TV screen together with the voice of an external speaker. The other VR systems the users feel the VE in a 2 dimensional flat screen and they focus on total body movement that controlled a mouse, joystick, keyboard [22].

Tactile feedback can be provided by a feedback glove, and force feedback by providing resistance with the joystick in the virtual environment (VE). Systems connected to the internet (tele-rehabilitation) have the potential to reach out to children who are in distant areas where healthcare services are limited [30]. (Figure III-IV)

Those that were not specially designed for use in rehabilitation

The cost of simple VR game systems like Nintendo Wii is low and they are available in the physiotherapy departments of many third world countries. The use of Nintendo Wii may be

http://dx.doi.org/10.5772/57486

beneficial in the rehabilitation process for physiotherapists who work in an area with limited resources [18]. Current virtual reality systems like Interactive Rehabilitation Exercise System (IREX) are expensive and inaccessible for the majority of the population. Therefore, the use of mainstream game consoles for treatment has become popular [46].

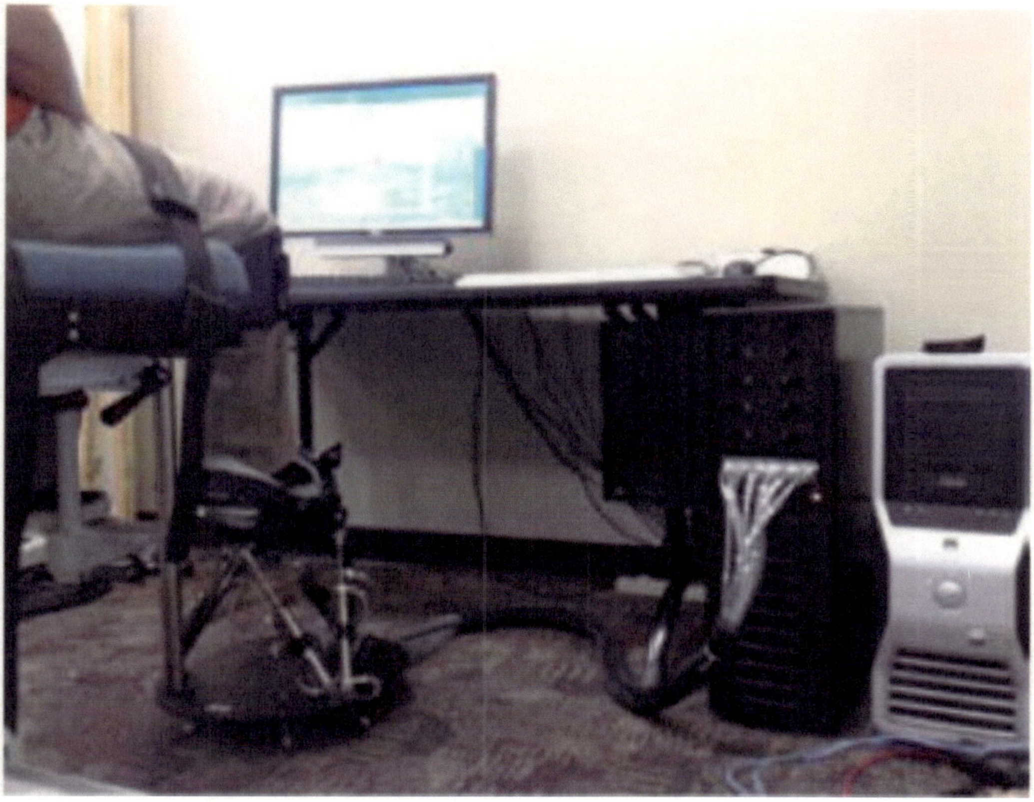

Figure 3. VR application with The Rutgers Ankle CP. Copyright Rutgers University Tele-Rehabilitation Institute and Washington University in St. Louis. Reprinted by permission.

These products (Nintendo Wii, Wii sport games, Wii fit) use motion-sensing technology. The technology perceives speed and orientation with the manual remote control. The player mimics the physical movements of games such as baseball and skiing with the remote control. A pressure-sensitive balance board can be used. Physical activities in yoga and similar balance games can be mimicked [47, 48]. Sony Eye-Toy is another video game and used with PlayStation 2. This system contains a small camera and perceives the body of the person and then transfers the appearance to the imaginary system [23]. Dance Dance Revolution (DDR) can be used with Nintendo, PlayStation and Xbox game. It contains a pressure-sensitive mat. It follows dance movements and transfers them to a virtual environment with e-dimensions [47].

Figure 4. The Rutgers Ankle CP: Game starting screens. Copyright Rutgers University Tele-Rehabilitation Institute and Washington University in St. Louis. Reprinted by permission.

5. Virtual reality systems designed for rehabilitation

Marketed games are not appropriate for patients with "severe spasticity". Pressing on a button in an extremely difficult way or holding the remote control for a long time can be required during the game. The failure of patients to finish the game is due to games being designed for young healthy adults and this decreases self-respect while leading to depression. Games that can automatically adapt to the decreased functions of each patient and that can provide the repetition necessary for neural change are necessary for amusing game-based hand therapy in children with CP who have severe spasticity [49]. Examples are Sony PlayStation 3-supported sensory glove and the Pediatric Intensive Therapy System (PITS); contains a sensory glove and games appropriate for rehabilitation. PITS has been developed for children with upper extremity dysfunction and decreases the dependence of patients, increases self-sufficiency in exercise control and decreases the therapy cost [50]. The Interactive Rehabilitation Exercise System (IREX); uses motion-sensing technology and video capture [46].

6. The advantages and disadvantages of VR systems

Knowing the advantages and disadvantages of the systems is important in determining appropriate virtual reality applications for clinical use and research. Galvin et al gathered and

http://dx.doi.org/10.5772/57486

classified the VR systems used in the field of pediatric rehabilitation in a very detailed and explanatory manner for clinicians and academicians in their review. This is valuable explanation for researchers, clinicians, master and doctorate students to support, help and make easier in choosing VR systems according to their target functions (Table II) [47].

Classification of Virtual Reality Systems for Rehabilitation		
VR systems focused on whole body movement, upper-lower extremities	Upper Extremity	PITS
		PS3 Glove
	Whole Body	Eye/Eye Toy
		IREX
		Wii/Wii Fit
	Lower Extremity	DDR
Cognitive–Motor relationship	Those with cognitive-motor relationship	IREX
		PITS
	Those without cognitive motor relationship	DDR
		Eye/Eye Toy
		Wii/Wii Fit
		PS3 Glove
Those with the ability to focus on movement quality	IREX	
	PITS	
	PS3 Glove	
	Eye/Eye Toy	
	Wii/Wii Fit	
Those without the ability to focus on movement quality	DDR	
Those that require the ability to maintain the straight posture of the body	Standing independently	Wii Fit
		DDR
	Sitting and others	IREX
		PITS
		PS3 Glove
		Eye/Eye Toy
		Wii

Galvin and Levac with kind permission [47].

Table 2. Classification of Virtual Reality Systems for Rehabilitation

7. Virtual Reality studies in rehabilitation

Virtual reality studies are promising for clinical use in pediatric rehabilitation, especially in CP. Parsons et al reported using VR treatment as a rehabilitation approach in children with CP to be more effective than in children with autism and attention deficit [51]. Researches focused on using VR systems in CP interventions, aimed to prove effectiveness of VR on the body structures and functions, activity and participation according to the ICF components. Most of them used applications to reduce impairments of body structures and functions [29]. These researches can be separated subgroups according to their goal as upper extremity, lower extremity, postural control, physical-cardiovascular fitness, and education.

8. Studies focused on upper extremity

All studies aimed to improve function or quality of movement in upper limbs in order to achieve better performance in daily life activities and increased social participation. Researches about this issue are rapidly increasing and then question that "Does VR applications really improve upper limb function?" came up. Accordingly, Galvin et al reviewed 5 studies that used VR to develop upper extremity skills in children with CP. The findings were reported to be limited due to the inconsistencies in result measurements [52]. Also, Wang et al reported that study designs, result measurements, and therapy intensities were heterogeneous and sampling groups small in their review on VR applications and supported the use of VR for upper extremity training in children with CP [53]. However, for CP treatment, it is difficult to demonstrate changes in quality of movement. For instance, Reid et al treated 4 spastic quadriplegia and diplegia patients between the ages of 8 and 12 years. Treatment was implemented for 1.5 hours per week for 8 weeks. They showed that the BOTMP scores increased but the quality of upper extremity movements did not change [31]. In the other study of Reid et al divided 31 children aged 8-12 years as 19 VR subjects and 12 control subjects in their randomized controlled study. Treatment for 1.5 hours per week for 8 weeks was implemented for the children. Canadian Occupational Performance Measurement (COPM) and Quality of *Upper Extremity* Skills Test (QUEST) were used to evaluate the effect of the treatment. While the results of all participants improved after the treatment, no significant functional difference was found between the groups but there was significantly increased social acceptance and motivation in the treatment group [54]. Small sample size or inability of assessment tools to measure quality of movement may induce these results. On the other hand, there are evidence-based studies showed improvement quality of movement. One of those, a study of You et al conducted on children with hemiparetic CP to investigate VR-based cortical reorganization and functional motor development. They treated the child with IREX for 60 minutes 5 times a week for one month. The Bruininks–Oseretsky Test of Motor Proficiency (BOTMP) score was shown to increase from 1 to 5, the Modified Pediatric Motor Activity Log (PMAL) questionnaire from 0 to 3 and the Fugl-Meyer (FMA) score from 39 to 52. The authors reported functional motor skills and amount of use in affected upper extremities and the quality of the motion, active movement control and coordination of upper extremity motor performance

http://dx.doi.org/10.5772/57486

increased. According to the functional magnetic resonance imaging (fMRI) results, the increased abnormal activities in the ipsilateral and contralateral primary sensorimotor cortex before the treatment decreased while the primary motor cortex and primary sensorimotor cortex were found to be more active. The authors showed that internalization of the motor process in target motor behavior is facilitated during visual sensory feedback and VR therapy. This internalization leads to the formation of new motor pathways especially around the sensorimotor cortices (SMC). They demonstrated that VR therapy stimulates neural motor pathways that were not previously used and might develop neuroplasticity. Thus, motor skills of the affected extremity of the child have developed and cortical reorganization was similar to that of a normally developing child after VR application [55]. Similarly, Chen et al investigated the benefit of VR use for reaching activity in 4 children with CP between the ages of 4 and 8. They used the Sony Eye-Toy system. Children were treated for 4 weeks at 2 hours per week. Reach kinematics and PDMS-2 results of the children were found to increase. The quality of reaching was shown to improve after individual training for 4 weeks [56]. Most of the researches showed improvement upper limb functions via VR interventions. Jannink et al included 12 children with CP for upper extremity training with a Sony Eye-Toy in a weakly randomized controlled study. A treatment of 30 minutes was implemented twice a week for 6 weeks between the ages of 7 and 16. Upper extremity functions were evaluated with Melbourne. The results of two children in the treatment group were shown to have increased from 9% to 13%. The authors reported the Eye-Toy to be a motivational education tool that developed upper extremity function in children with CP [57]. Also, home-based treatment approaches are important to integrate effectiveness of VR intervention in daily life and to increase the time of intervention. Huber et al developed a PS3 home hand rehabilitation system for hemiparetic young people with CP and Burdea et al reported pre-treatment hand values of 45% for extension, 70% for flexion and post-treatment values of 70% and 90% after using this system for treatment for 2 months. Grip strength increased by 50% and the Jebsen hand function test results increased. The bone mineral density also increased [49]. Then again, Winkels et al included 15 children with CP between the age of 6 and 15 in the study where they aimed to evaluate the effect of upper extremity function training in children with CP by using Wii games. Children at level I and II according to the manual ability classification system (MACS) were included in the study. The children were evaluated with the Melbourne Assessment of Upper Limb Function and ABILHAND-Kids before and after the treatment. A marked increase was found in the performance of daily living activities. Those who received low scores in the Melbourne evaluation at the beginning showed great improvement in upper extremity function. It has been said that treatment with games is more suitable for children with a more severe disorder. Health professionals and children have been satisfied with the Wii boxing and tennis game [41]. As result, these studies proved VR systems are motivational, familiar and useful for children with CP and can be used for improving upper limb function on the other hand, participants were included, had mild level of MACS or gross motor function classification system (GMFCS), future researches should investigate benefits of VR applications in moderately and severely affected children with CP.

9. Studies focused on lower extremity

The goal of studies associated with lower extremity generally is to enhance walking ability or strength. Another current review, Meyer-Heim et al focused on the treatment of motor impairments with new complementary technologies such as robot-assisted and computer-based rehabilitation systems. Previously, robot-assisted gait training (RAGT) was developed for adults and researched demonstrated its functional benefits such as increased gait velocity, improve balance, reduction of muscular hypertonia and gait endurance. RAGT is based on neuroplasticity of central nervous system within spinal cord to stimulate a basic locomotor pattern via central pattern generators [5]. Mutlu et al, declared partial weight bearing treadmill training can improve walking capacity in children with CP via motor learning [58]. Besides that, a pediatric model of the Locomat designed for RAGT of children starting from age 5 years old due to beneficial outcomes in adults [59]. But one of the limitations of conventional locomotor training is that walking on a treadmill prevents optic flow. Optic flow is important for arrange gait speed and stride length during walking, which visual motion sensed by the eyes as the body moves through the environment. For that reason, combination between VE and treadmill training provide the optic flow of forward motion and may improve walking patterns and also increase motivation and immersion. Integrated virtual environment rehabilitation treatment gained visual and proprioceptive feedback that are necessary for gait training. Because movement patterns can be modified using visual, auditory and proprioceptive feedback. In addition, it is essential to match proprioceptive feedback from the limbs [60]. Other limitation is that walking monotonously for 30-45 min during Locomat or treadmill can be boring for children because motivation is essential for rehabilitation process. Thus, some pediatric rehabilitation centers play music during the treatment [57]. For this reason, recent studies focused on combination with VR system and RAGT. Brütsch et al investigated the effect of VR and Locomat and created a virtual football scenario for the patients. The therapy of 10 children with a neurological gait disorder and 8 healthy children was conducted with the Locomat alone, with therapist support, with VR, and with VR and the therapist together. Results were obtained in the posture and swing phase from the knee and hip joints. No difference was found between VR and a therapist in cases with a neurological walking disorder. Walking with any motor support in children, whether healthy or with a neurological disorder, significantly increased motor results compared to walking without any motivational support. In other words, active participation increased with the verbal support of the therapist, with VR, or with the support of VR and the therapist at the same time. The VR and therapist combination was found to be more effective. VR games provide the necessary motivation for gait training in children. These studies showed that electromyography activity output was significantly higher during task with VR and physiotherapist motivation than during normal walking conditions when walking on the Locomat [34] (Figure V).

Last 20 years, interventions focused on strength training in CP rehabilitation due to understanding important of muscle weakness. In children with CP, muscle weakness influence negatively daily life activities and social participation and decreased functional capacity [61]. Chen et al identified that cycling is an applicable, effective and easy approach for improving muscle strength and developed home-based virtual cycling training (hVCT) program. They included 27 children with CP with GMFCS levels I-II at age of 6-12 years. Children with CP

http://dx.doi.org/10.5772/57486

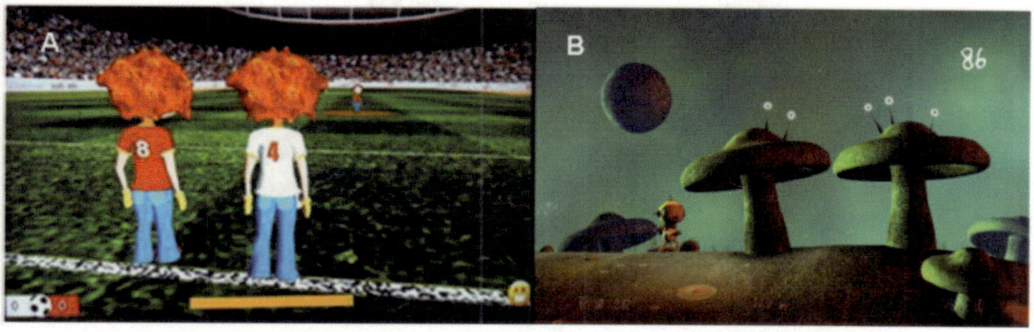

Figure 5. Examples of Pediatric Locomat virtual reality games. This game has been developed in a close collaboration among the Rehabilitation centre Affoltern, the ETH Zurich and the University of the Arts Zurich. (Color version of the figure is available online.) [5]

performed hVCT program 40 min/day 3 times per week for 12 weeks, that consisted of a 5 min warm-up exercise, 20 repetition of sitting to standing movements, cycling 20 min and a cool-down exercise for 5 min. They assessed gross motor function with BOTMP and muscle strength with isokinetic dynamometer. The results showed significant effect on muscle strength at post treatment. The hVCT group had greater peak torque of knee extensor and flexor at 60^0/s and 120^0/s angular velocities than control group. Changing in strength indices of knee extensor and knee flexor at 60^0/s post treatment were 19-41% in the hVCT group while those were -2 to 1 % in the control group. Also, at 120^0/s at post treatment were 30-36% in hVCT group while -6 to -19% in control group. They suggested that these findings might be lead clinicians to improve muscle strength more effectively [62]. Bryanton et al investigated ankle dorsiflexion kinematics in observational studies they conducted on 10 children with CP and 6 healthy children. The difficulty in voluntary muscle contradiction results in weak selective motor control in children with CP. Children with CP completed their selective motor control exercises with the VR exercise system and conventional exercises. Ankle movements were recorded with the electro goniometer. VR has been shown to provide more repetition than conventional exercises. The joint range of motion and duration of holding in a stretched position were found to be higher after VR exercises. They proved VR use increased compliance with exercise and its usefulness [63]. Burdea et al investigated ankle strength, motor control, gait, function and quality of life development in children with CP while playing VR games in their study. Plantar flexion strength of the children increased 0.15 Nm/kg and the quality of life increased by 2.8% according to the Pediatric Quality of Life Inventory (PedsQL). 400 repetitions were performed for each ankle with the Rutgers Ankle CP system in a game session [64]. Therefore, clinicians can be used VR systems to increase lower extremity strength in children with CP.

10. Studies focused on postural control

Postural and balance control is one of the key factors that affect performance of most functional skills such as walking and reaching. When children with CP have poor postural control, they may fall in walking or may not regulate the velocity of the reaching arm or to initial pelvis position. The main reasons of dysfunctional postural control are enhanced antagonistic co-activation, reduced capacity to modulate the degree of postural muscle contraction to the specifics of the situation [65]. In last decade, some interventions focused on impaired postural control and balance in children with CP to improve daily life activities [66]. However, there are few evidence-based studies showed effect of VR intervention on postural muscle activity. One of them is a case report by Deutch et al. which is the first study conducted with Wii. A 13-year-old spastic diplegic child was provided VR treatment with the Wii game console for 4 weeks at 11 sessions of 60-90 minutes. They showed that the visual perception process, postural control and functional mobility had increased. They emphasized that stretching behaviors developed with cortical reorganization in the rehabilitation of the upper extremity movements with VR [48]. In another study, Gordon et al included 6 patients between the ages of 6 and 12 years in treatment with Wii twice a week for 6 weeks. Total Gross Motor Function Measure (GMFM) score changed by 7%. The biggest change was seen in the sitting section (12%). The smallest change was seen in the turning section (2%). Two of them used balance assessments to demonstrate differences in postural control [18]. Sharan et al included 16 children (8 study - 8 control) in treatment 3 days a week for 3 weeks with Nintendo Wii sports in a study where they investigated the effect of VR application after surgery in children with CP. An increase was seen after the treatment with PBS. While VR had an important effect on the development of balance, no difference was shown between the control group regarding manual skills. The investigators proved that Wii-Fit use developed balance in the child and balance training decreased the swing of the children with this study [24]. Additionally, Brien et al investigated the functional balance and mobility of adolescents with CP level I according to Gross Motor Function Classification System (GMFCS) after intensive short-term VR application. Four children with CP between the ages of 13 and 18 were treated for 90 minutes per day for 5 days with IREX. Timed Up and Down Stairs (TUDS), 6-Minute Walk Test (6MWT) and Community Balance and Mobility Scale (CB&M) and GMFM E were evaluated. Functional balance and mobility were shown to develop with short-term, intensive VR implementation. The improvement was found to be significant with CB&M and 6MWT in the follow-up period. The development in CB&M is reported to be associated with the development in coordination, time and speed necessary for ambulatory performance of complex motor skills and especially within the society, and this effect was preserved for at least 1 month. The walking endurance necessary for daily life and social participation was proven to be increased [67]. Walking is also influenced of active control of pelvis and trunk. Balance perturbation responses in healthy individuals are formed with the simultaneous contradiction of the neck, body and hip muscles and are seen even before the activation of muscles. Distortion of proprioception in the core environment (body and pelvis) and decreased strength has been found to be associated especially with increased injury risk to the knee in prospective studies. Good control of core movement is therefore a prerequisite for better use of the legs. The interaction between the body and the pelvis is necessary for good control and performance of daily living activities. This interaction during walking depends on the walking speed. While the pelvis and body interact as when the body is standing in the transverse plane at slow speed, the interaction becomes that seen in the swing phase as the rate increases. The protraction of the pelvis together with the retraction of the body in the swing phase increases the step length of the leg.

Thus, the walk productivity improves. One of the primary problems in CP is the decrease in selective motor control. Poor selective motor control of the pelvis and body distorts walking and negatively affects daily living activities VR training leads to decreased combination of body and pelvis and increased selective control. Co-contraction and combination decreases with increase of selective control of the muscles in the body and pelvis region, while selective pelvis control increases and pelvis rotation-body rotation is facilitated. Co-contraction is the simultaneous contraction of the agonist and antagonist muscle and is used to prevent errors and increase stability when unaccustomed tasks are being performed. The co-contraction level decreases with increased practice [68, 69]. Due to these reasons, Barton et al treated a child with spastic CP 2 times a week for 30 minutes, for 6 weeks. The combination of pelvis and body increased after the treatment [70].

11. Studies focused on physical and cardiovascular fitness

Physical activity consists of body movements performed by using the skeletal muscles and results in spending energy. According to ICF, activity is divided into 2 areas as performance and capacity. Physical activity is made up of the activities the individual undertakes in regular daily life. Capacity is how much people can achieve, such as walking distance [6]. Increasing spend time on watching television, playing electronic games and computers, are generally associated with decreased physical activity and obesity [71]. On the other hand, the children and adolescents active video games are considered interesting alternatives to passive games. There is gradually increasing evidence that internet-based applications and active games can increase physical activity in healthy children [72]. Physical activity and fitness is reduced in children with CP than their typically developing peers and also spend most of their time with sedentary activities, facing a screen [73]. Examples are watching television and playing video games. For this reasons the risk factors to develop obesity, osteoporosis, diabetes, CVS disease or musculoskeletal pain are increased. There is new evidence indicating that VR implementation and the use of motion interactive games increases physical activity in children with CP. Mitchell et al investigated the effects of VR application in children with CP on physical activity. VR implementations are more intense than one-to-one training. For example, "move it to improve it- (Mitii)" provides a total of 70 hours of therapy. Therefore, these systems may provide an increase in physical activity. Physical activity capacity increased with Wii Sports and Mitii and physical activity performance increased with Eye-Toy 2. Functional strength also increased with Mitii training. A few intervention studies investigated effects on physical activity in home use and/or long-term use of active video games. These researches reported that active games could improve physical activity in a moderate level and reduce sedentary screen time. Home based interventions have ranged from 10-28 weeks in duration so the effects of long term use of active games are uncertain. Several studies indicated playtime was reduced during the intervention [74].

In recent review, Fehlings et al pointed that effect of VR on cardiovascular fitness (CVI) in children with CP. For active video games, it's necessary to appropriate physical activity. Active video games have great potential to promote increased physical activity and enhanced CVI fitness for children with CP [26]. Hurkmans et al investigated effects on energy expenditure (EE) among adults with CP when playing Wii sports. Several researches compared EE measured with indirect calorimetry or by an activity monitor during play of different motion interactive video games to EE during other activities of various physical exertions [75]. Generally activity levels compare with metabolic equivalent units (MET). MET is a physiological concept expressing the energy cost of physical activities and is determined as the ratio of metabolic rate during a specific physical activity to a reference rate of the metabolic rate at complete rest. Therefore 1 MET corresponds to the metabolic rate while at complete rest and 2 METs represent a doubling of the energy consumption. A common grading of physical activity according to METs is: *sedentary* (<3METs); *moderate* (3-6 METs); *vigorous* (6-9 METs); and *very vigorous* (>9 METs). Playing motion interactive games found to increase EE to light or moderate levels around 3 METs, same as brisk walking, skipping, jogging or stair climbing. Sustained vigorous activity above 6 METs is generally not obtained during play. Energy expenditure during play depends on engagement and type of game played. Games that playing on upper body movements compared to games with playing lower EE and higher values are achieved when all body movements are achieved [76]. As result studies demonstrated that active video games generates higher EE compared to rest and sedentary screen time activities but not as high EE values as performing the corresponding real activity in itself.

12. Studies focused on other conditions associated with cerebral palsy

Most of children with CP suffer from several comorbidities in addition to motor handicap such as behavioral, cognitive, and learning disabilities, further impeding their overall functional capacity. Approximately 40 percent of them have learning disabilities and common behavioral symptoms in high-functioning children with cerebral palsy are attention deficits and impulsivity, especially premature birth compatible with the diagnosis of attention-deficit hyperactivity disorder (ADHD). ADHD is one of the major causes of behavioral, friendship and school problems [77]. Pollak et al investigated effects of VR intervention on children with ADHD and produced virtual reality schoolroom environment to motivate children, to support active participation, to evaluate attention and motor behaviors in challenging. Their VR classroom designed according to Rizzo et al Digital Media Works and has head mounted display and gives visual-auditory stimuli with in the VE. In experimental group had 20 boys with ADHD, the control group consisted of 17 boys without ADHD. They assessed with Test of Variables of Attention (TOVA) and virtual reality continuous performance tasks (VR-CPT). According to VR-CPT findings, children with ADHD had slower reaction time, higher variability in RT is more errors of omission and commission than control group. They demonstrated VE provide test and training situations that are ecologically valid, motivating and dynamic. These findings consistent with literature that VR-CPT is a user-friendly method for children with ADHD, autism and intellectual disability [78]. Future researches may focus on VR applications on children with ADHD and CP. Akhutina et al administered VR therapy to the treatment group for 30-60 minutes, 6-8 times for 1 month in a semi-experimental study they conducted with 12 treatment and 9 control subjects. They showed that the visual-spatial abilities developed more

http://dx.doi.org/10.5772/57486

in the treatment group. They emphasized that VE-based spatial education was effective in children with complex disabilities [79]. Rosenbaum et al originally reported agency is essential component for our self-consciousness and the ability to control movements and interact appropriately with the environment, also the computer model help to investigate sense of agency in our experiment. Rosenbaum et al investigated that CP children's ability to correctly perceiving their own movement by training with cognitive process and motor control. The study consisted of 20 CP children in training group and 20 CP children in control group trained for at least 30 min daily in the 20 weeks period using the internet based home training system 'move it to improve it (Mi Tii)" and CP training group continued their routine daily life activities. Their results proved that children with CP improve their ability to determine whether they themselves or a computer are responsible for the movement of an observed object following 20 weeks of an inter active computer training designed to increased sensory-motor interaction [80].

Children with CP also generally occur pain and this affect to daily life activities, participation in rehabilitation, social life. Pain is also important for children to reduce motivation that induce human action. Distraction is one of the successful methods to reduce pain and behavioral distress for children suffered by pain. Pillay reported that interactive distraction that child attend activity the distraction task continuously require central attention resources much more effective than passive distraction activities that do not necessary management of central attentional functions such as watching cartoons. These interactive distraction activities consist of videogames, interactive musical storybooks, interactive toy robots and educational electronic games that reduce acute pain in children [81]. Law et al aimed to investigate effect of passive and active distraction task. They assessed pain by cold pressure tolerance. Participants separated two groups according to their age (6-9 years and 10-15 years). Because children react distraction differently associated to the age. Researches explained that attentional control improve rapidly during early childhood and develop greatly between 7-9 years and is relatively mature by 12 years. They used Nintendo Wii game system. Law et al found that there was significant improvement in pain tolerance all the interactive and passive distraction. But interactive distraction by increasing attentional load resulted higher improvements in pain tolerance. Their results are supported to the other researches in literature. In addition their findings showed that there were developmental differences on response to interactive distraction. Both of children have benefits on passive and interactive distraction but especially interactive distraction is much better. Also, older children response pain interactive distraction better than younger [82]. Current literature is required to research relationship between pain and VR interventions in CP.

13. Conclusion

We believe that VR provide new possibilities for the rehabilitation team of CP as the effect of active treatment and motivation together with functional use enable minimizing motor problems although the effect has not been completely shown in studies on VR applications in children with cerebral palsy due to heterogeneity of studies, sample size, outcome measures and etc. In addition to many physical benefits the concept of "cerebral plasticity" are important for independent activity perception, especially for the treatment of motor problems from the perspective of physiotherapists, as well as therapy including play, fun and enjoy from the perspective of the child encourage us to complement the use of the VR systems in the rehabilitation of children.

Author details

Mintaze Kerem Gunel, Ozgun Kaya Kara, Cemil Ozal* and Duygu Turker

Hacettepe University, Faculty of Health Sciences, Department of Physiotherapy and Rehabilitation, Ankara, Turkey

References

[1] Rosenbaum P, Paneth N, Leviton A, Goldstein M, Bax M, Damiano D, et al. A report: the definition and classification of cerebral palsy April 2006. Developmental medicine and child neurology Supplement. 2007 Feb;109:8-14. PubMed PMID: 17370477.

[2] Surveillance of Cerebral Palsy in E. Surveillance of cerebral palsy in Europe: a collaboration of cerebral palsy surveys and registers. Surveillance of Cerebral Palsy in Europe (SCPE). Developmental medicine and child neurology. 2000 Dec;42 (12):816-24. PubMed PMID: 11132255.

[3] Ashwal S, Russman BS, Blasco PA, Miller G, Sandler A, Shevell M, et al. Practice parameter: diagnostic assessment of the child with cerebral palsy: report of the Quality Standards Subcommittee of the American Academy of Neurology and the Practice Committee of the Child Neurology Society. Neurology. 2004 Mar 23;62 (6):851-63. PubMed PMID: 15037681.

[4] Rosenbaum P, Stewart D. The World Health Organization International Classification of Functioning, Disability, and Health: a model to guide clinical thinking, practice and research in the field of cerebral palsy. Seminars in pediatric neurology. 2004 Mar;11 (1):5-10. PubMed PMID: 15132248.

[5] Meyer-Heim A, van Hedel HJ. Robot-assisted and computer-enhanced therapies for children with cerebral palsy: current state and clinical implementation. Seminars in pediatric neurology. 2013 Jun;20 (2):139-45. PubMed PMID: 23948688.

[6] (WHO) WHO. International Classification of Functioning, Disability and Health. Geneva. 2001; World Health Organization.

[7] Kantak SS, Stinear JW, Buch ER, Cohen LG. Rewiring the brain: potential role of the premotor cortex in motor control, learning, and recovery of function following brain

injury. Neurorehabilitation and neural repair. 2012 Mar-Apr;26 (3):282-92. PubMed PMID: 21926382.

[8] Lollar DJ, Simeonsson RJ. Diagnosis to function: classification for children and youths. Journal of developmental and behavioral pediatrics : JDBP. 2005 Aug;26 (4): 323-30. PubMed PMID: 16100508.

[9] Vargus-Adams J. Understanding function and other outcomes in cerebral palsy. Physical medicine and rehabilitation clinics of North America. 2009 Aug;20 (3): 567-75. PubMed PMID: 19643354. Pubmed Central PMCID: 2719719.

[10] Novak I, McIntyre S, Morgan C, Campbell L, Dark L, Morton N, et al. A systematic review of interventions for children with cerebral palsy: state of the evidence. Developmental medicine and child neurology. 2013 Oct;55 (10):885-910. PubMed PMID: 23962350.

[11] Beckung E, Hagberg G. Neuroimpairments, activity limitations, and participation restrictions in children with cerebral palsy. Developmental medicine and child neurology. 2002 May;44 (5):309-16. PubMed PMID: 12033716.

[12] Krageloh-Mann I, Cans C. Cerebral palsy update. Brain & development. 2009 Aug;31 (7):537-44. PubMed PMID: 19386453.

[13] Rosenbaum PL, Walter SD, Hanna SE, Palisano RJ, Russell DJ, Raina P, et al. Prognosis for gross motor function in cerebral palsy: creation of motor development curves. JAMA : the journal of the American Medical Association. 2002 Sep 18;288 (11): 1357-63. PubMed PMID: 12234229.

[14] Kerem Gunel M. [Rehabilitation of children with cerebral palsy from a physiotherapist's perspective]. Acta orthopaedica et traumatologica turcica. 2009 Mar-Apr;43 (2): 173-80. PubMed PMID: 19448358. Fizyoterapist bakis acisiyla beyin felcli cocuklarin rehabilitasyonu.

[15] Anttila H, Autti-Ramo I, Suoranta J, Makela M, Malmivaara A. Effectiveness of physical therapy interventions for children with cerebral palsy: a systematic review. BMC pediatrics. 2008;8:14. PubMed PMID: 18435840. Pubmed Central PMCID: 2390545.

[16] Weindling AM, Cunningham CC, Glenn SM, Edwards RT, Reeves DJ. Additional therapy for young children with spastic cerebral palsy: a randomised controlled trial. Health technology assessment. 2007 May;11 (16):iii-iv, ix-x, 1-71. PubMed PMID: 17462166.

[17] Butler C, Darrah J. Effects of neurodevelopmental treatment (NDT) for cerebral palsy: an AACPDM evidence report. Developmental medicine and child neurology. 2001 Nov;43 (11):778-90. PubMed PMID: 11730153.

[18] Gordon C, Roopchand-Martin S, Gregg A. Potential of the Nintendo Wii as a rehabilitation tool for children with cerebral palsy in a developing country: a pilot study. Physiotherapy. 2012 Sep;98 (3):238-42. PubMed PMID: 22898581.

[19] Lotze M, Braun C, Birbaumer N, Anders S, Cohen LG. Motor learning elicited by voluntary drive. Brain : a journal of neurology. 2003 Apr;126 (Pt 4):866-72. PubMed PMID: 12615644.

[20] Tatla SK, Sauve K, Virji-Babul N, Holsti L, Butler C, Van Der Loos HF. Evidence for outcomes of motivational rehabilitation interventions for children and adolescents with cerebral palsy: an American Academy for Cerebral Palsy and Developmental Medicine systematic review. Developmental medicine and child neurology. 2013 Jul; 55 (7):593-601. PubMed PMID: 23550896.

[21] Marshall SJ, Gorely T, Biddle SJ. A descriptive epidemiology of screen-based media use in youth: a review and critique. Journal of adolescence. 2006 Jun;29 (3):333-49. PubMed PMID: 16246411.

[22] Wilson PN, Foreman N, Stanton D. Virtual reality, disability and rehabilitation. Disability and rehabilitation. 1997 Jun;19 (6):213-20. PubMed PMID: 9195138.

[23] Weiss PL, Rand D, Katz N, Kizony R. Video capture virtual reality as a flexible and effective rehabilitation tool. Journal of neuroengineering and rehabilitation. 2004 Dec 20;1 (1):12. PubMed PMID: 15679949. Pubmed Central PMCID: 546410.

[24] Sharan D, Ajeesh PS, Rameshkumar R, Mathankumar M, Paulina RJ, Manjula M. Virtual reality based therapy for post operative rehabilitation of children with cerebral palsy. Work. 2012;41 Suppl 1:3612-5. PubMed PMID: 22317271.

[25] Sandlund M, McDonough S, Hager-Ross C. Interactive computer play in rehabilitation of children with sensorimotor disorders: a systematic review. Developmental medicine and child neurology. 2009 Mar;51 (3):173-9. PubMed PMID: 19191834.

[26] Fehlings D, Switzer L, Findlay B, Knights S. Interactive computer play as "motor therapy" for individuals with cerebral palsy. Seminars in pediatric neurology. 2013 Jun;20 (2):127-38. PubMed PMID: 23948687.

[27] Andolsek D. Virtual reality in education and training. International journal of instructional media. 1995;22 (2):145-51.

[28] Weiss PL, Katz N. The potential of virtual reality for rehabilitation. Journal of rehabilitation research and development. 2004 Sep;41 (5):vii-x. PubMed PMID: 15558392.

[29] Snider L, Majnemer A, Darsaklis V. Virtual reality as a therapeutic modality for children with cerebral palsy. Developmental neurorehabilitation. 2010;13 (2):120-8. PubMed PMID: 20222773.

[30] Holden MK. Virtual environments for motor rehabilitation: review. Cyberpsychology & behavior : the impact of the Internet, multimedia and virtual reality on behavior and society. 2005 Jun;8 (3):187-211; discussion 2-9. PubMed PMID: 15971970.

[31] Reid DT. Benefits of a virtual play rehabilitation environment for children with cerebral palsy on perceptions of self-efficacy: a pilot study. Pediatric rehabilitation. 2002 Jul-Sep;5 (3):141-8. PubMed PMID: 12581476.

[32] Geerdink Y, Aarts P, Geurts AC. Motor learning curve and long-term effectiveness of modified constraint-induced movement therapy in children with unilateral cerebral palsy: a randomized controlled trial. Research in developmental disabilities. 2013 Mar;34 (3):923-31. PubMed PMID: 23291509.

[33] Papavasiliou AS. Management of motor problems in cerebral palsy: a critical update for the clinician. European journal of paediatric neurology : EJPN : official journal of the European Paediatric Neurology Society. 2009 Sep;13 (5):387-96. PubMed PMID: 18778959.

[34] Brutsch K, Schuler T, Koenig A, Zimmerli L, Koeneke SM, Lunenburger L, et al. Influence of virtual reality soccer game on walking performance in robotic assisted gait training for children. Journal of neuroengineering and rehabilitation. 2010;7:15. PubMed PMID: 20412572. Pubmed Central PMCID: 2877051.

[35] Rizzolatti G, Craighero L. The mirror-neuron system. Annual review of neuroscience. 2004;27:169-92. PubMed PMID: 15217330.

[36] Pfeifer JH, Iacoboni M, Mazziotta JC, Dapretto M. Mirroring others' emotions relates to empathy and interpersonal competence in children. NeuroImage. 2008 Feb 15;39 (4):2076-85. PubMed PMID: 18082427. Pubmed Central PMCID: 3840169.

[37] Dancause N, Barbay S, Frost SB, Plautz EJ, Chen D, Zoubina EV, et al. Extensive cortical rewiring after brain injury. The Journal of neuroscience : the official journal of the Society for Neuroscience. 2005 Nov 2;25 (44):10167-79. PubMed PMID: 16267224.

[38] Garrison KA, Winstein CJ, Aziz-Zadeh L. The mirror neuron system: a neural substrate for methods in stroke rehabilitation. Neurorehabilitation and neural repair. 2010 Jun;24 (5):404-12. PubMed PMID: 20207851.

[39] Cross ES, Kraemer DJ, Hamilton AF, Kelley WM, Grafton ST. Sensitivity of the action observation network to physical and observational learning. Cerebral cortex. 2009 Feb;19 (2):315-26. PubMed PMID: 18515297. Pubmed Central PMCID: 2638791.

[40] Sveistrup H. Motor rehabilitation using virtual reality. Journal of neuroengineering and rehabilitation. 2004 Dec 10;1 (1):10. PubMed PMID: 15679945. Pubmed Central PMCID: 546406.

[41] Winkels DG, Kottink AI, Temmink RA, Nijlant JM, Buurke JH. Wii-habilitation of upper extremity function in children with cerebral palsy. An explorative study. Developmental neurorehabilitation. 2013;16 (1):44-51. PubMed PMID: 23030054.

[42] Ketelaar M, Vermeer A, Hart H, van Petegem-van Beek E, Helders PJ. Effects of a functional therapy program on motor abilities of children with cerebral palsy. Physical therapy. 2001 Sep;81 (9):1534-45. PubMed PMID: 11688590.

[43] Thorpe DE, Valvano J. The effects of knowledge of performance and cognitive strategies on motor skill learning in children with cerebral palsy. Pediatric physical therapy : the official publication of the Section on Pediatrics of the American Physical Therapy Association. 2002 Spring;14 (1):2-15. PubMed PMID: 17053676.

[44] Chen KB, Savage AB, Chourasia AO, Wiegmann DA, Sesto ME. Touch screen performance by individuals with and without motor control disabilities. Applied ergonomics. 2013 Mar;44 (2):297-302. PubMed PMID: 23021630. Pubmed Central PMCID: 3572909.

[45] Lee D. The state of the touch-screen panel market in 2011. Inf Disp. 2011;27:12-6.

[46] Reid D. Virtual reality and the person-environment experience. Cyberpsychology & behavior : the impact of the Internet, multimedia and virtual reality on behavior and society. 2002 Dec;5 (6):559-64. PubMed PMID: 12556119.

[47] Galvin J, Levac D. Facilitating clinical decision-making about the use of virtual reality within paediatric motor rehabilitation: describing and classifying virtual reality systems. Developmental neurorehabilitation. 2011;14 (2):112-22. PubMed PMID: 21410403.

[48] Deutsch JE, Borbely M, Filler J, Huhn K, Guarrera-Bowlby P. Use of a low-cost, commercially available gaming console (Wii) for rehabilitation of an adolescent with cerebral palsy. Physical therapy. 2008 Oct;88 (10):1196-207. PubMed PMID: 18689607.

[49] Burdea GC, Jain A, Rabin B, Pellosie R, Golomb M. Long-term hand tele-rehabilitation on the PlayStation 3: benefits and challenges. Conference proceedings : Annual International Conference of the IEEE Engineering in Medicine and Biology Society IEEE Engineering in Medicine and Biology Society Conference. 2011;2011:1835-8. PubMed PMID: 22254686.

[50] Wille D, Eng K, Holper L, Chevrier E, Hauser Y, Kiper D, et al. Virtual reality-based paediatric interactive therapy system (PITS) for improvement of arm and hand function in children with motor impairment--a pilot study. Developmental neurorehabilitation. 2009 Feb;12 (1):44-52. PubMed PMID: 19283533.

[51] Parsons TD, Rizzo AA, Rogers S, York P. Virtual reality in paediatric rehabilitation: a review. Developmental neurorehabilitation. 2009 Aug;12 (4):224-38. PubMed PMID: 19842822.

[52] Galvin J, McDonald R, Catroppa C, Anderson V. Does intervention using virtual reality improve upper limb function in children with neurological impairment: a systematic review of the evidence. Brain injury : [BI]. 2011;25 (5):435-42. PubMed PMID: 21401370.

[53] Wang M, Reid D. Virtual reality in pediatric neurorehabilitation: attention deficit hyperactivity disorder, autism and cerebral palsy. Neuroepidemiology. 2011;36 (1):2-18. PubMed PMID: 21088430.

[54] Reid D. The use of virtual reality with children with cerebral palsy:A pilot randomized trial. Therapeutic Recreation Journal. 2006;40:255-68.

[55] You SH, Jang SH, Kim YH, Kwon YH, Barrow I, Hallett M. Cortical reorganization induced by virtual reality therapy in a child with hemiparetic cerebral palsy. Developmental medicine and child neurology. 2005 Sep;47 (9):628-35. PubMed PMID: 16138671.

[56] Chen YP, Kang LJ, Chuang TY, Doong JL, Lee SJ, Tsai MW, et al. Use of virtual reality to improve upper-extremity control in children with cerebral palsy: a single-subject design. Physical therapy. 2007 Nov;87 (11):1441-57. PubMed PMID: 17895352.

[57] Jannink MJ, van der Wilden GJ, Navis DW, Visser G, Gussinklo J, Ijzerman M. A low-cost video game applied for training of upper extremity function in children with cerebral palsy: a pilot study. Cyberpsychology & behavior : the impact of the Internet, multimedia and virtual reality on behavior and society. 2008 Feb;11 (1): 27-32. PubMed PMID: 18275309.

[58] Mutlu A, Krosschell K, Spira DG. Treadmill training with partial body-weight support in children with cerebral palsy: a systematic review. Developmental medicine and child neurology. 2009 Apr;51 (4):268-75. PubMed PMID: 19207302.

[59] Meyer-Heim A, Borggraefe I, Ammann-Reiffer C, Berweck S, Sennhauser FH, Colombo G, et al. Feasibility of robotic-assisted locomotor training in children with central gait impairment. Developmental medicine and child neurology. 2007 Dec;49 (12): 900-6. PubMed PMID: 18039236.

[60] Lamontagne A, Fung J, McFadyen BJ, Faubert J. Modulation of walking speed by changing optic flow in persons with stroke. Journal of neuroengineering and rehabilitation. 2007;4:22. PubMed PMID: 17594501. Pubmed Central PMCID: 1913055.

[61] Mockford M, Caulton JM. The pathophysiological basis of weakness in children with cerebral palsy. Pediatric physical therapy : the official publication of the Section on Pediatrics of the American Physical Therapy Association. 2010 Summer;22 (2):222-33. PubMed PMID: 20473109.

[62] Chen CL, Chen CY, Liaw MY, Chung CY, Wang CJ, Hong WH. Efficacy of home-based virtual cycling training on bone mineral density in ambulatory children with cerebral palsy. Osteoporosis international : a journal established as result of cooperation between the European Foundation for Osteoporosis and the National Osteoporosis Foundation of the USA. 2013 Apr;24 (4):1399-406. PubMed PMID: 23052930.

[63] Bryanton C, Bosse J, Brien M, McLean J, McCormick A, Sveistrup H. Feasibility, motivation, and selective motor control: virtual reality compared to conventional home exercise in children with cerebral palsy. Cyberpsychology & behavior : the impact of the Internet, multimedia and virtual reality on behavior and society. 2006 Apr;9 (2): 123-8. PubMed PMID: 16640463.

[64] Burdea GC, Cioi D, Kale A, Janes WE, Ross SA, Engsberg JR. Robotics and gaming to improve ankle strength, motor control, and function in children with cerebral palsy--a case study series. IEEE transactions on neural systems and rehabilitation engineering : a publication of the IEEE Engineering in Medicine and Biology Society. 2013 Mar;21 (2):165-73. PubMed PMID: 22773059.

[65] Peng YC, Lu TW, Wang TH, Chen YL, Liao HF, Lin KH, et al. Immediate effects of therapeutic music on loaded sit-to-stand movement in children with spastic diplegia. Gait & posture. 2011 Feb;33 (2):274-8. PubMed PMID: 21185725.

[66] El-Shamy SM, Abd El Kafy EM. Effect of balance training on postural balance control and risk of fall in children with diplegic cerebral palsy. Disability and rehabilitation. 2013 Sep 13. PubMed PMID: 24032716.

[67] Brien M, Sveistrup H. An intensive virtual reality program improves functional balance and mobility of adolescents with cerebral palsy. Pediatric physical therapy : the official publication of the Section on Pediatrics of the American Physical Therapy Association. 2011 Fall;23 (3):258-66. PubMed PMID: 21829120.

[68] Bruijn SM, Meijer OG, van Dieen JH, Kingma I, Lamoth CJ. Coordination of leg swing, thorax rotations, and pelvis rotations during gait: the organisation of total body angular momentum. Gait & posture. 2008 Apr;27 (3):455-62. PubMed PMID: 17669652.

[69] Allum JH, Bloem BR, Carpenter MG, Hulliger M, Hadders-Algra M. Proprioceptive control of posture: a review of new concepts. Gait & posture. 1998 Dec 1;8 (3):214-42. PubMed PMID: 10200410.

[70] Barton GJ, Hawken MB, Foster RJ, Holmes G, Butler PB. The effects of virtual reality game training on trunk to pelvis coupling in a child with cerebral palsy. Journal of neuroengineering and rehabilitation. 2013;10:15. PubMed PMID: 23391156. Pubmed Central PMCID: 3571979.

[71] Yamaki K, Rimmer JH, Lowry BD, Vogel LC. Prevalence of obesity-related chronic health conditions in overweight adolescents with disabilities. Research in developmental disabilities. 2011 Jan-Feb;32 (1):280-8. PubMed PMID: 21115323.

[72] Biddiss E, Irwin J. Active video games to promote physical activity in children and youth: a systematic review. Archives of pediatrics & adolescent medicine. 2010 Jul; 164 (7):664-72. PubMed PMID: 20603468.

[73] Maher CA, Williams MT, Olds T, Lane AE. Physical and sedentary activity in adolescents with cerebral palsy. Developmental medicine and child neurology. 2007 Jun;49 (6):450-7. PubMed PMID: 17518932.

[74] Mitchell L, Ziviani J, Oftedal S, Boyd R. The effect of virtual reality interventions on physical activity in children and adolescents with early brain injuries including cere-

bral palsy. Developmental medicine and child neurology. 2012 Jul;54 (7):667-71. PubMed PMID: 22283557.

[75] Hurkmans HL, van den Berg-Emons RJ, Stam HJ. Energy expenditure in adults with cerebral palsy playing Wii Sports. Archives of physical medicine and rehabilitation. 2010 Oct;91 (10):1577-81. PubMed PMID: 20875517.

[76] Ainsworth BE, Haskell WL, Herrmann SD, Meckes N, Bassett DR, Jr., Tudor-Locke C, et al. 2011 Compendium of Physical Activities: a second update of codes and MET values. Medicine and science in sports and exercise. 2011 Aug;43 (8):1575-81. PubMed PMID: 21681120.

[77] Spencer T, Biederman J, Wilens T. Attention-deficit/hyperactivity disorder and comorbidity. Pediatric clinics of North America. 1999 Oct;46 (5):915-27, vii. PubMed PMID: 10570696.

[78] Pollak Y, Weiss PL, Rizzo AA, Weizer M, Shriki L, Shalev RS, et al. The utility of a continuous performance test embedded in virtual reality in measuring ADHD-related deficits. Journal of developmental and behavioral pediatrics : JDBP. 2009 Feb;30 (1):2-6. PubMed PMID: 19194324.

[79] Akhutina T, Foreman N, Krichevets A, Matikka L, Narhi V, Pylaeva N, et al. Improving spatial functioning in children with cerebral palsy using computerized and traditional game tasks. Disability and rehabilitation. 2003 Dec 16;25 (24):1361-71. PubMed PMID: 14660204.

[80] Ritterband-Rosenbaum A, Christensen MS, Nielsen JB. Twenty weeks of computer-training improves sense of agency in children with spastic cerebral palsy. Research in developmental disabilities. 2012 Jul-Aug;33 (4):1227-34. PubMed PMID: 22502849.

[81] Pillay H. An investigation of cognitive processes engaged in by recreational computer game players: Implications for skills of the future.. Journal ofnResearch on Technology in Education. 2003; 34, :336–50.

[82] Law EF, Dahlquist LM, Sil S, Weiss KE, Herbert LJ, Wohlheiter K, et al. Videogame distraction using virtual reality technology for children experiencing cold pressor pain: the role of cognitive processing. Journal of pediatric psychology. 2011 Jan;36 (1):84-94. PubMed PMID: 20656761. Pubmed Central PMCID: 3107585.